Handbook of Trait-Based Ecology

From Theory to R Tools

Functional ecology is the branch of ecology that focuses on various functions that species play in the community or ecosystem in which they occur. This accessible guide offers the main concepts and tools in trait-based ecology, and their tricks, covering different trophic levels and organism types. It is designed for students, researchers and practitioners who wish to get a handy synthesis of existing concepts, tools and trends in trait-based ecology, and wish to apply it to their own field of interest. Where relevant, exercises specifically designed to be run in R are included, along with accompanying online resources including solutions for exercises and R functions, and updates reflecting current developments in this fast-changing field. Based on more than a decade of teaching experience, the authors developed and improved the way theoretical aspects and analytical tools of trait-based ecology are introduced and explained to readers.

Francesco de Bello is a researcher at the Spanish National Research Council (CSIC) and Assistant Professor at the University of South Bohemia, Czech Republic. He has developed and applied several widely used tools to compute indices of functional diversity and assess community assembly mechanisms.

Carlos P. Carmona is an Associate Professor at the University of Tartu, Estonia. Currently he is working on the development of tools to assess functional diversity across scales while incorporating within-species trait variability.

André T. C. Dias is an Assistant Professor at the Federal University of Rio de Janeiro (UFRJ), Brazil. His research uses trait-based approaches to understand both species response to environmental conditions and their effects on ecosystem processes.

Lars Götzenberger is a researcher at the Institute of Botany at the Czech Academy of Sciences and the University of South Bohemia, Czech Republic. His research focuses on how traits evolve and affect species coexistence at different spatial and temporal scales.

Marco Moretti is a Senior Scientist at the Swiss Federal Research Institute (WSL) in Birmensdorf, Switzerland. He uses trait-based approaches to assess the response of communities to global stressors and the effects on ecosystem processes within and across trophic levels.

Matty P. Berg is an Associate Professor at the Vrije Universiteit Amsterdam, the Netherlands and an Affiliate Professor at the Rijksuniversiteit Groningen, the Netherlands. He uses trait-based approaches to study the assembly of soil fauna communities, amongst other related aspects.

Handbook of Trait-Based Ecology

From Theory to R Tools

FRANCESCO DE BELLO
University of South Bohemia, Czech Republic and Spanish National Research Council (CSIC)

CARLOS P. CARMONA
University of Tartu, Estonia

ANDRÉ T. C. DIAS
Federal University of Rio de Janeiro (UFRJ), Brazil

LARS GÖTZENBERGER
Institute of Botany of the Czech Academy of Sciences, Czech Republic and University of South Bohemia, Czech Republic

MARCO MORETTI
Swiss Federal Research Institute (WSL), Switzerland

MATTY P. BERG
Vrije Universiteit Amsterdam, the Netherlands and Rijksuniversiteit Groningen, the Netherlands

CAMBRIDGE
UNIVERSITY PRESS

University Printing House, Cambridge CB2 8BS, United Kingdom

One Liberty Plaza, 20th Floor, New York, NY 10006, USA

477 Williamstown Road, Port Melbourne, VIC 3207, Australia

314-321, 3rd Floor, Plot 3, Splendor Forum, Jasola District Centre, New Delhi - 110025, India

103 Penang Road, #05-06/07, Visioncrest Commercial, Singapore 238467

Cambridge University Press is part of the University of Cambridge.

It furthers the University's mission by disseminating knowledge in the pursuit of education, learning and research at the highest international levels of excellence.

www.cambridge.org
Information on this title: www.cambridge.org/9781108460750
DOI: 10.1017/9781108628426

© Francesco de Bello, Carlos P. Carmona, André T. C. Dias, Lars Götzenberger, Marco Moretti, and Matty P. Berg 2021

This publication is in copyright. Subject to statutory exception and to the provisions of relevant collective licensing agreements, no reproduction of any part may take place without the written permission of Cambridge University Press.

First published 2021

A catalogue record for this publication is available from the British Library

Library of Congress Cataloging in Publication data
Names: Bello, Francesco de, 1974– author.
Title: Handbook of trait–based ecology : from theory to R tools / Francesco de Bello, University of South Bohemia, Czech Republic and Spanish National Research Council (CSIC), Carlos P. Carmona, University of Tartu, Estonia, André T. C.Dias, Universidade Federal do Rio de Janeiro (UFRJ), Lars Götzenberger, Institute of Botany of the Czech Academy of Sciences, Marco Moretti Marco Moretti, Swiss Federal Research Institute WSL, Matty P. Berg, Vrije Universiteit, Amsterdam and RijksUniversiteit Groningen.
Description: Cambridge, UK ; New York, NY : Cambridge University Press, 2021. | Includes bibliographical references and index.
Identifiers: LCCN 2020041965 (print) | LCCN 2020041966 (ebook) | ISBN 9781108472913 (hardback) | ISBN 9781108460750 (paperback) | ISBN 9781108628426 (epub)
Subjects: LCSH: Ecophysiology. | Ecophysiology—Statistical methods. | R (Computer program language)
Classification: LCC QH541.15.E26 B45 2021 (print) | LCC QH541.15.E26 (ebook) | DDC 571.2—dc23
LC record available at https://lccn.loc.gov/2020041965
LC ebook record available at https://lccn.loc.gov/2020041966

ISBN 978-1-108-47291-3 Hardback
ISBN 978-1-108-46075-0 Paperback

Cambridge University Press has no responsibility for the persistence or accuracy of URLs for external or third-party internet websites referred to in this publication, and does not guarantee that any content on such websites is, or will remain, accurate or appropriate.

The truth of the principle, that the greatest amount of life can be supported by great diversification of structure, is seen under many natural circumstances.

Charles Darwin, 1859

Contents

	Preface		*page* xi
1	**General Introduction**		1
	1.1	General Definitions	2
	1.2	From Species to Functions	4
	1.3	What Is a Functional Trait?	7
	1.4	Response and Effect Traits	12
	1.5	Open Challenges	15
2	**Trait Selection and Standardization**		17
	2.1	Which Traits to Select?	18
	2.2	How Many Traits?	21
	2.3	Where to Get Trait Values From?	22
		2.3.1 Traits from the Literature and Databases	22
		2.3.2 Measuring Traits	28
	2.4	How to Express Trait Values?	29
	2.5	Missing Trait Data	31
	2.6	Trait Standards	32
		2.6.1 Why and How to Standardize Trait Measurements?	32
		2.6.2 How to Work with Trait Protocols?	34
3	**The Ecology of Differences: Groups vs Continuum**		36
	3.1	Historical and Conceptual Overview	37
		3.1.1 Functional Groups	37
		3.1.2 Trait Trade-offs and r/K Selection	39
		3.1.3 C-S-R, L-H-S and 'Spectra' of Differentiation	41
	3.2	From Theory to Numbers	44
		3.2.1 The Species × Trait Matrix	44
		3.2.2 Computing the Dissimilarity between Species	46
		3.2.3 Overlooked Issues with the Gower Distance	50
		3.2.4 Grouping Species	53

4	**Response Traits and the Filtering Metaphor**	57
	4.1 From Early Biogeography to Trait–Environment Relationships	57
	4.1.1 Trait Variation within Species	58
	4.1.2 Trait Variation across Species	60
	4.2 Environmental Gradients	62
	4.3 The Environmental Filtering Metaphor	64
	4.4 From Species to Communities and Back	67
	4.5 Relating Functional Traits to Fitness	70
	4.6 Traits and Species-Distribution Models	72
5	**Community Metrics**	75
	5.1 Community Functional Trait Structure	76
	5.2 Community Weighted Mean (CWM)	78
	5.2.1 Computing CWM	78
	5.2.2 Considerations on Species Abundance	81
	5.3 Functional Diversity Indices	86
	5.3.1 Range and Convex Hull	86
	5.3.2 Sum of Distances	88
	5.3.3 Variance	89
	5.3.4 Mean Dissimilarity	91
	5.3.5 Regularity	94
	5.4 Functional Diversity Components	96
	5.5 Partitioning Functional Diversity	98
	5.6 R Tools to Compute Functional Diversity	103
6	**Intraspecific Trait Variability**	105
	6.1 The Source of Intraspecific Trait Variability	106
	6.2 The Importance of Intraspecific Trait Variability	110
	6.2.1 Speciation	111
	6.2.2 Population Size and Genetic Diversity	112
	6.2.3 Adaptation	112
	6.2.4 Distribution	113
	6.2.5 Invasion Predictability	114
	6.2.6 Community Assembly	115
	6.2.7 Trait-Mediated Species Interactions	115
	6.2.8 Ecosystem Processes	117
	6.3 Assessing Intraspecific Trait Variability	119
	6.3.1 Quantifying Intra- vs Interspecific Trait Variability	119
	6.3.2 Intraspecific Trait Adjustments vs Species Turnover	121
	6.3.3 Trait Variability across Ecological Scales	124
	6.3.4 Including Intraspecific Variability into Functional Diversity	124

7	**Community Assembly Rules**		129
	7.1 Community Assembly Mechanisms		130
		7.1.1 Defining the Reference Species Pool	131
	7.2 Biotic Interactions and Species Coexistence		132
		7.2.1 Historical Perspective	132
		7.2.2 Coexistence Theory	134
		7.2.3 Moving Past (Pairwise) Competition	137
	7.3 Trait-Based Community Assembly		138
	7.4 Assessing Assembly Rules		140
	7.5 Null Models		143
	7.6 Community Assembly Applications		147
		7.6.1 Predicting Species Abundances and Trait Structure	147
		7.6.2 Invasive Species	148
8	**Traits and Phylogenies**		151
	8.1 What Is a Phylogenetic Tree?		152
	8.2 Brownian Motion and Why Related Species Should Be Similar		153
	8.3 Linking Evolution and Trait Filtering		156
	8.4 Phylogenetic Signal		157
	8.5 Are Traits Brownian?		159
	8.6 Phylogenetic Comparative Methods		160
		8.6.1 The Comparative Method and Evolution	160
		8.6.2 Independent Contrasts	162
		8.6.3 Moving Past PICs	164
		8.6.4 Imputing Data with Phylogeny	166
	8.7 Phylogenetic Diversity and Community Assembly		167
	8.8 Combining Phylogenetic and Functional Diversities		170
	8.9 Evolutionary Niche Modelling		173
9	**Effects of Traits on Ecosystem Processes and Services**		177
	9.1 Links between Effect Traits and Ecosystem Processes		179
	9.2 Assessing Biodiversity and Ecosystem Functions (BEF) Relationships		181
	9.3 Disentangling Functional Traits Effects on Ecosystem Functioning		185
		9.3.1 Designing Experiments Disentangling CWM and FD	185
		9.3.2 Analysing Biodiversity Experiments	187
	9.4 The Response–Effect Trait Framework		190
10	**Response and Effect Traits across Trophic Levels**		194
	10.1 Multitrophic Controls on Ecosystem Functioning		195
	10.2 The Multitrophic Response–Effect Trait Framework		197
		10.2.1 Adding Numbers to the Framework	199

	10.2.2 Intraspecific Trait Variability in Trophic Cascades	201
10.3	Response and Effect Trophic Traits in Interaction Networks	202
10.4	Perspectives	207

11 Trait Sampling Strategies 210

11.1	The 'Ambitious Supervisor' Exercise	211
11.2	Accuracy and Precision	216
11.3	Trait Variation at Different Scales	218
11.4	Sampling Strategies	219
	11.4.1 Starting Point: Setting Limits	219
	11.4.2 Literature on Sampling Strategies	220
	11.4.3 'Length' of the Environmental Gradient	221
11.5	Species' Abundances and Missing Values	226
11.6	A Visual Guide to Choosing Sampling Strategies	230

12 Applied Trait-Based Ecology 231

12.1	Biomonitoring: Biodiversity and Ecosystem Health	232
12.2	Traits in Agricultural Systems	237
12.3	Ecosystem-Based Solutions	240
	12.3.1 Green Urban Infrastructure	240
	12.3.2 Restoration with Functional Targets	244
12.4	Exotic Species' Impacts on Ecosystems	246
12.5	Ecological Literacy	248

References 250
Index 292

Preface

When did it all commence? Both of us were invited to the beautiful medieval city of Coimbra, Portugal, by Prof. Dr Paulo Sousa, in 2007. The occasion was a European Union–funded project meeting (Rubicode) at which we met each other for the first time. During the meeting we entered into a passionate discussion about the application of trait-based methods. We soon realized that the ecological questions in which we were interested and the methods we applied to analyse our data were quite similar, even though we were using different organisms as model systems, i.e. plants (Francesco) and soil invertebrates (Matty). By the Mondego River, we discussed many aspects of the use of functional traits. Based on these discussions, Paulo invited us to develop a new course on the use of 'Traits in Ecology' for MSc and PhD students at the University of Coimbra, which started in September 2009. This course has been running biennially at the University of Coimbra ever since.

A few years later, Lars Götzenberger joined us as a third lecturer, adding his experience in community assembly and phylogenetic analyses to the course. Meanwhile, Francesco has been teaching similar courses elsewhere with Lars, Carlos Carmona, Marco Moretti and André Dias, among many other good colleagues. It therefore made sense to combine our shared experiences and join together in an adventure to assemble this book.

While teaching we realized there was an obvious lack of a handbook connecting the large body of theory and practical methods in trait-based ecology that we were discussing with the students. In 2017, after a day spent teaching in Coimbra, the idea to write this book was conceived. It was a sunny day, and Francesco, Matty and Lars were sitting on a bench under a large cork oak tree, next to an old fountain in the botanical garden. We started discussing existing works and felt that there was a clear gap on the bookshelves. We thought it would be useful to prepare a book addressed to MSc and PhD students either approaching trait-based ecology for the first time, or seeking a deeper understanding of concepts and methods they had already encountered. The book would also be intended for researchers and practitioners needing to apply functional ecology to their field of interest. The present book is the final result of that discussion.

The content of this book is structured following our experience in teaching theoretical and methodological aspects of trait ecology, and is based on several tools and algorithms, particularly using R, several of which were developed in our labs. The courses have helped us to improve the way theoretical aspects and analytical tools of trait-based ecology are, in our view, best explained and introduced, to both basic and advanced students. In particular, through our teaching experiences, we were able to identify which are the trickier aspects of trait ecology and deserve special care in communication and practice. The aim of the book is thus to provide a comprehensive toolbox for functional ecologists, with guidance on how to apply the tools both in experimental approaches and data analysis.

This is not the first, and most probably not the last, trait-based ecology book. We believe, though, that this is the first book to fully link trait-based approaches to ecological concepts and to R tools, particularly across multiple trophic levels. A recent book by Garnier, Navas and Grigulis (2016) on *Plant Functional Diversity* (Oxford University Press) nicely synthesizes many key aspects of the trait-based ecology of plants with a firm theoretical background. As a matter of fact, Francesco, Marco and Carlos have been involved in an itinerant course on plant traits organized by Alison Munson, Eric Garnier and Bill Shipley. In several chapters we refer to this book for further useful reading. At the same time, some increasingly important trait-based themes have not yet been addressed virtually anywhere in textbooks: for instance, trait analyses across trophic levels, phenotypic plasticity and tools to quantify components of intraspecific trait variability. Our book addresses such topics and also deals with practical issues that everyone faces when working with traits, such as trait sampling strategies, trait measurements, missing trait data, data handling and analyses.

Most importantly, we provide and explain many R tools to assess different aspects of trait-based ecology. For almost every chapter of the book we have developed R materials to show the tricks associated with the different analyses, or solving problems often not mentioned in the published literature. For this reason, the elevator-pitch title we had for this book was 'Tricks of the traits: a hitchhike-R guide to functional ecology'. Although this was a great title, we finally opted for a more formal one. In terms of trait-based analyses using R, we have also built our work on the seminal book of Swenson (2014a) on *Functional and phylogenetic ecology in R* (Springer), which provides the first synthesis of a few basic R tools to account for functional and phylogenetic diversity in ecology. We discuss these tools, but we have also expanded on and updated them. Another seminal book on methods in trait-based ecology is the one by Bill Shipley (2012), *From plant traits to vegetation structure* (Cambridge University Press), which provides various important tools to connect vegetation changes along gradients to ecosystem functions. Several ideas were borrowed from this fundamental book, dealing with detailed mechanisms of plant adaptations and coexistence along environmental gradients.

The present book is organized into 12 chapters. Most chapters are accompanied by specifically designed explanations of R tools available on a dedicated and stable website (https://digital.csic.es/handle/10261/221270) with the title "Trait-based ecology tools in R" also searcheable online. The material is organized in an intuitive way to connect it easily to the reference chapter in this book (e.g. 'R material Ch3' is the material accompayining Chapter 3, organized with R markdown step-by-step explanation). We 'outsourced' the R-code and related content so as not to overload the conceptual and theoretical parts of the book with methodological and technical aspects. The R material includes examples on how to use existing R functions and packages, while providing some ad hoc designed functions, and includes updates reflecting current developments in this fast-changing field. The collection of R material is, we think, the most encompassing and comprehensive summary that exists currently. While we acknowledge that the material cannot cover the entire range of trait-based methodological approaches, we do think it provides a solid basis on which further developments can be grounded.

For writing the book, each author took the lead on two chapters while contributing considerably to the other chapters as well. Matty led the introductory chapter (Chapter 1) and the one on intraspecific trait variability (Chapter 6). Francesco dealt with trait differentiation between species, trait dissimilarity and functional diversity (Chapters 3 and 5). Marco focused on trait selection, standardized trait measurement, and databases (Chapter 2) and trait effects on ecosystems (Chapter 9). Lars assessed how traits can help in predicting species responses to environmental change (Chapter 4) and how to combine functional and phylogenetic concepts and tools (Chapter 8). Carlos focused on community assembly theory and tests (Chapter 7) and the design of sampling campaigns (Chapter 11). Finally, André linked traits across trophic levels (Chapter 10) and dealt with how traits can be used in an applied ecological context (Chapter 12). All the chapters are preceded and accompanied by beautiful drawings to illustrate some key aspect of the chapter, skilfully made by Janine Mariën (Vrije Universiteit Amsterdam) in her leisure time. She also designed the cover of the book, and many of the figures. We are also grateful to Luís Gustavo Barretto for several other beautiful drawings made for various chapters and to Javier Puy for a nice drawing in Chapter 6.

Writing this book has been a pleasant adventure, which peaked in a retreat at the field station 'het Groene Glop' (Dutch for a 'wet deciduous small forest') of the Vrije Universiteit, Amsterdam, located on the barrier island of Schiermonnikoog, the most northerly part of the Netherlands. Here, during a hot summer's week, in the presence of many mosquitos, and surrounded by beautiful nature, the book was largely finalized after a long period. A pleasant interaction between us was possible because we share so many similar 'traits' ourselves – above all, the appreciation of a good meal accompanied by nice wine. At the same time, each author has particular traits and skills, which make this book a typical case of both trait filtering and trait complementarity, resulting, we hope, in some useful material for its readers.

Composing this handbook has been made possible by the support of many institutions, including the Spanish National Research Council (CSIC), the University of South Bohemia (USB), the Institute of Botany of the Czech Academy of Sciences (IBOT), the Vrije Universiteit Amsterdam, the Groningen Institute of Evolutionary Life Sciences, the University of Tartu (UT), the Swiss Federal Institute for Forest, Snow and Landscape Research (WSL) and the Federal University of Rio de Janeiro (UFRJ). Our families and colleagues have supported us in so many ways. Above all, we thank Paulo Sousa for initiating all this in 2007 and keeping the trait course in Coimbra active. Aleksandra Serocka, Vinithan Sethumadhavan, Jenny van der Meijden and Ken Moxham from Cambridge University Press, provided a very precious feedback on the assembly of the book. Many colleagues generously provided friendly reviews of individual chapters, including Eric Garnier, Norman Mason, Martin Zobel, Margaret Mayfield, Sandra Lavorel, Cecile Albert, Jan Lepš, Daniel García, Jacintha Ellers, Miguel Verdú, Paulo R. Guimarães Jr, Nagore Garcia Medina, Maria Majeková, Fabio R. Scarano, Martin Gossner, Yoann Le Bagousse-Pinguet and Simon Pierce. We offer our heartfelt thanks to you all. Robert Davis, Conor Redmon and Nichola Plowman helped us very professionally to improve the English in the different chapters. We are very grateful for their constructive criticism and very useful comments that helped us greatly to improve the clarity of the text. We also sincerely thank all the students who have attended our courses over the years, for raising so many interesting questions, testing our R scripts and, with their doubtful remarks and lost expressions, showing where explanations have needed improving.

Finally, we do hope you will enjoy this book and find it useful in your work.

Matty P. Berg and Francesco de Bello
Schiermonnikoog, the Netherlands, 29 August 2019

1 General Introduction

The last few decades have witnessed an explosion of studies using functional traits of organisms to answer long-standing and pressing ecological questions, such as how biodiversity varies along environmental gradients or what are the consequences of species loss for ecosystem processes. Functional traits are also used to develop applied tools for nature conservation, such as bioindicators for managing multiple ecosystem services. This rising interest in trait-based ecology (McGill et al. 2006; Cadotte et al. 2011) is partially a response to a growing demand for methods to predict the consequences of global-change drivers on biodiversity and the feedback of traits on ecosystem functions. Traits can ideally provide an essential tool to uncover and generalize biodiversity (dis)assembly mechanisms and allow the prediction of species and community effects on ecosystem functions and services. For this reason, ecologists are increasingly looking at traits, rather than species alone, in something that has been referred to as 'the biodiversity revolution' (Cernansky 2017).

In order to deal with the different theoretical and practical aspects of trait-based ecology presented in the following chapters, it is essential, firstly, to prepare the ground. In this chapter, we will introduce key definitions and concepts that are essential for most applications of the rapidly evolving field of trait-based ecology.

1.1 General Definitions

In this book we will borrow definitions and concepts already widely used and discussed in the literature. These definitions will allow us to set the scene for the development of the following chapters. *Biodiversity*, or biological diversity, is defined as the variety of life on Earth at all its levels, from genes to ecosystems, and the ecological and evolutionary processes that sustain it (Gaston 1996). Biodiversity thus includes different components, such as the variety in genotypes, phenotypes, populations, communities and ecosystems.

Because of the multiplicity of biodiversity components, it is difficult to quantify it in full. One of the big debates in ecology is what aspects of biodiversity are important to answer different questions. Is species richness per se important, or are species traits more critical if we want to understand community (dis)assembly, or how shifts in species composition affect ecosystem processes (Hooper et al. 2005)? The main premise of trait-based ecology is that one way to generalize patterns beyond taxa and locations is the use of species traits. It is thus assumed that two species with similar traits will have similar behaviour. In this direction, Cernansky (2017) recently referred to a 'biodiversity revolution', arguing that ecologists are increasingly focusing on traits – rather than species names and numbers alone – to measure the health of ecosystems. This raises the question of what the role of biodiversity is, expressed in terms of traits, in explaining observed patterns in community assembly and ecosystem processes, and how and what we should measure when species-based approaches are combined with trait-based approaches. Some consensus on the relative importance of different components of biodiversity based on traits has been reached, but before we dive into that we must provide some definitions that have been adopted throughout the book.

A *community* is an ensemble of all populations of all species that co-occur in the same place and at the same time. Usually it relates to species that interact with each other, via competition, trophic interactions or positive feedback, such as facilitation and symbiosis. These interactions may occur on a variety of temporal and spatial scales, i.e. from small, discrete and relatively stable food webs in tropical epiphytic tank bromeliads (Armbruster et al. 2002), to large, less distinct savanna communities with interactions between vertebrate grazers and their predators with a large home range. Although communities are often envisioned as delineated or 'closed' systems with a clearly defined boundary, most of them are dynamic and open to immigration or emigration of individuals, energy and nutrients. The term 'community' is often used to refer to a subset of the whole community, e.g. vascular plants, detritivores, pollinators, bird communities etc., as researchers usually focus on a few trophic levels and a few groups of organisms.

A *population* is composed of all the organisms, i.e. 'individuals', of a single species, which have the capability of interbreeding when it concerns a sexual species, and that live in the same area. In a local population, interbreeding is potentially possible between any pair of individuals and is more likely than cross-breeding with individuals from other regions. As individuals can disperse, populations and communities are connected with each other, forming so-called meta-populations (Hanski & Gaggiotti 2004) and meta-communities (Holyoak et al. 2005). Therefore, a distinction can be made between within- and between-population characteristics.

Species abundance is the local representation of the population size of a species in a particular ecosystem. In some cases, the abundance can refer to a larger taxonomic unit such as a feeding guild, or molecular operational taxonomic units (MOTUs; using molecular data to classify groups of closely related species). It is often measured as the number of individuals found in a sample, particularly in studies on fauna. When the number of individuals is not a feasible measure, other measures, such as biomass, cover, or frequency are often considered as well, particularly for plants. Very often, many indicators described in this book refer to the relative abundance of species with respect to other species in the community. The *species relative abundance* expresses the proportional abundance represented by a species in a community. Depending on the situation or environmental condition, species abundances can vary considerably across time and space. Information on species abundance provides insight into how species are distributed across communities and ecosystems, and which species are adapted to certain environmental conditions.

Species richness is the actual number of species (or, for example, MOTUs) in a community, without any reference to the abundance of species. This latter aspect is taken into account by indices of species diversity such as *dominance* or *evenness*, which mathematically combine species richness with species relative abundance. These components altogether define *species diversity*. The more even the abundance across species, the higher the evenness. The basic assumption behind species richness is that all species have an equal importance within communities. In this book we refer to species as taxa, and therefore taxonomic diversity is generally intended to be a synonym of species diversity.

Species composition is defined as the identity and contribution of multiple species to the community or ecosystem. Species composition is an important property of communities, provides an essential description of the characteristics of a group of species at a particular site and is central to community (dis)assembly (see Chapters 4 and 7). Species composition changes over time and space as a result of a multitude of factors. It is an important indicator of ecological processes in ecosystems, as species composition often drives the rates of multiple ecosystem processes. Species composition is usually based on inventories, which describe the abundance of different species in a community (see also the 'species × community' matrix in Chapter 4 and elsewhere).

In this book, an *ecosystem function* is defined as a biological, physical or geochemical process that occurs or takes place within an ecosystem. Ecosystem functions are also referred to as ecosystem processes. Typical ecosystem functions that are studied extensively are primary productivity, litter decomposition and plant pollination, but there are many more. Another common definition focuses more on beneficial aspects of these processes for humans. In this case, ecosystem function is defined as the capacity of natural processes and components to provide goods and services that satisfy human needs, either directly or indirectly (de Groot et al. 2002), and is then referred to as an ecosystem service.

Species effects refer to the role a species performs in a community or ecosystem process. In most cases, the belonging of a species to a functional group or a feeding guild (Chapter 3) defines its role. For example, isopods feed on leaf litter and belong to the feeding guild of macro-detritivores. As litter feeders, they play an important role in litter decomposition, which is a key process in soil fertility and the cycling of nutrients through ecosystems. Species effects could matter because different isopod species, as well as other macro-detritivores, differ in their consumption rates (Zimmer et al. 2002; Vos et al. 2011) and, therefore, in the individual effect they have on litter decomposition. The effect of species composition, which we will see corresponds to the composition in terms of species traits, is often larger than the effect of species richness. For example, the effect of mixtures of isopods, millipedes and/or earthworms on litter decomposition is not explained by species richness, nor taxonomic group richness but by the presence or absence of particular species or species combinations, hence species effects (Heemsbergen et al. 2004).

Functional diversity is a component of biodiversity and is defined as the functional trait differences between organisms present in a community or ecosystem (see also Chapter 5). Functional diversity can be of great importance to many ecological processes as it may have a strong impact on community dynamics and stability as well as ecosystem processes (Chapters 7 and 9). As we will see in Chapter 5, there are different components to functional diversity, mostly *functional richness, evenness* and *divergence*, which are complementary as they describe different features of the distribution of trait values across individuals and species in a community. The degree to which community and ecosystem features are influenced by functional diversity, as well as the relative importance of its components, is much debated.

1.2 From Species to Functions

An increasing number of studies suggest that a trait-based approach has some advantages over taxonomy-based research. While trait-based ecology has a long tradition (Chapter 3), the main focus of ecological research since the 1990s has often been centred on species taxonomy, for example on changes in species diversity and composition along gradients, or on how these factors affect ecosystem functions (Hooper et al. 2005). We have gone far with this taxonomical approach but it has its limitations. In special cases, such as ecosystems with a low species richness, one could argue that a taxonomic approach is extremely valid. With a limited number of species, it is potentially feasible to study the response of each species to changes in the environment and to unravel all, or the majority of, possible species interactions. However, ecosystems are generally complex and often species-rich, making it impossible to study all these species and their feedback in concert.

1.2 From Species to Functions

Figure 1.1 A contemporary agricultural landscape in which the former continuous forest is fragmented by agricultural fields into many forest patches that differ in size, shape and connectivity. Although it concerns the same habitat type (forest), the exact species composition cannot be predicted because size, shape and surrounding habitat affect community composition. Hence, species composition within a patch depends on the context. Figure created by Janine Mariën.

Another problem with taxonomic approaches is that even in rather simple systems there is *context dependency* (Fig. 1.1). Take, for example, a common contemporary landscape, in which former large forested areas are fragmented into small forest habitat patches, surrounded by agricultural fields or managed semi-natural grasslands. Assume the focus of your study is on plants, or animals, living in these forest fragments. Even without disturbances, and under equal regional geological and climatic conditions, these forest fragments will differ in species composition. Dissimilarity in species composition occurs due to fragments differing in size (think about the influence of the theory of island biogeography on species richness; Whittaker & Fernández-Palacios 2007), in connectivity due to the presence or absence of corridors that connect a patch to larger forested areas, or in the habitat surrounding these patches (e.g. small forest patches surrounded by agricultural fields are different in species composition compared to patches of the same size and connectivity but surrounded by semi-natural grasslands). If forest habitat patches under similar conditions are already different in species composition, how can we hope to predict changes in community composition when the environment will change? Trait-based approaches assume that changes in trait compositions in these forest patches can be more predictable than the species compositions (Fukami et al. 2005).

Imagine another case in which we want to compare the response of plant communities to climate change in two similar grasslands, but from different regions of the world. Without doubt, we cannot easily generalize results only in terms of species composition from one region to another, because the two grasslands will greatly differ in species, given the difference in regional species pools. We can of course say that, for example, species richness will decrease with drought in both grasslands. We can even go further and might conclude, in the case of an oversimplified example, that smaller species will become more abundant under increased drought, which could lead to a reduction in overall herbivory or lower herbivory damage on plants with tougher leaves (Ibanez et al. 2013). As such, one solution to the problem of context dependency might be to combine information about species ecological preferences (ecological niches) and traits of species in our studies when we want to forecast community responses to environmental stress.

A species' *ecological niche* is defined by all the resources needed to sustain healthy populations, i.e. for a species to successfully complete their whole life cycle. It can be envisioned as an n-dimensional hypervolume, in which each axis represents an abiotic or biotic factor that affects the fitness, hence presence and distribution, of an individual (Chase & Leibold 2003). If we take soil moisture as an example of a resource axis, we can observe that a particular species, let's say the sand sedge *Carex arenaria*, will occur only at the drier end of the moisture gradient in its typical climatic conditions. When soil moisture conditions become too wet, this species will not survive or will be outcompeted by other species that are more adapted to wet conditions. The sand sedge is a dry-soil-adapted species. Under these conditions it shows the highest fitness. It can do so because it has acquired physiological traits that allow optimal growth in warm, dry, sandy soil. One could say its traits allow it to cope with these environmental conditions. Interestingly, when we study other species that grow next to sand sedge, we will find that other species of this vegetation type all have some sort of physiological adaptations to cope with warm, dry, sandy soil. These types of adaptations can either be the same or different (see Chapter 4; Pistón et al. 2019). They grow in a resource-poor environment and have to develop specific strategies and functional traits that enable them to acquire enough nutrients. This becomes quite evident when we compare their traits with the functional traits of species that live in marshes. Marsh species have to cope with regular flooding, anoxic conditions in the rhizosphere and, in general, nutrient-rich soils. This asks for completely different physiological adaptations compared to plants that belong to sandy-soil vegetation.

Convergence in traits, or trait value, between organisms that are exposed to the same environmental conditions (see Chapters 4 and 7) may help us to predict which species will increase or decrease in abundance when the environment changes (Fukami et al. 2005). It may be difficult to predict, for a particular type of habitat, which species will increase or decrease based on taxonomic identity, but given a predictable environment–trait relationship we might be able to forecast which trait values will become prominent after a change in the environment. If we know the species traits for a great number of plant species, for example, we should be able to predict, based on such traits, which species will increase or decrease under a given stress. The same applies to the effect of species on ecosystems. If we know the traits that affect primary productivity for a great number of plant species, we should be able to predict, based on these traits, the impact of a shift in plant composition on primary productivity. This is the general premise of trait-based ecology.

1.3 What Is a Functional Trait?

A unified terminology is the first step toward a common understanding of traits and trait-based approaches. This is a far from trivial issue as ecologists have been using a variety of definitions and approaches when referring to traits, possibly due to the different applications of traits, ranging from individuals to species and ecosystems. It was only with the seminal work by Violle et al. (2007) that the great diversity of approaches and definitions achieved a certain coherence to make the entire field more 'functional' and less equivocal. Here, building on that study, we introduce and discuss important aspects of the definitions of traits.

According to Violle and colleagues, a *trait* is defined as '*any morphological, physiological or phenological feature measurable at the individual level, from the cell to the whole organism*'. It is important to realize that in their paper the authors generally refer to plants, although not exclusively so. This definition can also be applied to animals, with one additional feature, namely 'behavioural', as this reflects an important component of animal ecology. In practice, the definition states that whatever we can measure on a single organism is a trait. For example, we can, with more or less effort, evaluate plant height, body length, age at maturity, ability to fix nitrogen or mouthpart morphology of an individual. In addition, it has been stressed that traits are heritable, genetically or epigenetically, so that some authors also explicitly include this aspect in the definition (Garnier et al. 2016), while others assume it is implicit in the definition. We follow the latter point of view for simplicity, while keeping in mind the importance of the genetic basis of traits.

In the courses we teach, we always stress that the reference to a single individual is a very important point because it allows communication between ecologists and evolutionary biologists, who use a similar definition of traits, as heritability acts on individuals via phenotype. The measurement of traits on an individual can, strictly speaking, sometimes be a challenge. For example, functional traits of microbes often cannot be measured on a single cell ('individual'), while social insects operate as a colony.

Let's continue focusing on the definition of traits. According to Violle and colleagues, a trait, to be truly functional, needs to '*impact fitness (of an individual) indirectly via its effects on growth, reproduction and survival*'. This reference to fitness is important as it determines whether a trait is truly functional or not. Let's assume a situation in which spiders are exposed to drought stress because of a long period without any precipitation, while the temperature is within their tolerance range (no heat stress in this case). One important functional trait to include in our study would be water-loss rate (Fig. 1.2), expressed as the amount of water lost per unit of body mass per unit of time (Moretti et al. 2017). The lower the water-loss rate the higher the survival, hence

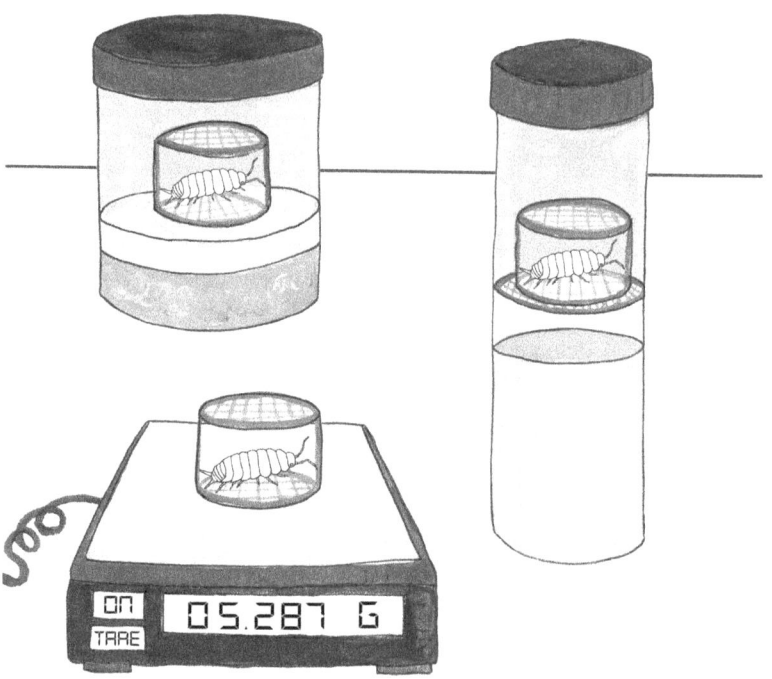

Figure 1.2 Measuring water-loss rate in invertebrates. Specimens are kept in an open tube placed within a larger vial with a heat space of a certain relative humidity (due to a specific glycerine–water mixture at the bottom of the vial). Water loss is measured by mass loss of specimen over time when the tube with specimen is placed on a balance. Figure created by Janine Mariën.

fitness, when individual spiders are exposed to drought. Here, measurements of lower thermal limit, a trait that defines variation in tolerance to cold stress, are not very useful in the context of the system studied. To our understanding, although the lower thermal limit is a trait (can be measured on a single individual), and is functional in cold habitats, in this particular case it is not a functional trait as it cannot be associated with individual fitness when spiders are exposed to extremely dry conditions. The lower thermal limit, on the contrary, can inform us about survival of spiders when they are exposed to frost, and would then in turn influence their individual fitness, while water-loss rate would not affect fitness under cold stress. In other words, almost anything one can measure on an individual is a trait, but the example here indicates that the functionality of a trait largely depends on both the environmental situation and the research question, hence on the context of the study. We recognize that some researchers choose not to include the word 'functional' when they refer to traits, to avoid the conflict of deciding whether (or when) a trait is functional or not.

Following the arguments above, it is important to discuss the final important component of the definition of a trait by Violle and colleagues, i.e. the part related to '*without reference to the environment or any other level of organization*'. Although in practice the trait definition gives great freedom in the choice of traits we can measure, this part adds some constraints and, by our own experience, is often found to be the most obscure for many students and ecologists in general. To our understanding, this part of the definition intends to clarify the specific scale at which traits are defined, i.e. the individual, and excludes information that can only be defined at larger levels of aggregation, such as populations or species. Examples of the latter are habitat width, range size, or anything that relies on data obtained from species distributions across environmental gradients (such as Ellenberg indicator values; Box 1.1). A typical example is defining a species as 'forest specialist'. These should not be considered functional traits because values are based on multiple individuals (i.e. you need to observe the distribution of many individuals) and have a clear reference to the environmental preference, rather than anything else. We agree entirely with Violle and colleagues that using the observed distribution of species along environmental gradients (also referred to as the niche of species or sometimes named *ecological traits*; see above) is certainly not good practice to define a trait. This is because the observed niche of a species reflects a complex response to a combination of abiotic and biotic factors and might include facilitation of species enlarging the niche of species or competition reducing niche space. For this reason, we discourage the use of the term 'traits' when referring to habitat preference. Again, this happens because habitat preference is a population or species feature determined by the observed distribution of species rather than an individual characteristic, thus without selection on individual fitness. Moreover, the question of which functional traits determine species distribution along gradients

---

Box 1.1 Environment–species associations, such as Ellenberg's indicator values for plants.

Can we use environment–species associations, such as Ellenberg's indicator values for plants, as traits? Basically, no. Ellenberg's indicator values (Ellenberg 1974) indicate the species' preferred environmental conditions, generally based on species' observed distribution along natural gradients. These plant indicator values, or similar indices for animals (e.g. the Species Temperature Index, which indicates the average temperature of the species range (Thuiller et al. 2005; Hijmans & Graham 2006; Devictor et al. 2008)) can only be inferred from aggregated data and cannot be measured on a single individual. Although there might be merit in using variables derived from the living environment of a given species, they cannot be called traits, as they are species properties and not a characteristic of an individual. Properties inferred from aggregated data are called 'ecological features' or 'environmental associations' in Garnier et al. (2017). They are often derived from the environmental conditions at the place where the individuals of a given species were observed. For a critical analysis of the reliability of Ellenberg's indicator values as traits, see e.g. Schaffers and Sykora (2000) and Zelený and Schaffers (2012), but see Klaus et al. (2012) and Wildi (2016) who show circularity in predicting species distributions based on these ecological traits.

(Chapter 4) cannot be answered using habitat preferences derived by the observed distribution of species, given the circularity in reasoning.

There is one additional note necessary. Existing trait protocols advocate measuring individuals in optimal environmental conditions, e.g. leaves in well-lit environments (see Cornelissen et al. 2003), or water-loss rate of animals after acclimatization and under standardized conditions (Moretti et al. 2017), and field measurements that require recording information describing the local environment in which the measures were taken (Garnier et al. 2016). This standardized way of measuring traits is important to be able to compare trait values between species or organisms. For instance, if you want to compare trait values across species, you need to compare similar types of organisms, e.g. mature individuals growing in good conditions (for example, but not necessarily, in those conditions in which the species grows well). As such, when measuring such traits, this is always done in a given 'environment' and requires, therefore, some reference to the abiotic conditions. However, these are true traits when they can be measured on an individual and become functional when related to individual fitness once placed in an ecological context (see examples above on drought and cold adaptations). As such, strictly speaking, the reference to the environment is important when measuring traits.

As explained above, for a trait to be considered 'functional', it needs to influence, to some degree, the performance of the individual. Individual performance depends on three properties – growth rate, reproductive output and survival – which are the three main components of fitness (see below). Violle and colleagues call these '*performance traits*', while the impact of performance traits on fitness is sometimes called 'elasticity' (Adler et al. 2014). As mentioned above, one could argue that most traits, if not all, can directly or indirectly influence an organism's fitness, at least in some conditions. For example, in the case of terrestrial isopods, smaller individuals lose water faster when exposed to dry spells than large individuals, making body size (next to water-loss rate) a functional trait (Dias et al. 2013a). While both traits are functional, as they affect individual fitness under dry conditions, water-loss rate is more strongly related to survival under drought conditions than body size is, making body size a '*proxy*' (sometimes called a '*functional marker*'; see Garnier et al. 2016) for the physiological trait water-loss rate. Therefore, some traits might be more 'functional' than others (Fig. 1.3). Furthermore, in isopods, small and large species also differ in the morphology of respiratory tissue, also making the morphology of the pleopodal lungs a proxy for water-loss rate (which is directly related to drought tolerance; Dias et al. 2013a). Often, we do not have the information about more complex traits, or it is too hard or labour-intensive to measure them on many or even a few individuals, and therefore we have to rely on proxies (Ackerly & Monson 2003). Some proxies can act as a substitute for functional traits that directly affect species performance but for which data are missing. Using proxies can be very informative, but as they are more loosely related to performance than more integrative functional traits (Adler et al. 2014). Before use they should be tested for their actual correlation with the functional or performance trait of interest (Rosado et al. 2013). In all cases, the merit of a trait as a functional marker of fitness depends on the research question, study organisms and environmental context.

1.3 What Is a Functional Trait?

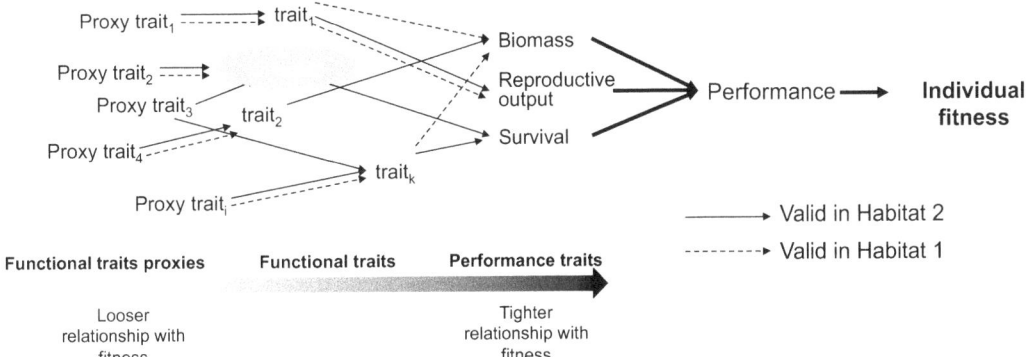

Figure 1.3 Arnold's (1983) framework linking functional traits to individual fitness, generalized for all organisms for which morphological, phenological, physiological and behavioural (especially for animals) traits can be measured on individuals. Functional traits (from 1 to k) can directly affect species performance, i.e. growth, reproduction and/or survival, which are named performance traits. Individuals' performance traits ultimately affect their fitness. We include proxy traits (1 to i), i.e. traits not directly linked to performance traits but that are correlated with those functional traits more closely related to performance traits. Note that a marked delineation between what is a functional trait and what is a functional trait proxy does not exist. Adapted from Violle et al. (2007), with permission from Wiley. © OIKOS. Published by John Wiley & Sons Ltd.

Figure 1.4 Plant–herbivore interactions. Examples of organisms' functions, in this case caterpillar growth and plant survival, both defined at the levels of an individual. Figure created by Janine Mariën.

To assess which traits can act as a proxy of performance it can be necessary to measure changes in performance traits, i.e. to quantify survival, growth and/or reproduction directly, across environmental gradients (Figs. 1.3 and 1.4). For example, when species are exposed to frost, the variation in survival (number of individuals still alive at a given temperature and after a given exposure time) can be defined as 'ecological performance' (Violle et al. 2007). Once measured under standardized conditions, such 'performances' in standardized experiments can be linked to simple traits (in the search for the best proxy). An alternative is to relate such performances to species distribution in nature. This has, therefore, a high potential for predictive power. In this particular example, the ranking of species survival along temperature or altitude gradients could be compared with the ranking of species based on freezing tolerance (lowest temperature a species can withstand under standardized conditions). Freezing tolerance in controlled conditions is thus a potentially good proxy of performance if this trait strongly correlates with cold sensitivity based on survival measurements. For these reasons, some authors (e.g. Moretti et al. 2017) consider the variation in a performance trait in standardized conditions as, potentially, a functional trait in its own right.

1.4 Response and Effect Traits

Traits do not only affect the fitness of organisms. At least some traits can also explain the effect of organisms on the environment, which includes different ecosystem processes (Chapter 9) and other trophic levels (Chapter 10, Fig. 1.4). For example, fixing nitrogen facilitates plant survival in nitrogen-poor soils, but by doing so it also increases nitrogen content in the ecosystem, which may facilitate the establishment, and increase the fitness, of other plant species. Because of these two distinct facets of traits, they have been further classified into *response traits* and *effect traits* (Lavorel & Garnier 2002; Fig. 1.5). Below we explain the important distinction between the two types, together with some important considerations.

Response traits are those that allow organisms to cope with a particular stress, i.e. to be able to survive, grow and reproduce under the experienced stress level or under different environmental conditions (both biotic and abiotic factors). For example, if smaller plants, or plants with secondary compounds that make them unpalatable, can better survive in grazed pastures, plant height and the presence of secondary compounds are, in this sense, important response traits along a grazing-intensity gradient. Or, in the case of a significant heatwave, the critical upper thermal limit (CT_{max}) of a species determines if that species can cope with extreme high temperatures by controlling whether the species (or individual) will survive or not when exposed to heat stress. Obviously, species differ in their CT_{max} (Franken et al. 2018), making species with a low value more sensitive to heatwaves compared to species with a high CT_{max}. If a strong heatwave occurs, the average CT_{max} of the community will shift to a higher value, as species with a low CT_{max} will be 'filtered out' (see Chapter 4) and

Figure 1.5 An example of response and effect traits. Grazing induces the formation of plant secondary compounds (response trait) that increase fitness (survive or not) of the plant. Presence of plant secondary compounds in litter, the afterlife of fresh leaves, affects the palatability of litter to detritivores. In this particular case, plant secondary compounds are both a response trait and an effect trait. Figure created by Luís Gustavo Barretto.

species with a high CT_{max} survive. This means that if we know the CT_{max} values of community members, or at least the (sub)dominant species, we should, in principle, be able to predict which species will be at risk when exposed to a heatwave, and CT_{max} can be used as a predictor of community response. It is important to note that a functional trait is, by definition, a response trait, but the reverse is not necessarily true. Every trait that varies in response to a specific stress can be used as a predictor, even proxies of functional traits. However, one should strive for selection of functional traits and not trait proxies, when possible, because the former provides a causal mechanistic link to the stress.

Effect traits are traits that either have an effect on other trophic levels, as in the case of predator–prey interactions or mutualistic relationships, or on ecological processes, such as nutrient cycling, pollination or primary productivity (Lavorel & Garnier 2002). Basically, any trait can act as an effect trait as long as it significantly affects another species or an ecosystem process of interest. As an example, let's look at an important soil process: litter decomposition, a key process for nutrient cycling through ecosystems. Macro-detritivores, such as termites, isopods and millipedes, influence the speed of litter degradation. The consumption rate of litter-feeding, terrestrial Isopoda will affect litter decomposition (Hättenschwiler & Gasser 2005) because species with a high consumption rate assimilate more litter, generally produce more faeces, and enhance litter surface area for microbes, the principal decomposer organisms, via litter fragmentation. All these combined actions of macro-detritivores and microbes result in high litter decomposition. If the average consumption rate in the community shifts to higher values because there are more species (or individuals!) with a high intake rate, litter decomposition will increase. This means that a shift in species composition in the community, hence a change in the average effect trait value in the community, has consequences for ecosystem processes, either indirectly via an impact on the next trophic level or directly as it affects process rates. Hence, litter consumption is a key effect trait, and information on the trait values of the species in the community can be used to predict in which direction an ecosystem process will change across an environmental gradient.

Not all traits that have an effect on the ecosystem necessarily translate into a positive effect on the performance or fitness of the species, especially in animals. The opposite is also true: not all functional traits (traits affecting fitness) affect ecosystem functions. So, not all response traits are effect traits, and vice versa, although in practice things will never be so black and white. For instance, while water-loss rate of an animal affects its fitness in an environment with fluctuating soil moisture conditions, it does not affect ecosystem processes such as litter decomposition and nutrient mineralization. However, litter consumption rate does affect decomposition (i.e. it acts as effect trait) but does not help detritivores to overcome drought (i.e. it is not a response trait) or other environmental stressful conditions. Nevertheless, some traits do affect fitness and can also have an effect on the processes maintained by ecosystems, although this latter aspect is not part of the formal definition of a functional trait. However, we do acknowledge that some authors include the effect part of species in the definition of a functional trait, which we believe, from a conceptual point of view, could be avoided. As a matter of fact, the traits an organism possesses can have an effect on several ecosystem processes, and related ecosystem services (de Bello et al. 2010a; Chapter 9). In this particular case, a functional trait also has an effect on the ecosystem. Several plant functional traits that have been shown to allow species to respond to environmental change, such as those pertaining to the cycling of elements (carbon, nutrients, water), e.g. leaf nitrogen, leaf dry matter content etc., also impact the ecosystem (e.g. net primary productivity; de Bello et al. 2010a). Therefore, these traits are response and effect traits at the same time (Suding et al. 2008). However, by no means can it be taken for granted that all functional traits act as effect traits; this should be properly tested. Most importantly, the potential link between response and effect traits in organisms is central to various applications in functional ecology (Lavorel & Garnier 2002; Lavorel et al. 2013). Further details about response and effect traits and their interactions are given in Chapters 4 and 9, respectively, while insights on these aspects in a multitrophic context are available in Chapter 10. For a phylogenetic perspective on response versus effect traits see Chapter 8.

1.5 Open Challenges

While the basis of the definition of functional traits includes a link to species fitness, interestingly enough not many studies have assessed how strong single traits, and their interactions, do actually affect individual fitness (but see Adler et al. 2014; Ellers et al. 2018 for a review on soil fauna; Pistón et al. 2019). So, often it is assumed that many traits are 'functional', but this is not often explicitly tested (Ackerly & Monson 2003). Regarding the correct use of the term functional trait, only when the link (either direct, or indirect for proxies; Fig. 1.3) between a given trait and the individual's performance is known or tested can one use the term functional trait with confidence. In all other cases, we suggest referring to the general term 'trait' or acknowledging that there is an assumption that a given trait is a proxy of a functional trait.

One open challenge in trait-based ecology is the selection, and proper validation, of which combinations of traits affect species fitness. It is possible that traits do not affect fitness components in isolation but only in combination (Fig. 1.3). For many organisms, there are guidelines as to which traits can be prioritized (Chapter 2). An important avenue for further research is the search for interactions between, and combinations of, various traits in defining fitness components (Pistón et al. 2019). Moreover, it is possible that different combinations of traits could result in similar levels of fitness for organisms in a given environment – so-called alternative-designs (Dias et al. 2020). For example, in dry areas, different types of plant species can coexist (Fig. 2.4), including shrubs, succulents and ephemeral species, each with a different combination of traits. Taking these effects into account with proper statistical tools represents an open scientific challenge (Pistón et al. 2019).

It is also important to note that the assumption of a link between traits and performance, either directly or indirectly via a single trait or a combination of traits, reflects the general philosophy of comparative functional ecology (Shipley et al. 2016). This field of research was born from the premise that it is a Herculean task to know the performance of each single species for the multitude of environmental conditions in which that species occurs, and that, ideally, it is possible to measure one or a few characteristics (functional traits, or their proxies) that can help us to predict the performance of other species with similar trait values. As such, it is very hard to measure, for example, species performance on all vascular plants on Earth. But, ideally, if we validate the relationship between functional traits and performance for a few species, we can predict performance components for other species, based only on their traits. As we will see in this book, we are actually far from reaching this ideal goal, but nonetheless we hope to show that attempts are worthwhile.

How to measure functional traits on different types of organisms is another open question. In the cases of several taxa, trophic levels, or other levels of aggregation, one might face the problem that certain traits cannot easily be measured on individuals or on all community members or be expressed in the same unit. For instance, litter decomposition is performed by a large array of species, from bacteria and fungi to nematodes, springtails and macro-detritivores. In most invertebrates, body size is related to consumption rate, which impacts litter decomposition. This trait is easy to measure in most arthropods but how do we measure the body size of a fungal species? Similarly, mouthpart morphology might be a trait underlying the variation in consumption rate of taxa, but the mouthparts of earthworms are so differently constructed to those of isopods or millipedes that it is very difficult to find a common unit of expression for a trait that is present in all taxa. Similarly, in large arthropods, physiological traits, such as those underlying tolerance to environmental conditions, can more or less be easily quantified, but this is not the case in species groups with very small body sizes such as nematodes. As we advocate the adoption of standardized protocols for trait measurements (Chapter 2), the quantification of trait differences between organisms should be approached with care (Chapter 3).

Finally, one of the most crucial challenges we face as ecologists is to predict how environmental change impacts ecosystem processes via changes in community composition. Can we predict shifts in community composition from those functional traits that directly relate to the stress experienced by component species (Chapters 4 and 7)? And if we know through which trait a community has an effect on a particular ecosystem process, and we have information on this trait from all community members, can we then predict the consequences of a shift in community composition on an ecosystem process (Chapter 9)? Lavorel and Garnier (2002) published a seminal paper in which the authors introduced a response–effect trait framework that might provide an answer to these questions. Central to the response–effect trait framework is the notion that we might be able to predict how ecosystem functions change under a given environmental stress if we can identify the group of species that regulates or controls that particular process, via traits, together with the traits they possess to withstand a given stress (Lavorel et al. 2013). In the various chapters of this book we try to illustrate these and other aspects of trait-based ecology and guide ecologists through their potential uses and existing tools to incorporate them into research questions.

Take-Home Messages

- The last few decades have witnessed a shift in ecological research from a focus on species composition and species richness to a focus on the functional traits of species.
- Functional traits are any of the characteristics of organisms, generally measurable at the individual level, that affect their fitness. These are called 'response traits'. At the same time, some traits have strong effects on ecosystem functions or other trophic levels, regardless of whether they affect fitness or not. These are called 'effect traits'.
- We lack studies validating which traits, and combinations of traits, are functional, i.e. which traits and up to what degree affect species fitness in which habitats.
- Functional traits can enhance our ability to provide a mechanistic understanding of observed ecological patterns in the field and may facilitate the formulation of generalizations of these patterns across species and ecosystems.

2 Trait Selection and Standardization

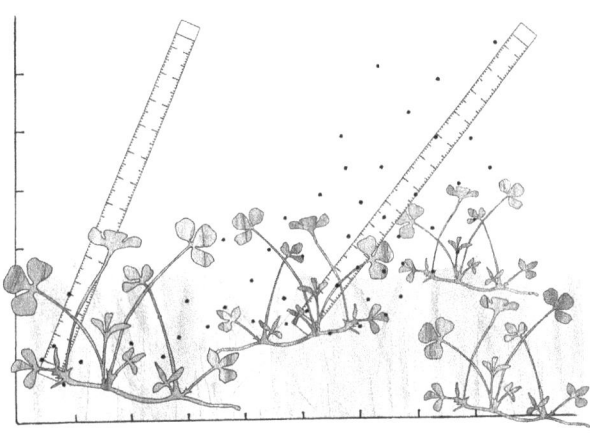

As we saw in the previous chapter, ecologists have long adopted a taxonomic-based approach to understand species distribution, community structure and ecosystem functioning. However, non-taxonomic approaches have also been used to investigate a wide range of topics in ecology. These include adaptation of individuals to stress (Bijlsma & Loeschcke 2005), population dynamics (Umana et al. 2017) and community responses to biotic and abiotic factors (Violle et al. 2007). These trait-based approaches have also been used to understand community assembly processes and predict changes in species and community distribution patterns at different spatial scales (Messier et al. 2010), and to quantify the influence of community composition on ecosystem processes and underlying services (Lavorel 2013; Deraison et al. 2015). Finally, traits are used as indicators to assess if the adoption of management strategies has beneficial effects on conservation goals (Vandewalle et al. 2010; Chown 2012).

These examples, together with many others (see Garnier et al. 2016 for a larger synthesis on plants), show that trait-based approaches can be very powerful in generating predictive models. It is therefore not surprising that they are widely used nowadays. However, in our experience, using traits for the first time raises a lot of questions. Now that we have explained what a trait is, and when a trait is functional or not (see Chapter 1), we will move on to other frequently asked questions: How to select the right trait(s) and how many traits should I select? Where can I find reliable trait values? Are the trait values provided in the literature or databases appropriate for my study system or should I measure traits myself? How should I deal with missing trait values? This chapter attempts to answer these questions although, for some of them, we will provide a more complete answer in other chapters of the book as well.

2.1 Which Traits to Select?

Some of the most important methodological decisions to make in trait-based ecology are the selection of particular traits, the number of these traits, and whether traits should be analysed separately or combined. Trait selection is crucial as it will have a strong impact on the outcome of trait-based studies, therefore it should not be taken lightly, and should be well justified. Most often ecologists tend to select the few traits available in the literature, or those easiest to measure. This could lead to poor results and a general scepticism about trait-based approaches.

In general, we highly recommend clearly stating hypotheses and expectations about the causal link between candidate traits and focal environmental gradients, stress factors or ecological processes of interest (Lefcheck et al. 2015). Expectations could also be based on previous observations or patterns tested in other studies. For example, inter-tegula distance in bees (Fig. 2.1) is known to reflect flying distance, with this being related to key functions such as dispersal and foraging (Greenleaf et al. 2007). Therefore we can take this trait into account for studies on pollination.

For guidelines on appropriate trait selection, we suggest adopting a three-step hypotheses-based approach proposed by Brousseau et al. (2018) and shown in Fig. 2.2. This framework, built on the distinction between response and effect traits (Chapter 1), provides some simple steps to follow when selecting traits.

The very first step is to define, as precisely as possible, the stressor and/or ecosystem process under study. This strongly determines the potential functional trait(s) under

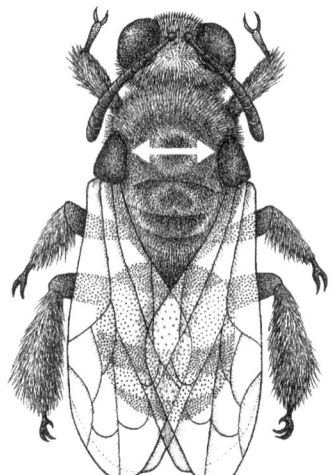

Figure 2.1 The 'inter-tegula distance' in bees is the distance between the points of connection of the wings to the thorax. Below this area is the musculature necessary to move the wings. In general, the bigger the inter-tegula distance, the bigger the muscles and the longer the bee can fly. Figure created by Luís Gustavo Barretto.

2.1 Which Traits to Select?

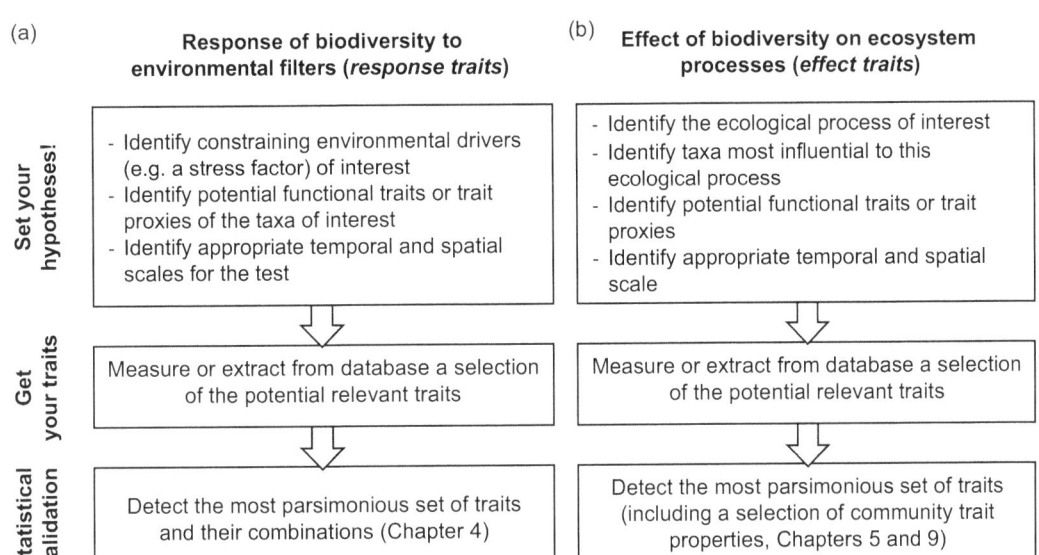

Figure 2.2 A framework showing the steps for selecting appropriate traits depending on your research aim. The hypothesis can be based on either (a) the response of an organism to a given stress (environmental driver) or (b) the effect of a species on some given ecological processes. The hypotheses-based screening shown in this figure implies a detailed ecological and functional knowledge of the focal model system. Based on these hypotheses you can measure, or extract trait information if available, for both traits and species. Once you have the trait information, you can use statistics to detect the most parsimonious number of traits needed to optimize predictions. Based on Brousseau et al. (2018) and Díaz et al. (2007). © 2018 The Authors. Journal of Animal Ecology © 2018 British Ecological Society.

selection that relates to the fitness of individuals. If your hypothesis on the relationship between the candidate trait/s and the stressor/s or ecosystem process/es involved is sharp, then it is easier to make the right decision as to which trait to select and how to interpret the results. As such, the selected response traits (Fig. 2.2 on the left and Chapter 1) should represent a potentially relevant link between the environmental conditions (e.g. humidity level, management practice; the possibilities are endless) and the performance (e.g. growth, survival, fecundity) of individuals exposed to those conditions. Thus, species with a suitable trait value in a given environmental condition, will be more likely to increase their population size through higher performance (McGill et al. 2006). In other words, these species have a strong trait-value–environment linkage.

If you are, instead, interested in testing hypotheses on how species effect ecosystems (Fig. 2.2 on the right), then trait selection should focus on the impact effect traits have on the ecosystem process under consideration (Díaz & Cabido 2001). For instance, if you are interested in plant productivity, then relative growth rate and/or nitrogen-fixing ability are important traits to consider. Again, the more precisely an ecosystem process of interest is defined within its corresponding model system, the easier it will be to select the appropriate effect traits. For example, if plant pollination

Figure 2.3 Two examples of traits involved in plant pollination: tongue length (on the left) and head hairiness (on the right). Figure created by Luís Gustavo Barretto.

is the focal ecosystem process, the hairiness of a bee's head and thorax are key traits reflecting the amount of pollen potentially available for pollination (Stavert et al. 2016), while proboscis length is the key trait reflecting which floral types are likely to be pollinated (Ibanez 2012) (Fig. 2.3).

Similarly, if litter decomposition is the process of interest, then isopods or millipedes are key organisms as they are a strong determinant of litter decomposition (e.g. Petersen & Luxton 1982), and variability in individual 'litter consumption rate' is an appropriate effect trait as the more leaf litter they consume in a given time, the higher the decomposition rate will be (Bílá et al. 2014). In Chapter 10 we expand on these ideas for the selection of traits when we address the interaction between trophic levels in more depth.

The second step in the selection of traits is often, largely, a pragmatic one. Ideally, in the first hypothesis-based step, we come up with a number of very useful traits and a number of potentially interesting ones. However, most of the time, the specific trait information we need is either not available for the species in our sample, for example in the literature or an existing database (see later in this chapter), or is laborious to measure for many species. Also, in some cases we end up with a list of potentially interesting traits which are correlated. So, in this second step while remaining ambitious we need to be realistic and try to get trait information (either measuring or extracting from existing databases) which is both relevant and feasible (see also Chapter 11).

The third step is relating the selected traits to the function (response) or process (effect) of interest. This can be done in different ways, depending on whether the focus is on response traits (Chapter 4) or effect traits (Chapter 9). Most of the time, the selection procedure implies using some sort of step-wise approach which detects the most parsimonious set of traits which serve our purpose. In some cases formulating *a priori* hypotheses on environment–trait or trait–process interactions might not be an easy task, due for example to the lack of information for given taxa, regions or ecosystem functions. When proper hypotheses cannot be identified, then a more explorative approach is surely justified. Following this approach, several potentially relevant traits are selected and their responses to environmental gradients or effects on ecosystem processes is evaluated. The obvious risk with this approach is the possibility of finding spurious or non-mechanistic relationships, where a trait (often a proxy) is detected as important, for example due to a correlation with the trait really affecting fitness or to an effect trait driving ecosystem functions. However, explorative approaches are fully justified for many taxonomic groups when little is known about which traits to consider and how to measure them.

Finally, after we compile a list of potentially important traits for the question being asked, further decisions including the type of traits (proxy or functional; Chapter 1), their unit of expression (continuous, categorical (either ordinal or nominal); Chapter 3) and the final number of traits (single or multiple) to consider (see section 2.2), will have consequences for the functional metrics (Chapter 5) and analytical techniques (Chapters 6 and 7) used, as well as for the interpretation of the results.

2.2 How Many Traits?

Defining, *a priori*, how many traits are needed to answer your ecological question is a tricky task. Exploring the functional role of single traits can be a first step in the search for underlying mechanisms. However, most of the time traits do not act in isolation and single traits do not act independently of each other. The combined effect of multiple traits can be, generally, of two types: (1) syndromes of either positively or negatively correlated traits, and (2) independent effects. While we discuss these issues in Chapter 3 in detail, together with practical ways to face these challenges, some general ideas need to be introduced here.

The functional role of traits might emerge from trait combinations. For example, trait–environmental relationships could be affected by the interaction of independent traits (de Bello et al. 2005), when multiple combinations of traits provide different strategies with similar fitness (Pistón et al. 2019). For example, plant species can adapt to drought by either being drought-tolerant (i.e. being either succulent or having small, thick leaves) (Fig. 2.4) or by being drought-avoidant (annual).

Hence, multiple solutions (i.e. trait combinations) can exist that provide a similar function. In all cases, a multiple-trait approach should provide more informative outcomes compared to single-trait methods (Lefcheck & Duffy 2015). Highlighting the underlying mechanisms in such a multi-trait approach is challenging, as we will discuss several times in this book. However, as a general rule in your study, it is better not to include more traits than the one for which you have a clear hypothesis. Sometimes ecologists use multiple traits hoping that a complex multiple-trait analysis will somehow compensate for the lack of a clear hypothesis (Lefcheck & Duffy 2015). This can often lead to a suboptimal situation, particularly when we include traits that are not 'functional', as is often the case with multi-trait analyses (Chapter 1). For example, computing functional diversity indices based on multi-traits can produce misleading, confusing outcomes when traits that are not of interest are included (Swenson & Enquist

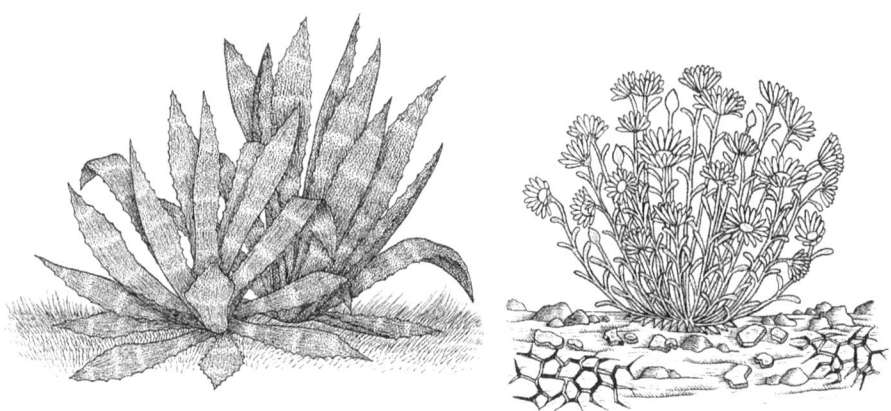

Figure 2.4 Examples of two strategies to cope with drought conditions: drought-tolerant plants, such as succulent species (*Agave americana*, on the left) and drought-avoidant plants, such as annual and ephemeral species growing in the desert regions (*Xylorhiza tortifolia*, on the right). Figure created by Luís Gustavo Barretto.

2009; see Chapter 5 for more details). Similarly, phylogenetic diversity is often advocated as a proxy for multi-trait diversity or when there is no information on trait values (Chapter 8), but this approach may mask the effect of important traits at play (Lepš et al. 2006; Swenson & Enquist 2009). In this context, there are also recent approaches that try to establish which particular traits, among larger sets of traits, are the ones that explain the majority of variation in the organisms' general functional 'strategies', so that one can potentially discard traits that do not contribute to this variation (e.g. for plants see Laughlin 2014a,b; Pistón et al. 2019).

2.3 Where to Get Trait Values From?

After having established which trait(s) are of potential interest (Fig. 2.2), there are two basic options to obtain trait values: (1) a dive into the literature and/or into existing trait databases, or (2) measure the trait(s) yourself. Under which conditions can you reliably use database values and when should you measure traits? How do you measure traits so that they are comparable to other studies? Below we will discuss the strengths and limitations of the different approaches.

2.3.1 Traits from the Literature and Databases

Depending on the focal taxonomic group, the particular traits under investigation and the spatial scale of interest, both published literature and databases can be a rich source of trait information. More and more scientific journals require authors to publish their data in data repositories or as online supplementary material. Also, journals that exclusively publish biological datasets, including trait data, are on the rise (Klimešová et al. 2017). Nowadays extraction of data from these sources can be facilitated by existing tools (e.g. software that extracts tables from pdf documents). It is important to stress that retrieving trait data can be time-consuming when the data you need are scattered over large numbers of papers and books, each of them containing the data for only a single or few species. It can therefore be worthwhile to check if the data you require are already stored in some database, as often trait values are taken from the literature and databases. Several trait databases for different taxonomic groups already exist (see Table 2.1), either available free online, accessible after registration, or by sending a data request to database curators and/or owners (see below). We provide some R tools that may help you to get data from trait databases ('R material Ch2') and discuss issues in more detail below.

An important question is whether traits from literature or database, usually published in connection with a particular study, can be used for your own research. The answer will of course be very dependent on (i) what you need the trait data for, and (ii) how and especially where on the globe these data were obtained and in what environmental context. As a rule of thumb, the more localized and context-dependent your ecological question is, then the less you can rely on the usefulness of database-retrieved trait data, because they will likely be measured for and from other context-dependent conditions (Chapter 11). For example, if we are interested in leaf size changes in response to fertilization, we will generally be interested in both trait variability within a species (Chapter 6) in response to fertilization and data that have been measured in fertilized conditions. The trait values given in existing databases, or published papers, on the contrary are often averaged trait values measured on individuals over multiple populations and habitats (Cordlandwehr et al. 2013). The habitat conditions in which these individuals were measured can be rather different from the conditions for which trait information is missing, i.e. your own study. As such, existing patterns in trait–environment relationships can be masked by using trait values that have been measured in habitats that do not reflect the conditions of your study system (Cordlandwehr et al. 2013; Kazakou et al. 2014). For instance, one might be worried about using a trait value for a cosmopolitan species that was measured in the arctic, but is used in a study that is carried out in a tropical habitat. Moreover, intraspecific trait variability (see Chapters 6 and 11) might be very important and this information is often not available in trait databases. From a statistical standpoint, using traits measured in other conditions can lead to low precision of the estimates (Chapter 11) leading to noise (i.e. unexplained variation) in the analyses. In biological terms, it can fail to capture a particular functional aspect of a focal species or lead to erroneous conclusions regarding underlying mechanisms.

Using traits from a database relies on the hypothesis that the ranking of species in terms of traits should be consistent (Garnier et al. 2001; Mudrák et al. 2019). In other

Table 2.1 List of existing trait databases and other sources of trait values (often species-based ones) of different taxonomic groups (Taxa) and details about the data source and the geographical extent. Note that source URLs might be subject to change (see other trait databases available in Schneider et al. (2019) and in http://scales.ckff.si/scaletool/?menu=6). For more information about when and how to extract traits from databases, see section 2.3.1.

Taxa	Database	Geographical extent	Source	Reference
Plants				
Vascular plants	TRY	Global	www.try-db.org/TryWeb/dp.php	Kattge et al. 2011
Vascular plants	LEDA	NW-Europe	www.uni-oldenburg.de/en/landeco/research/projects/LEDA	Kleyer et al. 2008
Vascular plants	BiolFlor	Central Europe	www.ufz.de/biolflor/index.jsp	Klotz et al. 2002
Vascular plants	TOPIC	Canada	www.nrcan.gc.ca/forests/research-centres/glfc/topic/20303#topic	Aubin et al. 2012
Vascular plants	USDA Plants Database	United States	https://plants.usda.gov/characteristics.html	USDA & NRCS 2018
Vascular plants	D^3 – Dispersal Diaspore Database	Europe	https://doi.org/10.1016/j.ppees.2013.02.001	Hintze et al. 2013
Vascular plants	BROT 2.0: A functional trait database for Mediterranean Basin plants	Mediterranean	https://doi.org/10.6084/m9.figshare.c.3843841.v1	Tavşanoğlu & Pausas 2018
Fine roots	FRED	Global	http://roots.ornl.gov/	Iversen et al. 2017
Clonal growth	CLO-PLA	Central Europe	http://clopla.butbn.cas.cz/	Klimešová et al. 2017
Fungi, Mosses and Lichens				
Fungi	FUNGuild	Global	https://doi.org/10.1016/j.funeco.2015.06.006	Nguyen et al. 2016
Fungi	Fungaltraits aka funtofun	Global	https://github.com/traitecoevo/fungaltraits	Cornwell & Habacuc 2018
Mosses and Lichens	–	Global	Publication	Cornelissen et al. 2007

Invertebrates				
Ants	ANT PROFILER	Europe	www.antprofiler.org/	Bertelsmeier et al. 2013
Ants	GlobalAnts	Global	http://globalants.org/	Parr et al. 2017
Bees	–	Europe	Contact the author	Stuart Robert <spmr@msn.com>
Beetles, True bugs, Orthopterans, Spiders	–	C-Europe	https://www.nature.com/articles/sdata201513	Gossner et al. 2015
Carabid beetles	Carabid.org	Europe	www.carabids.org	Homburg et al. 2014
Carabid beetles, Orthopterans, Dragonflies, Butterflies	Fauna Indicativa	Switzerland	www.wsl.ch/de/publikationen/fauna-indicativa.html	Klaiber et al. 2017
Cavity-nesting hymenopterans	–	Europe	http://scales.ckff.si/scaletool/?menu=6&submenu=3	Budrys et al. 2014
Copepods	A trait database for marine copepods	Global	https://doi.pangaea.de/10.1594/PANGAEA.862968	Brun et al. 2016
Corals	The Coral Trait Database	Global	https://coraltraits.org/	Madin et al. 2016
Freshwater organisms	freshwaterecology.info	Europe	www.freshwaterecology.info/	Schmidt-Kloiber & Hering 2015
Gastropods	Excluded aquatic species and slugs	Europe	Publication	Falkner et al. 2001
Homopterans	–	Europe	Publication	Nickel & Remane 2002
Orthopterans	–	Europe	Contact the author	Frank Dijoze <dziock@htw-dresden.de>
Saproxylic beetles	FRISBEE	France	Contact the author	Bouget et al. 2008; christophe.bouget@irstea.fr
Syrphids	Syrph The Net	Europe	Contact the author	Speight 2014; "Martin Speight" <speightm@gmail.com>
Diverse taxa of soil fauna	BETSI	Europe	http://betsi.cesab.org/	Pey et al. 2014
Diverse taxonomic groups	CRITTER	Canada	www.nrcan.gc.ca/forests/research-centres/glfc/topic/20303#critter	Handa et al. 2017

Table 2.1 (cont.)

Taxa	Database	Geographical extent	Source	Reference
Diverse taxa (microbes, plants, animals)	Global Biotraits Database (thermal responses of physiological and ecological traits)	Global	https://doi.org/10.1890/12-2060.1	Dell et al. 2013
Vertebrates				
Amphibians	AmphiBIO	Global	https://doi.org/10.6084/m9.figshare.4644424	Oliveira et al. 2017
Amphibians	A database of life-history traits of European amphibians	Europe	https://doi.org/10.3897/BDJ.2.e4123	Trochet et al. 2014
Bats	Bat Eco-Interactions	Global	www.batplant.org	info@batplant.org
Birds	Bird trait database	Europe	https://doi.org/10.1111/geb.12127	Pearman et al. 2014
Birds	Avian body size and life history	Global	www.esapubs.org/archive/ecol/E088/096/default.htm	Lislevand et al. 2007
Birds	Functional Traits in 99 Bird Specie	Germany	https://doi.org/10.3390/data2020012	Renner & van Hoesel 2017
Birds, Mammals	EltonTraits 1.0: Species-level foraging attributes	Global	www.esapubs.org/archive/ecol/E095/178/	Wilman et al. 2014
Birds, Mammals, Reptiles	Amniote Database	Global	http://esapubs.org/archive/ecol/E096/269/	Myhrvold et al. 2015
Fishes	FishBase	Global	www.fishbase.org	Froese & Pauly 2018
Mammals	YouTHERIA	Global	www.utheria.org/	Kate Jones or Nick Isaac <youtheria@gmail.com>
Mammals (incl. bats)	Atlantic Mammal Traits	S-America	https://doi.org/10.1002/ecy.2106/suppinfo	Goncalves et al. 2018
Reptiles	Reptile Trait Database	Europe	https://datadryad.org/stash/dataset/doi:10.5061/dryad.hb4ht	Grimm et al. 2014
Diverse taxonomic groups	AnAge		http://genomics.senescence.info/species/	Tacutu et al. 2018

words, if one species has a larger trait value than another species, then this should also apply across different environments. This will clearly be the case for a number of traits locally, but the problem occurs because traits are measured in different sites, so the consistency of ranking is not guaranteed. In general, the more fine-scale your ecological study is, as in the above-mentioned example of the response of traits to fertilization in a particular field, the more likely it is that the traits from the database will not cover all aspects of trait–environment linkage. This is potentially less problematic when the spatial scale of your study is broader, for example the effect of climate change on populations across the American continent (Lamanna et al. 2014), as intraspecific trait variation is less important in such cases (Chapter 6). Some databases (e.g. TRY) will allow the retrieval of intraspecific trait data for most of the traits, so that researchers can (at least potentially) filter the raw data according to the suitability to their own study.

It is important to stress that only a few trait databases contain trait values measured using standardized protocols (see below) and even fewer provide information about the actual source, location, abiotic conditions, trait variability within species, number of individuals measured, or details about the actual protocol used to measure traits. For these reasons, when selecting trait values from the literature or databases, we recommend that for each trait you have enough (metadata) information on the number of specimens measured, their spatial distribution, the environmental conditions at the sampling sites, and the protocol (method) used to measure the traits so that you may determine if the trait values can be used. To summarize, be aware of the limitation of using trait values from the literature and databases, and use them wisely. A strength of using easily available traits is, however, that they can be very useful for exploratory analyses, for instance to determine which traits really matter and then, in your own study, measure these for yourself in a more accurate way.

Another issue that can hamper the use of trait databases is the matter of species taxonomy. Species names change with time due to the revision of species groups, splitting of species and newly discovered species, often leading to a wealth of synonyms. In the ideal case, a database uses a taxonomic 'backbone', i.e. a clearly defined taxonomy that is applied to all species entered into the database. Problems can emerge when researchers bring their own species list to query a given database, following a different taxonomic concept, and therefore species names might not match the ones from the database. Moreover, a mismatch in species names can also occur due to simple misspellings or typos, which are certainly not uncommon when working with large numbers of species. While the awareness of problems stemming from the use of synonyms is not new, addressing this issue in global trait databases can be an enormous undertaking, at least for taxonomically diverse species groups. The global vascular plant list (www.theplantlist.org/) is an attempt to overcome this challenge for plant species names on a global level, and is already implemented by the TRY database as its taxonomical backbone reference. A similar effort has recently been made for amphibians (http://research.amnh.org/vz/herpetology/amphibia/). However, similar backbone reference lists currently do not exist for many other taxa, so zoologists must address this issue with extra care, which often means resolving synonymies and other taxonomical issues manually, requiring a detailed taxonomical knowledge of the study group.

2.3.2 Measuring Traits

Although the use of trait values from databases can be useful for preliminary analyses of environment–trait linkages or trait–process interactions, measuring traits yourself is often the ideal option. Measuring traits is recommended when you study processes acting at the local scale (see above), e.g. when focusing on niche partitioning and competitive exclusion when comparing contrasting habitats (Cordlandwehr et al. 2013). There is, moreover, increasing evidence that intraspecific trait variability plays a significant role in demography and community assembly (see Chapter 6), which requires trait measurements to be done on-site. Specifically when the traits of interest are expected to vary within and between populations it might be wise to measure traits yourself, as the quality of the trait values will generally be higher (Chapter 11). This is particularly true, for instance, when the species of interest shows an important degree of polymorphism, sexual dimorphism or ontogenetic niche shift (Yang & Rudolf 2010; Violle et al. 2012) or if you evaluate the effect of trait plasticity and local adaptation, as shown in Fig. 2.5.

Last but not least, measuring traits yourself might be necessary when there are missing trait values in databases or the literature, as is often the case for functional traits in many animal groups and highly biodiverse ecosystems. If only a few trait values are missing, then imputations using a phylogenetic-based system are also possible (Penone et al. 2014). See Chapter 8 and the corresponding 'R material Ch8').

How many individuals need to be measured? Is there some better or more standardized way to measure a given trait? On the one hand you can try to capture the full variability of a given trait for a given population, or set of populations, in a species. In this case a proportional number of individuals should be measured from different populations, seasons, communities and ecosystems (Pakeman & Quested 2007; de Bello et al. 2011; Violle et al. 2012). The actual number depends on the source of intraspecific variation, such as polymorphism, sexual dimorphism and ontogenetic stages (Yang & Rudolf 2010; Violle et al. 2012), which are particularly important among animals. In general, the minimum number of individuals to be measured for a given species will depend on the amount of variation within a given trait, both within and across populations. The higher the variation, for example, in the case of behavioural traits, the higher the numbers of individuals that must be measured to achieve reliable estimates of the species' mean trait value. Luckily, for some organisms at least, there are standardized protocols for how, and how many, individuals should be measured (see section 2.6). For example, standardized protocols suggest measuring more individuals for traits expected to be more variable (for plants, see appendix 1 in Pérez-Harguindeguy et al. 2013). In Chapter 11 we make some suggestions on how to design your sampling strategies based on standardized protocols.

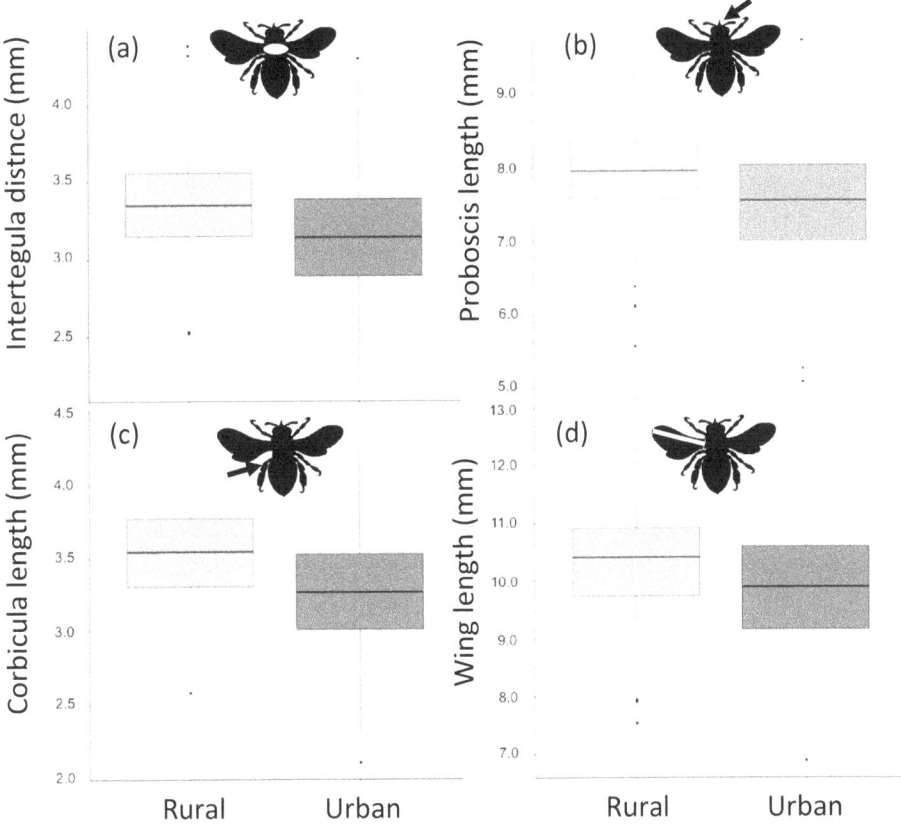

Figure 2.5 The woRkers of the common carder bee (*Bombus pascuorum*) living in rural areas tend to have larger intertegula distance (a), longer proboscis (b), longer corbicula (c) and longer wings (d) than their conspecifics living in urban areas, although there is great variability in these traits within both areas. Taken from Eggenberger et al. (2019), with permission from Wiley. © 2019 The Authors. Journal of Animal Ecology © 2019 British Ecological Society; Published by John Wiley & Sons Ltd.

2.4 How to Express Trait Values?

Traits can be expressed in different ways depending on the type of trait and the way it has been measured, research question, and the spatial scale of the study. For example, longevity can be expressed as a *quantitative* variable (the number of years) or as a *binary* variable (annual, perennial) or a *semi-quantitative* variable (categorical: 1, 2, 3 expressing less than two years, between two and five, and more than five, respectively). Of course the more precise the information, the better the resolution. Additionally, how we code traits will affect results (see, for example, Chapters 3 and 4). When investigating species distribution at large spatial scales, trait values can be coarser compared to studies on plant physiology conducted at a small spatial scale where more precise trait values are needed. More generally, as with any other variable, trait values can be of different types: categorical (nominal or ordinal), continuous, proportional, or circular.

Nominal traits are those referring to names, such as forms, which cannot be expressed on a quantitative scale. Typical examples of nomial/categorical traits are different type of clonal organs in plants, life form, type of diet, plant photosynthetic type (C3, C4, CAM), dispersal vectors etc. Where there are just two categories, one typically refers to *binary* values, basically 'yes/no', such as flight ability (able/not-able) or gender (male/female). *Ordinal* traits have, instead, a clear ordering of the values: for example, when you have a trait, let's say body size, with three categories (small, medium, large) which can be converted meaningfully into the numbers 1, 2, 3. However, for this particular body size example, expressing it as a continuous variable would be best. With ordinal analyses one must be aware that when interpreting the results, even though you can order traits from lowest to highest value, with equal mathematical spacing between adjacent categories, the real spacing between the values may not be the same across all levels. For instance, you might assign tongue length of bees as 1, 2, 3 for short, intermediate and long. While the integer interval is always 1, in reality a long tongue (e.g. *Bombus hortorum* mean ± StDv 12.38 mm ± 1.787) can be up to 15 times longer than a short tongue (e.g. *Hylaeus pictipes* 0.82 mm ± 0.064; own data). *Discrete* traits, in this respect, are traits that can only take on a certain number of values, e.g. when you can count a set of items, such as the number of legs or leaves per branch. If, instead, your trait can have an infinite number of possible values, then it is a *continuous* trait (e.g. body size in mm or mass in g). Contrary to nominal and ordinal traits, which are unit-less, discrete and continuous traits are expressed with units (grams, millimetres, legs, leaves). In general, we recommend avoiding converting continuous traits into discrete trait values, categories or proportions, since you lose biological information, and depending on the type of analyses, also lose statistical power.

In some cases, trait values are expressed as *proportional data* (or percentage), for instance, when assessing a species diet as the composition of different food resources eaten summing up to 1 (i.e. 100%). For example, the diet of the Eurasian jay, *Garrulus glandarius*, is composed (on average) of 60% seeds, 20% arthropods and 20% vertebrates (Bezzel 1985), and this can be expressed either as using three different columns for the three different food items (seeds, arthropods, vertebrates) or assigning the values 0.6, 0.2 and 0.2 respectively (see examples in Chapter 3). To express traits in that way can also be helpful when different individuals of the same species belong to different categories that make up a trait. This is referred to as *fuzzy coding*. We provide more detail and quantitative examples of this in Chapter 3.

Other sets of traits, such as *phenological* traits, are often expressed in number of days, weeks or months when some phonological event generally happens for a species, for example the period in which a plant species is generally flowering. These values should be used with caution since these are circular variables, where, e.g., October (10) cannot be considered 10 times bigger than January (1). Also, it is important to note that when calculating distance-based diversity measures (see Chapter 3), species occurring in January (1) is closer to December (12) than to, e.g., April (4).

For multiple-trait approaches, traits with different units and types of values can be used together, but this requires some caution and normalization. In Chapter 3, we will show how their information can be combined and in Chapter 5 we discuss how to use different traits to compute multi-trait functional diversity.

2.5 Missing Trait Data

Despite your best efforts collecting trait data from multiple sources, you might still miss trait values for some species. These missing values typically are not distributed randomly across your species pool, meaning that there can be an underrepresentation of trait data for rare species, species from a certain clade, or when taking a multiple-trait approach, for certain traits that you have more information on than others. A common practice, for several types of analyses, is to remove species with missing values (NAs) from your data sheet. This practice unfortunately not only reduces sample size, but also introduces bias that can lead to incorrect conclusions (Penone et al. 2014; Borgy et al. 2017). For instance, if the species removed is very abundant, removing it can affect the results of the analyses (Pakeman 2014; Majeková et al. 2016a for some examples).

Another common practice for handling NAs is to use trait values from phylogenetically close species, for example using a trait average for other species of the same genus. A potential solution to this problem involves more sophisticated approaches to impute missing trait values based on shared evolutionary history with species for which these traits are available (Penone et al. 2014; Taugourdeau et al. 2014). This method assumes that phylogenetically similar species will share similar trait values to the one for which we are missing trait information (see Chapter 8 also for corresponding R material). There are different ways to proceed in this direction, mainly depending on the type of trait data and their distribution over the phylogenetic tree, as well as on the functional trait metrics and the spatial scales you want to investigate (Penone et al. 2014; Taugourdeau et al. 2014; Majeková et al. 2016a). Some of these techniques are explained in Chapters 8 and 11. This phylogeny-based imputation of missing trait values may be a reasonable solution for traits that are not too plastic in space or time and that are phylogenetically conserved. Nevertheless, it is important to point out that, depending on the analyses being conducted, imputed traits based on phylogeny might only be spuriously correlated to other factors, thus generating circular relationships. For these reasons, the imputation of missing trait values using phylogenetic relatedness has to be applied with caution. Instead, we encourage trait data collection and, in particular, measuring traits of the species under investigation for yourself, when possible.

2.6 Trait Standards

2.6.1 Why and How to Standardize Trait Measurements?

The potential of traits calls for the availability of high-quality trait data (e.g. McGill et al. 2006). For many species we lack reliable functional trait data, and this calls for an effort to measure these in a comparative manner, by a standardization of trait measurements within and across taxonomic groups. Using standardized protocols we can ensure that comparisons are also meaningful across studies and can minimize sources of noise occurring from the use of trait databases (see above). Trait standardization is indeed one of the most important aspects in comparative functional ecology (Grime et al. 1988), which aims to determine the factors that lead to the observed differences between biomes (Garnier et al. 2016). Therefore, standardization is important for all types of traits, independent of whether they are expressed using continuous or categorical variables. It allows you to compare traits across different types of organisms within biomes or within organisms between different biomes or across large spatial scales.

A coherent, unified trait approach first requires a consensus on the basic set of traits to be considered as a priority, their definition and, most importantly, on how these traits should be measured based on standardized protocols. Currently there are four major handbooks that aim to standardize the measurement of functional traits, detailing the definitions and methods for measuring key traits, applicable for species measurements all over the globe. There is one handbook of protocols for plants by Cornelissen et al. (2003) and further updated and extended by Pérez-Harguindeguy et al. (2013), one for invertebrates by Moretti et al. (2017), one for protists (Altermatt et al. 2015) and, most recently, one for fungi (Dawson et al. 2019). More initiatives for other groups of organisms are underway. These four handbooks propose a series of protocols of standardized measurement techniques, applied both in the field and/or under controlled laboratory conditions depending on the trait and method. Also the recent publications by Garnier et al. (2017) and Schneider et al. (2019) are important contributions to trait standardization and harmonization of concepts and terms across taxonomic groups.

The standardization of trait measurements starts at the moment the specimen is selected for sampling. For example, it is often recommended that fully developed and healthy individuals are measured preferentially, unless there are some special ecological questions under assessment. Quite often the main focus is on mature individuals (but see below), especially if comparisons between species and populations are the goal. It is also essential that the environmental conditions in which the trait is measured are reported. For this, details of the site where the specimen was sampled are recorded and published together with the trait data, such as latitude, longitude and elevation, as well as minimum, mean and maximum temperature and annual precipitation at the time of sampling (see Moretti et al. 2017). When dealing with plants and soil organisms, typically we also reference soil characteristics, such as pH, chemical and physical characteristics, mineral substrate or the surrounding vegetation type (see Garnier et al. 2016).

2.6 Trait Standards

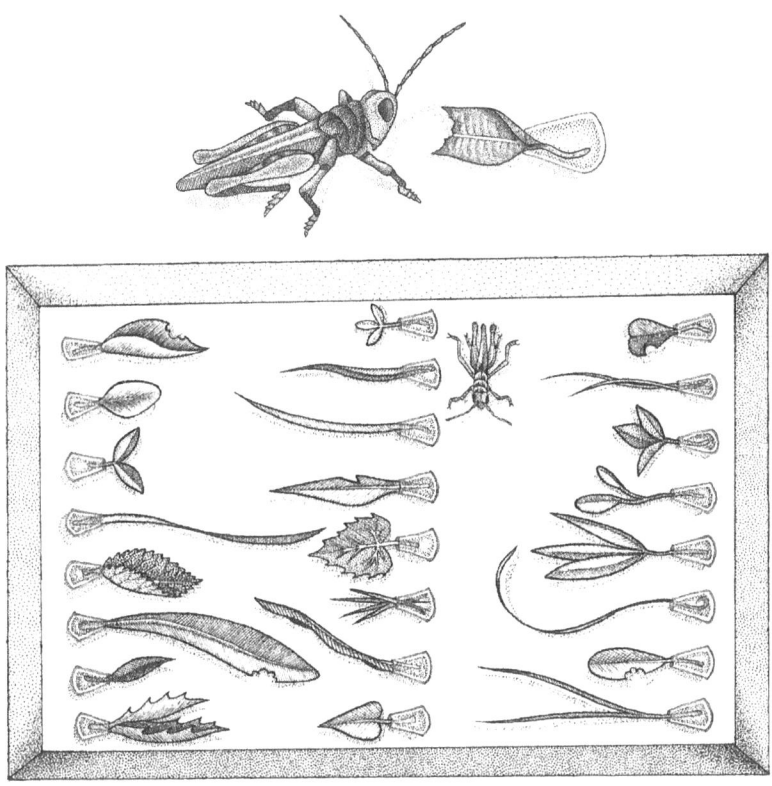

Figure 2.6 Standardized measurement of herbivore consumption rate in a cafeteria experiment, where the feeding preference of a target grasshopper species for different leaf species is tested under control conditions, based on the amount of leaf surface consumed. Figure created by Luís Gustavo Barretto.

In order to achieve true standardization of trait measurements, it is often deemed necessary to measure specimens under the same environmental conditions so that we measure the same 'type' of individuals. For example, feeding preferences of invertebrates are performed in standardized cafeteria experiments, as shown in Fig. 2.6. In plants the suggestion is to select individuals in full-light conditions, allowing for a comparison of individuals and species under similar conditions. For many traits, particularly in animals, this requires an acclimatization period under standardized conditions. This minimizes the effects that local conditions at the original sampling site have on the traits of interest, allowing for a true comparison of trait values between species obtained from different sites (Pérez-Harguindeguy et al. 2013; Moretti et al. 2017). For animals, the developmental stage at the time of measurement, and, if possible, sex and social caste (if applicable), should also be described, again to allow meaningful comparisons. Similarly for plants, the ontogenetic stage (seedling, sapling, mature plant), the time of sampling, location of an individual (i.e. position of leaves on a tree), and other properties are important to consider as they may determine trait values. The exact acclimatization conditions, such as its duration, and the abiotic conditions during the acclimatization, i.e. temperature, relative humidity (RH), light:dark (L:D) cycle, should also be reported together with the trait values. By doing this, the trait variability within species will more tightly reflect genetic rather than environmental effects, and information about intraspecific trait variability can become valuable. Additional information on the standardization of trait measurements, especially for plants, can be found in Chapter 9 of the book by Garnier et al. (2016).

2.6.2 How to Work with Trait Protocols?

Our research questions are changing continuously as plants, animals and microbes are exposed to new environmental stresses. This means that existing trait handbooks might not contain protocols for all the traits we need, so new protocols have to be written and incorporated into handbooks. A trait protocol should provide rationale as to why a certain trait is of importance, and describe how to measure this trait using a standard format aimed to facilitate comparisons among species and biomes. Protocols of the handbooks of plants and animals (Pérez-Harguindeguy et al. 2013; Moretti et al. 2017) are slightly different, but provide a similar structure in four sections: (1) Definition and relevance – providing a formal definition of the trait and a short, non-exhaustive justification why that particular trait is of ecological significance based on its role in responding to stressors and/or effecting trophic interactions or ecosystem processes. This section also describes the main approaches to measure a particular trait. (2) What and how to measure describes the standardized method and provides the units of measurement and, if applicable, mathematical formulas for trait value calculations. (3) Additional notes contains, if available, alternative techniques, which are often more expensive and challenging, and mainly used by more specialized research groups to answer deeper research questions. The rationale of standardized trait protocols is that these methods are applicable all over the globe, using techniques that are affordable by most research groups. This section may also list modifications of the methods for specific taxonomic groups and draws attention to potential caveats and improvements. Finally, (4) a References list gives a number of key papers which are cited in the protocol. We encourage young researchers that are planning to measure traits to adopt these approaches, which will improve the quality of trait databases and facilitate better trait–environment linkages.

Take-Home Messages

- Traits should be chosen based on clear ecological hypotheses with respect to the specific questions being asked. After defining a list of potentially useful traits, we need to act pragmatically as trait information is often either not available for the species in our sample or is laborious to measure for many species (see also Chapter 11).
- Often we need multiple traits to answer most existing ecological questions. There are different types of traits (quantitative, categorical, circular etc.), which can be expressed in different ways.

- Trait information can be extracted from trait databases or measured in the field. Trait databases are great, but measuring traits yourself is usually better, particularly when intraspecific trait variability and local site conditions are deemed important.
- Whether we use trait databases or traits measured in the field, we need to refer to standardized protocols of trait measurements.
- If you provide trait data that you measured yourself, do not forget to provide the corresponding metadata indicating how you performed your measurements, together with the associated environmental conditions.

3 The Ecology of Differences
Groups vs Continuum

The simple truth that organisms differ in their forms, life cycles and life strategies can be witnessed in any garden, meadow or forest. Even in a vegetable patch, the differences between a cabbage and a carrot are spectacular. However, the difference between species can also be subtle (e.g. in species complexes, such as for many fly and moth species only differentiated by their genitalia, or mimetic species, when unrelated species 'converge' in their traits). In this chapter we will learn how to conceptually and mathematically approach such differences between organisms in terms of traits.

While *trait differences* between species are often obvious, we should not forget that they also exist within species. It is enough to observe a room full of students to notice the importance of intraspecific trait variability, in terms of size, eye colour, diet requirements etc. (we present a trait game we usually play with students in the R material accompanying this chapter, 'R material Ch3'). At the same time, the differences within species, generally speaking, are smaller than those between species (see Chapter 6). This is true for a great variety of traits (Westoby et al. 2002; Siefert et al. 2015) and that is why in this chapter we will start by focusing on the dissimilarity between species. However, exceptions to this general pattern are actually not so rare, and many species exhibit great intraspecific trait variability. Even in the lifetime of a single organism, their traits can change dramatically. We will treat these within-species differences in more detail in Chapters 6 and 10.

In this chapter we will first provide an historical perspective on how researchers have tried to approach trait differences between species. To reduce the complexity of biological diversity, biologists have long categorized species into different 'types' based on their traits. This eventually led to the development of 'functional groups', or 'functional types', and the discovery of trait trade-offs between species. We will introduce these concepts first (section 3.1), then we will explain how to express, mathematically, the differences between species for different types of traits (section 3.2). The ancillary R material accompanying this book provides some practical analyses related to these challenges. In the following chapters we will further discuss the implications of between- and within-species differences for local adaptation, coexistence and ecosystem functioning. The title of this chapter was inspired by the study of Cadotte et al. (2013), further discussed in Chapter 8.

3.1 Historical and Conceptual Overview

3.1.1 Functional Groups

Attempting to classify organisms into groups of species with shared morphological features has a long tradition in ecology. As humans, we like to classify things, put them into 'boxes', and classifying organisms is surely a good example of this. Already three centuries before Christ, the Greek philosopher Theophrastus (*c.* 371–*c.* 287 BC), successor of Aristotle, proposed a scheme to group plant species into different 'types': trees, shrubs and herbs. Fast-forward to Alexander von Humboldt (1806) who classified species into groups according to different morphological features. Similarly Eugenius Warming attempted to classify plant life-forms into different groups. His pupil, Christen C. Raunkiær (1934), in disagreement with his mentor, proposed a more successful alternative scheme based on a single trait, the position of perennial buds during the most unfavourable season for plant growth. His scheme, known as the *plant life-form* system, showed a respectable correlation between species distributions and climate (Fig. 4.3). This finding (which we now call environmental filtering; see Chapter 4) highlighted the relationship between species forms and species adaptations (the so-called equivalence between *form and function*). His scheme has also proven useful to develop basic descriptions of plant distribution along disturbance gradients (de Bello et al. 2005). For these reasons, Raunkiær's classification is still widely used today and is the foundation of the field of modern functional ecology, together with Charles Darwin of course (Hortal et al. 2015).

Since Raunkiær's classification many other schemes have tried to allocate organisms into groups, based on specific functional traits. In 1997, the year of the Kyoto Protocol on climate change, a fundamental book called *Plant functional types* (Smith et al. 1997) made a clear call to define groups of organisms which have similar traits and similar responses to the drivers of global change. The authors remarked, 'We have very little information, in many cases none, about how plants will respond in the future. In order to circumvent this problem, and until more information on species accumulates, we reduce the diversity of species to a diversity of functions and structures.' This call, together with the seminal work by Lavorel et al. (1997), has resulted in the development of many different approaches and classification schemes. These approaches are generally based on the following principles: (1) we do not know the ecology of many species and (2) we can identify some specific functional traits that can distinguish species into groups with a comparable 'behaviour' (e.g. environmental preferences, response to global change drivers, effects on the ecosystem or other trophic levels etc.; see below); (3) we can thus predict the behaviour of species based on their traits. With these principles it is thus possible to assess, for example, how the distribution or abundance of poorly studied species might change in the face of environmental change.

As we will see below, and in Chapter 4, there are a variety of tools for categorizing and grouping species based on their traits. However, it is important from the beginning to identify the purpose of the functional groups. On the one hand, as discussed above, we can be interested in defining (i) species that share similar traits related to their response (i.e. adaptation) to environmental factors, leading to similar environmental preferences. These are often called '*functional response groups*'. On the other hand, we might instead be interested in defining (ii) groups of species with similar traits and similar effects on the ecosystem (often called 'functional effect groups'). For more on the distinction between 'functional response groups' and '*functional effect groups*' see also the description of functional response and effect traits in Chapter 1.

Let's see some examples. First, we might be interested in understanding how plants respond to environmental change. If a short life cycle in plants (i.e. annual plants, or therophytes according to Raunkiær) is associated with dry climates, or increased grazing (de Bello et al. 2005), then we can expect that annual plants will 'respond' positively to increased drought conditions under climate change or increased grazing pressure with land-use intensification. As such, we can say that annual plants are a 'functional response group', because they have the same trait (a short life cycle) and have similar environmental preferences, hence they can respond in the same way to environmental change.

Second, we might want to understand what effect plants will have in an ecosystem. We know that nitrogen-fixing plant species (i.e. those with symbiotic bacteria in the nodules of their root systems) can increase nitrogen availability in a field. Here, we would simply distinguish nitrogen-fixing vs non-fixing species and predict that species belonging to these two groups will have very different effects on soil nutrient content (Scherer-Lorenzen et al. 2003). Hence, species with nitrogen-fixing abilities can be grouped into a 'functional effect group', as they have the same trait and comparable effects on soil properties.

In practice when we try to group species into some 'types' we have to keep in mind the purpose of the grouping. Are we interested in detecting a trait (or set of traits) that determines species distribution in relation to given ecological conditions (example above with annuals adapted to dry areas grazing conditions)? This approach clearly requires 'response groups'. Of course, different response groups will be needed depending on the environmental conditions considered. Are we interested, on the contrary, in discovering which trait leads species to have a similar effect on the ecosystem? This clearly narrows down the group to the specific ecosystem function of interest ('effect groups'). This also means that, despite earlier attempts to create them, universal functional groups do not exist. Theophrastus' trees, shrubs and herbs have different traits, but this is not enough to define a functional group. If we create functional response groups, species grouping will depend on if we are interested in adaptations to drought, fire or contaminants, or on the traits which determine environmental 'preferences'. If we create functional effect groups, then we must clearly define the ecological function of interest, be it soil nutrient cycling, water fluxes, or erosion control, and the underlying traits of interest.

3.1.2 Trait Trade-offs and r/K Selection

The concept of trait trade-offs helps to explain why, during speciation and local adaptation, organisms evolve different traits and life strategies. A *trade-off* in biology is when adopting one strategy results in diminishing or losing another. For example, if plants invest in larger seeds they typically produce fewer than plants which produce small seeds (Garnier et al. 2016). Such trade-offs are very common in nature. They result from an 'evolutionary dilemma' in which species have to 'choose' where to invest resources to cope with environmental constraints and competition with other species. In other words, species cannot be good at everything, they need to compromise. Plants, for example, have many options of where to invest their resources: in allocation to defence, to seed production, to seed storage, to root growth, to leaf growth. In reality, no species can be perfect in all aspects, because resources are finite. Adopting a strategy is costly, and results in losing an advantage over other organisms in some other function. Such trade-offs apply also within species. In humans, for example, athletes who specialize in short distance vs long distance tend to have different type of muscles allowing them to run faster for shorter distances or longer but at lower speed.

Among the most well-known schemes based on trade-offs is *r/K selection* (Fig. 3.1). This theory relates to the selection of a combination of traits in an organism with trade-offs between quantity and quality of offspring. At one extreme of the trade-off there is a strategy of increased quantity of offspring at the expense of individual parental investment (r-strategists). At the other there is a strategy of a reduced quantity of offspring

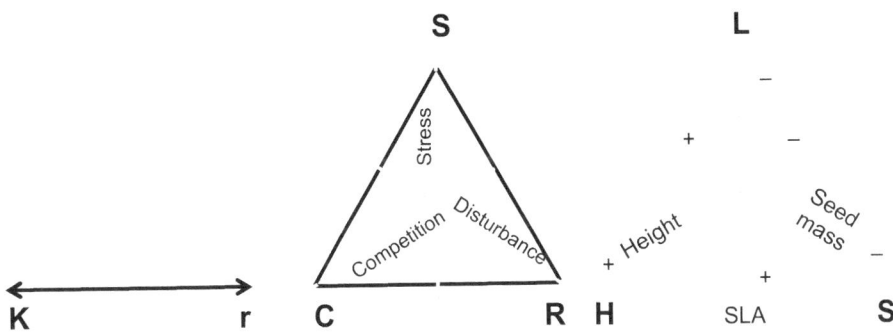

Figure 3.1 Three well-known systems proposed to distinguish species strategies, based on trade-offs between a set of traits: the continuum of r/K strategies, the C-S-R triangle and leaf-height-seed scheme of plant classification.

with a corresponding increased parental investment (K-strategists). The two opposing strategies are expected to promote success in particular environments: more unstable (more disturbed) should benefit from r strategy and more stable from K strategy. The terminology of r/K selection was coined by the ecologists Robert MacArthur and E. O. Wilson (1967) based on their work on island biogeography, and the terminology is drawn from population dynamics models, in which r is growth rate and K is carrying capacity.

The main trade-offs in r/K selection involve several associated traits. The *r-selected species* are those that emphasize high growth rates, typically exploit less-crowded ecological niches, and produce many offspring, which have a relatively low probability of surviving to adulthood. A typical r-species is the dandelion (genus *Taraxacum*). In unstable or unpredictable environments, r selection predominates due to specific traits, i.e. the ability to colonize (highly dispersible seeds), grow and reproduce quickly (high relative growth rate). In such situations, there is little advantage in adaptations that permit successful competition with other organisms, because the environment is likely to change again. Hence, r-species do not have the traits typical of competitive species (such as greater size). Traits that characterize r selection are, therefore, high fecundity, small body size, early maturity onset, short generation time and the ability to disperse widely.

By contrast, *K-selected species* display traits associated with living at densities close to carrying capacity where competition is expected to be intense. Therefore, K-selected species (for example, a tree) typically are strong competitors in crowded conditions, investing more heavily in fewer offspring, which have a relatively high probability of surviving to maturity. In stable or predictable environments K selection predominates, as the ability to compete successfully for limited resources is crucial and populations of K-selected organisms typically are more constant in number and close to the maximum that the environment can support. Traits that are characteristic of K selection include large body size, long life expectancy and the production of fewer offspring, which often require parental investments (or care, for animals). Organisms with K-selected traits include large organisms such as elephants, humans and whales, but also smaller, long-lived organisms such as arctic terns, parrots and eagles.

The r/K dichotomy can be explained, as an analogy, using economic concepts. The r-species use resources more quickly but also release them (return them to the ecosystem) more quickly, and vice versa. K-species invest and conserve resources for a longer time. Clearly trees in ancient forests retain carbon for longer than plants in more disturbed conditions. It is important to stress that, contrary to the approaches purely based on functional groups, the r/K dichotomy can also be expressed as a continuous *spectrum*, where species do not belong completely to one strategy but could have intermediate or mixed strategies. In this sense the focus of the r/K dichotomy is not strictly to create functional groups (see sections 3.1.1 and 3.2.3) but to characterize the prevalence of one or the other type along a continuum of differentiation (like a species that is halfway *r* and halfway *K*). In practice, the existence of trade-offs such as r/K strategies implies differentiation along a set of coordinated traits (e.g. production of a high number of seeds, low investment in defence structure and fast growth for the r strategy). As we will see, it is possible to detect the main axes of differentiation across multiple traits using multivariate analysis, and identify the specific trait trade-offs between species along different environments (see below and 'R material Ch3').

3.1.3 C-S-R, L-H-S and 'Spectra' of Differentiation

One of the most well-known schemes to characterize species ecological strategies is the *C-S-R triangle*, proposed originally by Phil Grime for plants (Grime 1979) and then expanded to other organisms (Grime & Pierce 2012). The scheme builds on the concept of trait trade-offs and borrows ideas from earlier developments, such as the r/K selection. In practice, for pedagogical purposes, we could say that this scheme builds on the r/K dichotomy in which r becomes R (for ruderal species) and the K strategy is split into either a C strategy (competition) or an S strategy (stress; Fig. 3.1). The R strategy corresponds largely to the r strategy described in section 3.1.2, describing species able to cope with unstable and disturbed conditions. The C strategy describes competitive superior species which are able to displace poorer competitors in productive and resource-rich conditions. However, in stressful conditions, i.e. when resources are limited and/or there are unfavourable conditions, like high or low temperatures, toxins etc., the S strategy dominates. Remember we commented above that small, long-lived organisms such as arctic terns could be considered as K strategy. Possibly, in the C-S-R scheme, they would be classified as S, because of their adaptation to stressed arctic conditions. The S strategy comprises organisms with traits associated with a slow growth rate, high investment in defences, and in plants small leaves, often evergreen. In general, these traits lead to a strategy of resource conservation, which can be advantageous when resources are scarce. This strategy should dominate under stressful conditions not as a result of competition, but because of better adaptation to a resource-poor environment.

The triangle proposed by Phil Grime has created a great deal of debate. While a useful framework, different types of disturbances and stressful conditions each favour different types of traits, at least to a certain extent. The adaptations to livestock grazing and fire, for example, are likely not to be completely the same, hence ruderal species in each of these conditions might show different traits (for example, rosette species in grazed conditions and resprouting shrubs in burnt conditions). The concept of competition is also often widely debated and it is assumed to be related to size (greater size implying greater competition) in the C-S-R scheme (Craine 2005). But the scheme undoubtedly has a great *communication power* because it immediately suggests the existence of three simple-to-understand types of organisms, which can be easily used to communicate scientific results to experts and non-experts alike. This is surely its greatest value, in our view. While the C-S-R scheme has been widely applied (Grime & Pierce 2012), originally it required many traits to be measured to define a species' position along the three 'types'. The validation, moreover, was originally applied to the British Flora. This approach has been simplified by later studies such as Pierce et al. (2017) by defining a lower set of traits (mostly leaf traits).

In 1998 Mark Westoby proposed an interesting alternative. The approach is based on a pre-selection of, ideally, the most essential traits to differentiate species.

A minimum set of three specific traits is proposed, with the idea that they should be measured on a great variety of species, across the globe, to characterize their ability to cope with competitive, disturbance and stressful conditions. The scheme, called the *L-H-S scheme*, proposed measuring a leaf trait (L) i.e. specific leaf area (SLA, leaf area divided by dry weight), and the size-related traits plant height (H) and seed mass (S). The aim of the approach was certainly not to limit research to only these three traits, acknowledging that three traits cannot cover all important differences between plants (see also Klimešova et al. 2017), but to allow general inferences to be made on the strategies of species from different communities, regions and continents. The L-H-S scheme is also based on the concept of trade-offs, where trade-offs along these three traits would indicate adaptations to different competitive, disturbance and stress conditions (Fig. 3.1). The L-H-S scheme is easy to apply, as it requires few trait measurements in each species, but as a result of this it is likely imprecise and oversimplified. Furthermore it might be more complicated to communicate to non-experts because it does not translate immediately to some easily understandable concepts of plant types.

Following the C-S-R and L-H-S approaches, many researchers have measured plant traits and extensive trait databases have begun to grow (see Chapter 2). This has allowed analysis of trait correlations and quantification of allometric changes and trait trade-offs (e.g. Díaz et al. 2004; Wright et al. 2004 and recently Díaz et al. 2016). Multivariate analyses, such as principal component analysis (PCA), are very useful to simultaneously compare multiple quantitative traits (for categorical traits the function *dudi.mix*, in the package *ade4* can be used). For example, PCAs have been used to characterize 'trait' differentiation among the main types of whiskies (Wishart 2009). In the whisky PCA based on 12 taste-related traits, the first two axes explained 50% of trait variability: the first axis distinguishing winey vs flavour, and the second axis distinguishing delicate vs rich intensity. This meant that the 12 traits, i.e. a 12-dimensional space, could be boiled down to an easy-to-interpret two-dimensional space (intensity and flavour), reflecting a great deal of variability between whiskies. Each of these two axes reflects a *'spectrum'* of differentiation along multiple 'traits'.

Hence we can use multivariate approaches such as PCAs to quantitatively explore which major trait trade-offs exist between organisms (Fig. 3.2). The PCA approach was used as evidence to support the emergence of C-S-R strategies (Grime & Pierce 2012), with some mixed results. Certainly trait-driven analyses can tell us which main trade-offs occur in nature and allow us to create hypothesis-driven groups of species. For instance, Díaz et al. (2004) and Wright et al. (2004) detected a fundamental trade-off between species in terms of traits associated with a rapid acquisition of resources (i.e. live fast and die young) and conservation of resources (the opposite, grow slowly but steadily), the so-called *'leaf economic spectrum'*. Díaz et al. (2004) specifically found that the main difference between a large set of plant species from three regions in the world was in terms of leaf traits, irrespective of the region where the species was sampled (Fig. 3.2). Garnier et al. (2016) explain well the mechanisms behind such trade-offs, echoing the economic concept of resource investment already described in the r/K dualism. The spectrum runs from plants showing a high rate of photosynthesis

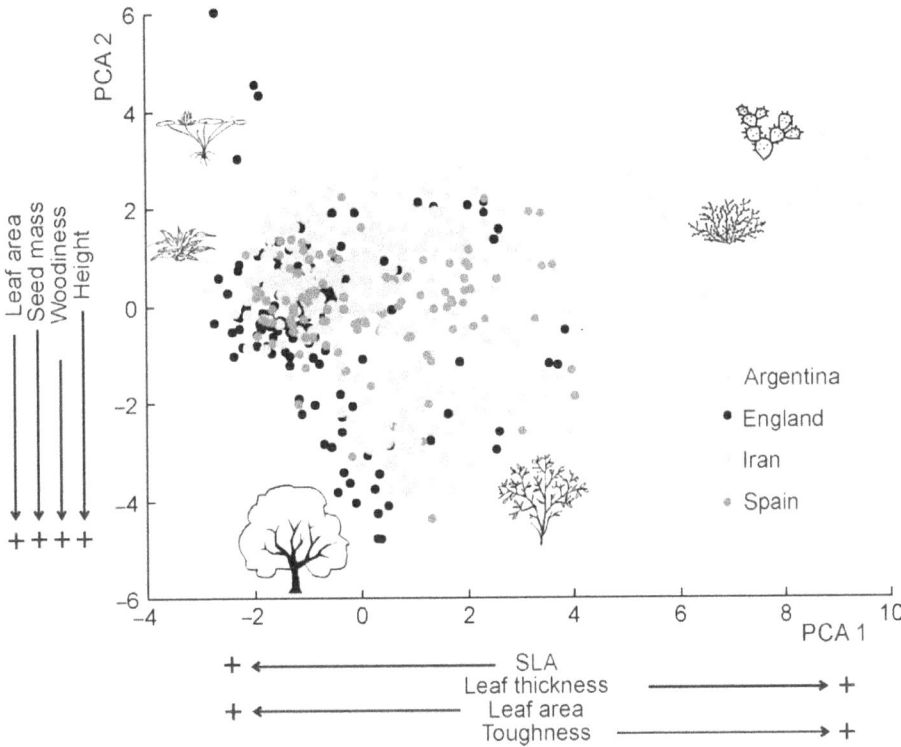

Figure 3.2 A PCA for traits of species sampled in four regions across the world. The traits indicated beside the two first PCA axes reflect the trait better correlated to the species scores on the axes (hence the traits being more important in species trait trade-offs) and in which direction the correlation was found. Taken from Díaz et al. (2004), with permission from Wiley. Published by Wiley; © 2004 IAVS – the International Association of Vegetation Science.

and a rapid return on investment in terms of mineral nutrients and leaf dry matter (acquisitive syndrome), without large investment in defence structure, to others characterized by low photosynthetic rates, with a much slower return on investment (conservative syndrome). This conservation strategy is also often associated with well-defended tissues against herbivores and slower afterlife effects on decomposition (Cornelissen et al. 1999; Díaz et al. 2004).

Of course, the main trade-offs detected among organisms will depend on which trait data are available and the type of species considered. For example, Díaz et al. (2016) found a 'global spectrum of plant form and function', which generally shows multiple traits trade-off between species in size-related traits, particularly on the first PCA axis, and then leaf-related traits on the second axis. Maybe not surprisingly the greatest differentiation was found between woody and herbaceous species, which generally made up two distinct clusters of species. As we know, Theophrastus reached a similar conclusion, a couple of years before (see above). Without any doubt, the detection of useful trait trade-offs, traits spectra and associated differentiation is an open task for a variety of species and organism types.

3.2 From Theory to Numbers

In the previous sections we discussed several conceptual approaches to characterize, understand and interpret the trait differences between species. In this section we will provide some tools and techniques, and some helpful tricks, to achieve these goals. The majority of these techniques begin with a type of data called a *'species × trait matrix'*. So, we will start by introducing this type of matrix, which can include different types of traits.

3.2.1 The Species × Trait Matrix

This type of data (see a basic example in Table 3.1) is crucial for many analyses in functional ecology and will be treated several times in this book. It generally includes information about different traits (columns) of multiple species (rows). In general, one line corresponds to one species (but see other options below and in Chapter 6), and one column corresponds to a particular trait (although sometimes we need more columns to define a single trait). In Table 3.1 we see a simple example of one such matrix. In the example we have data from seven species (species 1, species 2 etc.) and information about six of their traits (size, weight, life cycle, carnivory, growth form type, colour, and flowering period).

The first thing to observe in this type of table is that the information about functional traits can include different types of variables (Chapter 2). Two traits (size and weight) are defined by quantitative, continuous variables, each of them in different units. The trait 'life cycle' can be generally interpreted using a semi-quantitative (*ordinal*) variable in which a continuous scale is transformed into discrete classes (for example 0 = short, 1 = intermediate; 2 = long). Of course, here users can also include more levels within the scale (for example 0, 0.5, 1, 1.5 etc.) and the minimum and maximum are generally arbitrary. For these *quantitative* or semi-quantitative traits, very often the information in these columns reports some *average value* of each species or the most frequently found value. In this case, each column represents a single trait.

Table 3.1 A common type of a 'species × trait matrix', with a row for each different species and columns providing information on their different traits.

Species	Size	Weight	Life cycle	Carnivory	Type	Colour – type A	Colour – type B	Colour – type B	Flowering start	Flowering end
species 1	10	0.2	0	1	grass	1	0	0	120	180
species 2	20	0.3	1	1	forb	0	1	0	120	240
species 3	30	0.4	0	0	legume	0.5	0	0.5	200	220
species 4	40	0.5	1	1	grass	0	0	1	200	300
species 5	50	0.6	2	0	forb	0.2	0.3	0.5	100	300
species 6	60	0.7	0	1	legume	0	1	0	20	50
species 7	70	0.8	2	0	legume	1	0	0	300	20

Two other traits (carnivory and 'type') are *categorical* traits, so that each species belongs to a different category. Carnivory has only two 'levels', i.e. two categories (yes = 1, no = 0), so can be treated as a 'binary' trait. Notice that here users could also consider a case of the trait having some intermediate value, for example equal to 0.5, possibly indicating that the species is partially carnivorous. If the frequency of carnivory in the species is known, the value can even be set as 0.2 for example, indicating that the species is only carnivorous in 20% of cases (see below also for fuzzy coding).

The second categorical trait ('type') has three levels. In this case we cannot use the strategy we used for coding 'life cycle', because we cannot assume that there is a meaningful order among the three categories. Each category is unique, there is no intermediate category, there is nothing better, greater and so on. They are three mutually exclusive categories and cannot be put on a directional scale. So, the best that we can do is to use 'labels' (or *'factors'* in R terminology) to define the trait categories to which each species belongs.

So far, the traits considered are described only by a single column. For each of these traits we generally include either the most frequent value, for quantitative traits, or the most frequent type (for the categorical traits). In some cases, though, we might need to use *more columns* to provide information on a single trait. This is the case, for example, of a categorical trait which cannot be treated as semi-quantitative and when species can belong to more than one category. A typical example is colour, or food resource, when there are multiple categories and species can be composed of or utilise more than one. For instance, colours are always composed of three main 'types', the three primary colours. We can say that each species exhibits, to a different extent, all three types of primary colours. Each species will have a unique combination of the three colours. Similarly, in the C-S-R scheme, species could be partially C and partially S, so they could partially belong to two types.

Another example is diet. This is the case, for example, when the distinction between carnivorous and not carnivorous is not enough, and there are more subtle differences in food type, or prey (e.g. a generalist diet of fish, insects etc.). In this case, each of the categories within the trait has a single column. Then each species can be assigned to one type (e.g. species 1 belongs 100% to the type A in the trait colour) or to different types (e.g. species 3 belongs 50% to type A and 50% to type C; species 5 belongs 20% to type A, 30% to type B and 50% to type C). This type of trait coding is generally called *fuzzy coding* and it applied to a type of information called a *dummy variable*. The dummy variable is a categorical variable represented by multiple columns, in which each column represents a category of that variable. Notice that the values across the different columns of the dummy variable must sum 1 (i.e. 100%).

The last type of trait we consider here is the '*circular*' type of trait. This is the case of traits connected with angles (for example of leaf or bones position) or any phenotypic event, like flowering period. In these cases, it is impossible to use purely quantitative scales because values are placed in a circular scale. For example, we could say that species 8 in Table 3.1 starts flowering 300 days after 1 January in one year and stops flowering 20 days after 1 January in the following year. In these cases, as we will see below, computing dissimilarity between species cannot follow a simple quantitative scale, because values are circular. Moreover, we could be interested in the overlap of flowering time (see also Chapter 6), so we need both the beginning and ending of flowering to determine the dissimilarity between species, therefore two columns are required.

3.2.2 Computing the Dissimilarity between Species

After having introduced different types of traits and their coding into a 'species × trait matrix' we are almost ready to estimate the differences between species. Most analyses that we will introduce in this book, including in this chapter, require quantifying the differences between the whole set of *pairs of species* in an assemblage. In other words, quantifying how dissimilar species are in terms of a single trait or, more often, multiple traits. Existing R functions appear to solve this without difficultiy (but appearances can be deceiving!), by using a matrix like the one in Table 3.1. However, there are several potential pitfalls in the calculations. Here we show some cases, often unacknowledged in the literature, that can result in unwanted patterns in our results.

Starting from a 'species × trait matrix' (Table 3.1), we can estimate functional dissimilarity between pairs of species in different ways; the most common are summarized in Fig. 3.3, for three of the traits described in Table 3.1. These different methods depend to a large extent on the type of traits considered.

Let's start by computing dissimilarity for each trait separately, considering first a *binary* trait, e.g. carnivorous vs not carnivorous species (1 and 0 respectively in trait 'carnivory'). The dissimilarity between two species is, in this case, simply given by whether they share the same carnivory level/category or not. For example, species 1 and 2 are both carnivorous so that the dissimilarity between them is zero, i.e. $d_{1,2} = 0$ (where $d_{1,2}$ indicates the dissimilarity between species 1 and 2). This can be computed simply as the difference of the trait value of species 1 (which is 1) and species 2 (also 1), as $|1 - 1| = 0$ (notice that we express this as absolute difference because dissimilarities cannot be negative). On the contrary, species 1 and 3 are completely different: one is carnivorous and the other is not. In that case $|1 - 0| = 1$, so that $d_{1,3} = 1$, which implies that the two species are 100% different between them.

Computing dissimilarity produces an object composed by two specular/symmetric *triangular matrices* (Fig. 3.3). Each of the triangles provides a dissimilarity value for each pair of species in the dataset, and the two triangles contain the same information (so that many R functions just work with one of the two triangles). Now notice that by coding the trait 'carnivory' between 0 and 1 (i.e. no vs yes), all the dissimilarities between species are constrained between 0 and 1. This is quite conveniently indicating that species are either equal between them (as for example $d_{1,2} = 0$) or completely different (as for $d_{1,3} = 1$). It is important to notice that the '*diagonal*' of this object is, most frequently, assumed to be equal to zero. This means that in most approaches the dissimilarity of one species to itself is equal to zero, meaning that a species is equal (i.e. not different) to itself.

Let's now consider a *quantitative* trait, body size. Intuitively, the dissimilarity between one student 1.80 m tall and another 1.70 m tall is 10 cm. Similarly, in

3.2 From Theory to Numbers

	Body size	Carnivory	Col-red	Col-yellow	Col-blue
species 1	10	1	1	0	0
species 2	20	1	0	1	0
species 3	30	0	0.5	0	0.5
species 4	40	1	0	0	1
species 5	50	0	0.2	0.3	0.5
species 6	NA	1	0	1	0
species 7	70	0	1	0	0

'Species × traits' matrix

Body size
Quantitative

	sp1	sp2	sp3	sp4	sp5	sp6	sp7
sp1	0						
sp2	10	0					
sp3	20	10	0				
sp4	30	20	10	0			
sp5	40	30	20	10	0		
sp6	NA	NA	NA	NA	NA	0	
sp7	60	50	40	30	20	NA	0

standardization

	sp1	sp2	sp3	sp4	sp5	sp6	sp7
sp1	0.00						
sp2	0.17	0.00					
sp3	0.33	0.17	0.00				
sp4	0.50	0.33	0.17	0.00			
sp5	0.67	0.50	0.33	0.17	0.00		
sp6	NA	NA	NA	NA	NA	0.00	
sp7	1.00	0.83	0.67	0.50	0.33	0.17	0.00

Carnivory
Binary

	sp1	sp2	sp3	sp4	sp5	sp6	sp7
sp1	0						
sp2	0	0					
sp3	1	1	0				
sp4	0	0	1	0			
sp5	1	1	0	1	0		
sp6	0	0	1	0	1	0	
sp7	1	1	0	1	0	1	0

Colour
Fuzzy coded dummy variable

	sp1	sp2	sp3	sp4	sp5	sp6	sp7
sp1	0.0						
sp2	1.0	0.0					
sp3	0.5	1.0	0.0				
sp4	1.0	1.0	0.5	0.0			
sp5	0.8	0.7	0.3	0.5	0.0		
sp6	1.0	0.0	1.0	1.0	0.7	0.0	
sp7	0.0	1.0	0.5	1.0	0.8	1.0	0.0

Average

	sp1	sp2	sp3	sp4	sp5	sp6	sp7
sp1	0.00						
sp2	0.39	0.00					
sp3	0.61	0.72	0.00				
sp4	0.50	0.44	0.55	0.00			
sp5	0.82	0.73	0.21	0.55	0.00		
sp6	0.50	0.00	1.00	0.50	0.85	0.00	
sp7	0.67	0.94	0.39	0.83	0.37	0.67	0.00

Figure 3.3 Input and output data when calculating the dissimilarity between species using, in this case, Gower distance based on three different traits. See the text for more details.

Fig. 3.3 the dissimilarity between species in terms of body size is, to start with, simply the difference in their observed trait values. For example, between species 1 and 2 $d_{1,2} = |10-20| = 10$; between species 1 and 3 $d_{1,3} = |10-30| = 20$, and so on. However, a question arises: are these values big or small? How can we compare these values of dissimilarity with results from other traits, for example 'carnivory' which has a maximum dissimilarity of 1? How can we compare this trait to another trait expressed in another unit, for example grams? To allow such comparison all traits need to be *scaled* in similar ways. To do so, dissimilarities are most often expressed on a 0–1 scale, as in the case of 'carnivory', i.e. no difference (0) and maximum difference (1). To do this for quantitative traits we need to define the maximum possible dissimilarity, which will correspond to $d = 1$.

Using the *Gower distance* is the most common way to standardize dissimilarities for different traits (Gower 1971; Botta-Dukat 2005; Pavoine et al. 2009). To standardize quantitative traits, it uses the highest dissimilarity value in the dataset. In our case, the maximum difference in body size is given by the difference between the biggest and the smallest species, i.e. $|70-10| = 60$ (Fig. 3.3). Then, we can simply divide all the dissimilarities in the original scale by the maximum dissimilarity value in the dataset, i.e. $d_{1,2} = |10-20|/60 = 0.17$, and $d_{1,3} = |10-30|/60 = 0.33$ and so on. By doing this, the maximum dissimilarity between species is now 1.

Other approaches are certainly possible, of course. For example, quantitative trait values could be standardized before computing the dissimilarity between species, using the function *'scale'* in R. This centres and standardizes each column in a species × trait matrix, meaning that the mean trait value (across all species) is subtracted from each trait value; then the mean value equals 0 and becomes the 'centre'. These centred values are then rescaled by dividing by the standard deviation of the trait values. With this approach, traits are expressed in a common scale (standard deviations from the mean), and can then be easily compared across quantitative traits. The problem is that this approach does not work on categorical traits, so quantitative and categorical traits would still not be comparable. As such, scaling the dissimilarity between 0 and 1 is always necessary when we have different types of traits (mainly quantitative and categorical traits). As a matter of fact, using Gower distance is also often recommended.

As in the case of the binary traits, Gower distance for quantitative traits represents the relative difference between two species compared to the maximum possible value. For example $d_{1,3} = 0.33$ implies that species 1 and 3 differ by one-third of the possible maximum dissimilarity (which is 60 cm). This idea is advantageous in that we can compare among traits, regardless of the unit (e.g. metres vs grams). Similarly, we can compare quantitative traits to other type of traits ($d_{1,2} = 0.17$ for height and $d_{1,2} = 1$ for carnivory). This property makes it possible to combine these two dissimilarities. Using Gower dissimilarity, this is simply expressed as an *average* of the two dissimilarities, using the arithmetic mean [$(0.17 + 1)/2 = 0.58$] or the geometric mean (Euclidean distance; Pavoine et al. 2009) [$\sqrt{(0.17)^2 + (1)^2} = 1.01$]. The latter approach can be rescaled between 0 and 1 when needed.

Let's now consider the *fuzzy coded dummy variable* trait, species colour, that we introduced above. How can we transform a trait with multiple, unordered, levels into a 'quantitative trait'? Is red bigger than blue? In some cases we could use a quantitative approximation, such as the concentration of a pigment, to define the colour only in one column. But on most occasions we have trait information expressed over multiple levels and it is difficult to determine what is smaller or bigger. In the case of the trait colour, we could say that it has three main levels (red, blue, yellow, i.e. the three primary colours). Then the trait can be summarized by three columns in the species × trait matrix, one for each primary colour. In the columns, we can then add 1 to indicate the level to which a species belongs. For example, species 1 is red, then we will add 1 to the column referring to the red colour and 0 to the other columns. For each row, the sum must be equal to 1, because adding 1 into a column would reflect that the species is 100% that colour (as species 1 is 100% red).

Using dummy variables makes it possible to account for one species having an intermediate value among the different levels of traits. For example, a species of orange colour could be codified as half red and half yellow. We could then add 0.5 to the red column and 0.5 to the yellow one (i.e. the species is 50% red and 50% yellow, which produces orange). If the species is violet then it would be half blue and half red, like species 3 in the example in Fig. 3.3. These intermediate cases are quite frequent in nature. In fact, this strategy allows us to account for many types of plasticity. For instance, in the case of diet, we could account for intraspecific variation in food resources, where some individuals fall into one category while others belong to another. As we saw above, this fuzzy coding approach can also be used for binary traits; for example, some plant species are sometimes annual, sometimes perennial (or in the case of carnivory mentioned above, a species could be both carnivorous or not). This can be solved by using one binary trait going from 0 to 1 but also considering intermediate values, like 0.5 for species that are equally common in one form or the other. Imagine a hypothetical species which is sometimes carnivorous and sometimes not; we would code this species as 0.5 for the trait 'carnivory'. Then, the dissimilarity of this species to purely carnivore or not-carnivore species (either having 0 or 1 in 'carnivory'), would be $|1–0.5|$ or $|0–0.5|$, i.e. 0.5. This value of 0.5 implies that the species are 50% dissimilar.

Let's now see how dissimilarity should be computed when using dummy variables with more than two levels. How much is the colour dissimilarity between species 1 and 2? And between species 1 and 3? And between species 1 and 5? And between species 3 and 5? Let's see one possible solution. First let's assume, for the sake of the example, that yellow and red colours are completely dissimilar. This is an assumption we follow implicitly by using the dummy variables in Fig. 3.3. Then $d_{1,2} = 1$, because each species 'belongs' to different categories of the trait (they are included as 1 in the red and 1 in the yellow columns, respectively, i.e. they are 100% red and 100% yellow, respectively). On the other hand, $d_{1,3} = 0.5$ because species 3 is partially red (50%). Similarly, $d_{1,5} = 0.8$ because species 1 and 5 overlap only 20% of the time. Finally, $d_{3,5} = 0.3$ because both species are red and yellow but species 5 is also 30% yellow.

There are several possible ways to compute this type of distance in R, but we warn users that the mainstream algorithm (*gowdis*) does not solve this situation properly. As far as we are aware, the function *dist.ktab* in *ade4*; provides a solution, albeit slightly complicated. In 'R material Ch3' we show how to resolve this issue by hand, and how to combine the three types of traits together following Fig. 3.3 and other examples. For example, applying the commonly used function *gowdis* to the 'species × trait' matrix in Fig. 3.3 will not result in obtaining these expected dissimilarity values, because the function will not 'understand' that the three columns referring to species colour are actually reflecting the same trait information. Therefore, when combining the three traits to compute a combined dissimilarity, the function will actually 'think' that there are five traits, because there are five columns in the 'species × trait' matrix. When computing the average across the dissimilarities for the single trait it will thus give 1/5 of the weight to each column. By doing this it will wrongly weight the quantitative trait as 1/5, while it should be 1/3 (there are three traits only!) and weight the trait colour as 3/5 instead of 1/3. Of course, the function *gowdis* allows the user to change the weight of each trait. But unfortunately, in this case it is simply not possible to use this option of manually changing the weight of the columns, or at least we are not aware of a clear method to do so. Instead, in such cases you need to compute dissimilarities for each of the three traits individually and then average these dissimilarities (Fig. 3.3). This can be done in different ways using R (see 'R material Ch3'). Alternatively, you can apply our function *trova* (de Bello et al. 2013a) which can deal with this type of trait, as well as the above-mentioned 'circular' phenological traits that depend on the start and end of flowering (see 'R material Ch3'). The new function *gawdis* can be used as well.

Another important scenario illustrated by the example in Fig. 3.3 is the case of *missing trait values*. In this case, for the quantitative trait the dissimilarity between species 6 and the other species cannot be computed (i.e. '*NA*'). At the same time, the good news is that overall dissimilarity can be still computed using the other traits when, and only when, there is at least one trait value available for species 6, and the species to which species 6 is compared. The resulting dissimilarity for species 6 is simply the average dissimilarity for those traits without NA. In the case of the example in Fig. 3.3 this would correspond to the average dissimilarity for carnivory and for colour only, excluding body size (because there is no data for this trait). Then, we can still use the dissimilarity results from other traits and discard, for the comparisons with species 6, the missing information about body size. For example, the distance between species 1 and 6, based on both colour and carnivory, would be $d_{1,6} = 0.5$, since dissimilarity based on carnivory is equal to 0 and dissimilarity based on colour is equal to 1, and so on. Of course, we can also try to avoid NAs, for example by assigning some estimation of the possible trait value based on an educated guess or quantitative estimations based on species phylogeny. We will treat the filling of NA values in Chapters 8 and 11; however, we also generally recommend (a) avoiding NAs as much as possible and (b) excluding a species, particularly if not very abundant, when it has too many missing values for most traits.

3.2.3 Overlooked Issues with the Gower Distance

The idea that we can express dissimilarity using different traits is very appealing biologically because the general expectation is that species differ in more than one trait (Petchey & Gaston 2002). However, combining different traits can pose some problems, which are often not addressed in the literature (including the issue with the fuzzy coding already described in section 3.2.2).

The first issue is that traits could be correlated. We already saw that many trait trade-offs include a variety of correlated traits. For example, in Table 3.1 there are two quantitative traits that would be clearly correlated. Even biologically they are probably providing similar information, e.g. weight and size generally provide the same information about the 'dimension' of a species. If we treat these traits independently then it means that that 'dimension' will have a greater weight on the combined dissimilarity.

In the case of the L-H-S scheme described in section 3.1.3, it is often the case that ecologists measure several leaf traits together (also because it is simple to do). Imagine a case where you have data for Height, Seed Mass and five Leaf traits (a total of seven traits), with the leaf traits being intercorrelated. If we 'blindly' apply the Gower distance, each of the seven traits will have the same weight on the combined dissimilarity (i.e. 1/7), because Gower distance is doing an average of the dissimilarity computed for each single trait (see section 3.2.2). This means that, altogether, the weight of leaf traits will be 5/7, which is a bit unfair to the other type of traits. Here, we have two solutions. The first is that we can compute leaf trait dissimilarity separately from the other traits first (so only using the five leaf traits), while also independently computing dissimilarity for height and seed mass. Then, the three dissimilarity matrices can be averaged (in a similar way as for the three traits in Fig. 3.3).

The other approach, as proposed by Villéger et al. (2008), is to use multivariate analyses to account for the correlation between traits. If we have only quantitative traits, without missing values, we can simply use PCAs. When we do this, for example in the case of Díaz et al. (2016), we can conclude that the first couple of PCA axes already reflect a great deal of the differentiation between organisms. In this case we can use the position ('score') of species on these axes to estimate the dissimilarity between species (for example by computing a Euclidean distance based on the multivariate coordinates). For example, in the case of Díaz et al. (2016) we could conclude that the first axis reflects a combination of correlated traits related to plant size and woodiness. The second axis reflects the so-called 'leaf economic spectrum' described above, with a combination of multiple leaf traits. If we use species scores on these PCA axes we could summarize the main ecological differences between species, with all traits represented fairly. However, if we use too many correlated traits the first PCA axis will reflect these traits, even if biologically they are not the most important.

Another option, when we have both categorical and quantitative traits, and some NAs, is to use PCoAs. In this case we need to compute trait dissimilarity first, using the 'species × trait' matrix and calculating Gower distance before applying a PCoA. Then we can again use, as in the case of the PCA, the species scores on the PCoA axes, to summarize the main differences between species. In practice, the species scores are used as 'traits' and the Gower distance is computed on these scores. The function '*dbFD*' in the package '*FD*' mostly uses the PCoA approach to compute functional dissimilarity between species.

Using a PCA combining quantitative traits is certainly an elegant solution. However, there is generally an unresolved issue when combining quantitative and categorical traits into a dissimilarity matrix (which will also affect any possible PCoA). Imagine a case, summarized in 'R material Ch3', where we have 20 species and two traits, one quantitative trait (where values are randomly sorted between 1 and 100) and a binary trait with randomly attributed 0 and 1 values. If we compute the dissimilarity for each trait and then combine these two dissimilarities with a simple average (or even with Euclidean distance), the resulting dissimilarity will be affected much more by the binary trait than by the quantitative (see the corresponding R material for details). The reason for this is simply that the dissimilarities based on the binary trait are 'extreme' values (i.e. there are only 0s and 1s). In our example, around 50% of the distances between species pairs will be equal to 1, and 50% will be equal to 0 (Fig. 3.4). In the quantitative trait, however, only one species pair will have $d = 1$, the one which had the lowest and highest trait values respectively (in Fig. 3.3 this is $d_{1,7}$ as we discussed above). All other values will be smaller than 1 and, in general, increasing the number of species will produce lower mean dissimilarity values. Thus, the two dissimilarities will have a

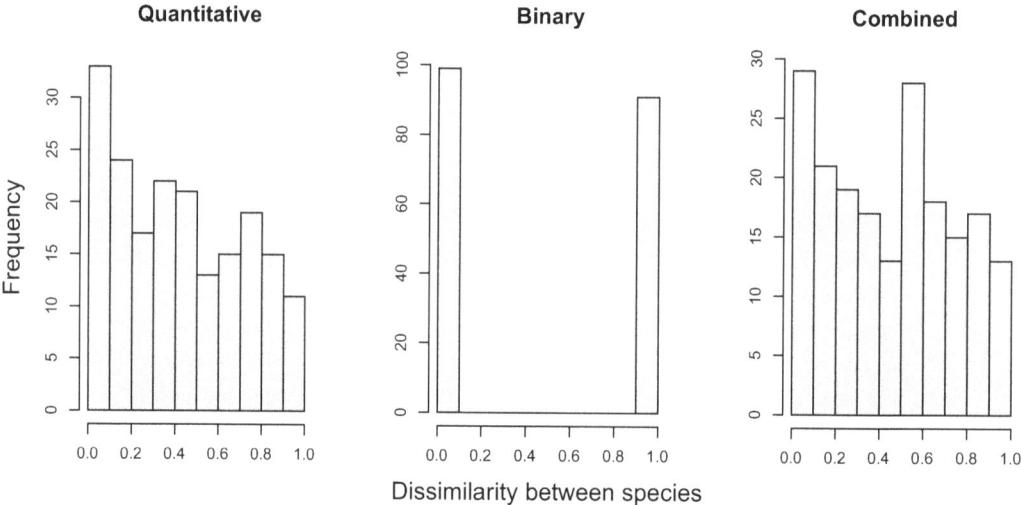

Figure 3.4 Distribution of the values of dissimilarity between 20 species following a simple R simulation with a quantitative and a categorical trait with two levels (binary trait). The dissimilarity is expressed for a quantitative trait, one binary and their combination. The contribution of the binary trait to the combined dissimilarity is much greater (approximately twice as big in terms of correlation, see text).

different distribution (see histograms in Fig. 3.4). This means that when averaging the dissimilarities over the two traits, the binary trait has a much greater effect on the mean than the quantitative trait. This issue was well discussed by Pavoine et al. (2009) but was basically overlooked in the literature.

The contribution of each single trait to the combined dissimilarity can be computed in simple way, for example in the function '*k.dist.cor*' in *ade4*. This is based on the correlations between the distances obtained for each trait and the global distances obtained by mixing all the traits. This can reflect mathematically the contribution of each trait to the global distances. When computing this index, one fair approach could be to produce a dissimilarity matrix which equally represents the different types of traits. One possible solution is to down-weight the contribution of the binary trait, or to increase the weight of the quantitative trait, for example by using a weighted average (see 'comb.dissim2<-(dissim.quant*0.7+dissim.bin*0.3)' in the R material). This option is available in several of the functions computing the Gower distance between species, the most used possibly being *gowdis* in the package *FD*. Notice that this down-weight should be applied before computing a PCoA, if using this approach.

A similar issue is when traits are not '*normally*' distributed (Pavoine et al. 2009). In traits such as height, leaf area or seed mass, the biological differences between species will be better characterized on a log scale (Westoby 1998). If we forget to log-transform such traits, the dissimilarity between species will be very much skewed, e.g. species with very big leaf size will be extremely different from the others while species with small-sized leaves, even if quite different between them in size (for example, double the size), will come out as very similar. Hence, it is generally recommended to log-transform not-normally distributed traits, before computing PCA or the dissimilarity needed for PCoA.

3.2.4 Grouping Species

Grouping things into 'boxes' represents a typical human approach to simplify reality. We mentioned above that grouping organisms into functional groups can be done with different aims (response vs effect groups), which mostly affects the decision of which traits to consider in your 'species × trait' matrix. Here we discuss several quantitative approaches that can be followed to make groups, already broadly discussed by Kleyer et al. (2012), based on a 'species × trait' matrix.

Following Fig. 3.5, and based on Kleyer et al. (2012), we identify two common approaches to define functional groups, the so called '*a priori*' and '*a posteriori*' approaches (we explain these strange names below; see also Lavorel et al. 1997). Let's start with the *a priori* approach, which is more common. Hypothesis-driven classification starts with a trait (or set of traits) that we assume to be important for the ecological question of interest. For instance, can we use the difference between C3 and C4 grasses as an indication of water use efficiency (Cornelissen et al. 2003)? This approach can be further divided into hypothesis-driven or data-driven classification.

The *data-driven a priori* approach for grouping species into functional groups requires the computation of a dissimilarity matrix, such as the one already produced in Fig. 3.3, from the 'species × trait' matrix. The second step generally involves the use of some clustering techniques to create groups (see Fig. 3.5). Much has been written on the best clustering approach required to 'transform' a dissimilarity matrix into a corresponding dendrogram (Fig. 3.5) and the following statistical definitions of discrete groups based on such dendrograms (Mouchet et al. 2008; Kleyer et al. 2012). In 'R material Ch3', we provide some solutions, although a full discussion of these techniques is outside the scope of this book (the R appendix of Kleyer et al. 2012 is also helpful).

In the *clustering* step there are several important issues to consider. First, it has been reported that slight changes in the clustering approach can drastically change the results of the grouping, even when using the same set of traits. In other words, the algorithms used for the clustering and the identification of distinct groups has a great effect on the output of the grouping. This was recognized by Mouchet et al. (2008), who proposed, as a possible solution, the idea of producing a consensus based on multiple possible dendrograms. The take-home message here is that the placement of species to one group, or another, is quite 'volatile', i.e. we can obtain quite different results depending on the methodological choice, rather than on the biological information. We thus think that grouping species (Fig. 3.5) should be approached with much care, and a healthy dose of scepticism. In this sense, we strongly recommend the use of the hypothesis-driven approach following its validation.

The second issue we want to discuss here is of the aim and strategy of grouping. The *a priori* approach illustrated in Fig. 3.5 is needed when authors want to define groups of species based on trait (dis)similarity so they can later test whether such groups have different environmental preferences, different effects on the ecosystem, or different

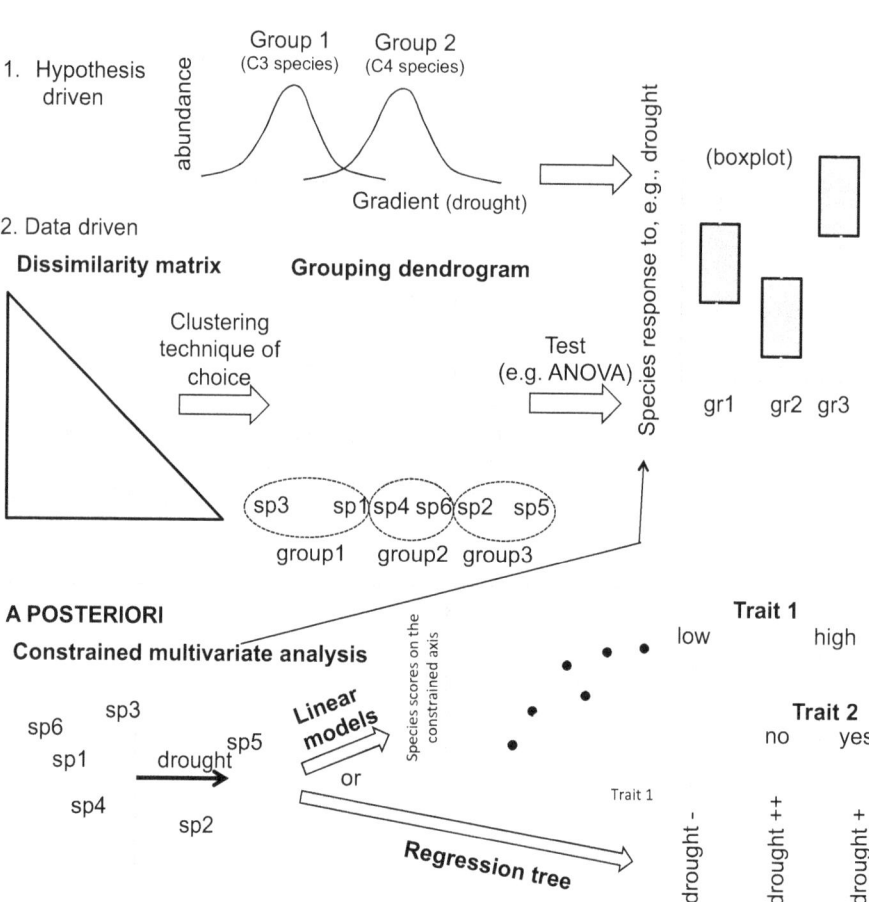

Figure 3.5 Schematic of the typical strategies used to define functional groups of species. In the 'a priori functional group' approach we start the analyses from groups based on either an ecological hypothesis (e.g. C4 species are more drought-tolerant than C3 species; hypothesis-driven approach) or based on a data-driven approach. In the latter, we begin with a trait dissimilarity matrix and use a clustering technique to build a dendrogram, and then define groups based on trait similarity. The second approach is to validate these groups, i.e. testing if the grouping predicts, for example, species' environmental preferences. In the 'a posteriori functional group' we must first define species' environmental preferences, for example using multivariate constrained analyses. Then we use traits to predict such species preferences, so the predictors are the traits directly (not the groups). To do this we can use linear models or regression trees (in the latter case, some groups can be identified). Obviously, species' environmental preferences can be computed for the 'a priori' approach in the same way as the 'a posteriori'. Species scores on a constrained axis are generally used as the dependent variable to define species' environmental responses. In the regression trees this is usually considered the response variable as well (summarized in the figure as 'drought' ++, or + or −).

fitness etc. These '*a priori* functional groups' are based only on trait similarity, and do not take into account whether species have similar/different environmental preferences or similar/different effects on the ecosystems. As a second step, authors can check if these '*a priori*' groups have also different environmental preferences, or even different effects on the ecosystem. This can be modelled with the groups as predictors of a given species function. This *a priori* grouping, obviously requires a good ad hoc choice of traits. If we want to focus on plant groups with different responses to either grazing, or fire, or drought, we would select the most relevant traits (e.g. palatability for grazing, resprouting ability for fire, succulence for drought etc.). If we focus on groups with different effects on the ecosystem, say nutrient cycling or water use, we might consider different traits for each (e.g. nitrogen fixing ability, water use efficiency etc.). In some analyses we need both response and effect traits, for instance, to define functional redundancy indices (see also Chapter 5), following Mouillot et al. (2013a) or Laliberté et al. (2010). In the approach by Laliberté et al. (2010) two groupings are needed, one based on response traits and one on effect traits, while distinguishing them might not always be straightforward (Chapters 1 and 2).

The alternative to the '*a priori* functional groups' is the '*a posteriori* functional groups'. In this case, the first step is not to look at species trait similarity/dissimilarity but, for example, at their environmental preferences. We can use species composition data in communities along either artificial or natural gradients and define species preferences along these gradients using constrained multivariate analyses (de Bello et al. 2005; Kleyer et al. 2012). This reflects earlier attempts to define, for example, species increases or decreases following grazing (Díaz et al. 2001), fire and other environmental gradients. Once the environmental preferences of species are defined (for example using species scores on a constrained multivariate axis), we can use traits to predict such preferences. Such predictions can be made with simple linear models, or with regression trees (de Bello et al. 2005; Kleyer et al. 2012) which allow for an *a posteriori* clustering (i.e. grouping) of species based on the combination of traits which predicts species behaviour. The advantage of regression trees (and booster regression trees; Pistón et al. 2019) is that they allow for non-additive effects between species. This allows also for taking into account the case in which different combinations of traits result in similar effects on species fitness, or species effects on the ecosystems, which is sometimes called 'alternative designs' (i.e. different types of species are adapted, each in a different way, to given environmental conditions; Fig. 2.4). For example, de Bello et al. (2005) showed that species distinction into annual and perennial species was important in some biomes to predict species preference for grazed or ungrazed conditions. However, and only within annual species, flowering period further distinguished groups of species with different grazing preferences. Such interactions between traits seem important to define trait syndromes that can be used to build robust trait predictions (Pistón et al. 2019).

We think that, in general terms, the *a posteriori* approach is a more neutral approach, because the selection of traits is based on which traits are statistically more important in distinguishing species behaviour, whereas the *a priori* choice of traits could be more arbitrary, if not based on strict hypotheses. However, it has one big disadvantage: the associations between traits and groups might not be causal. This is because we first characterize species according to their response (or effect) and then start 'mining' for traits that could explain such differences. In this sense, we can say that this is a more exploratory approach and should only be used if our information about the system is insufficient to generate hypotheses linking traits to species responses or functions.

Take-Home Messages

- It is generally helpful to attempt classifying, at least broadly, species into different 'types', particularly for interpreting ecological patterns and for communication with non-experts. On most occasions functional differentiation occurs along continuum axes of differentiation, but in some cases groups with contrasted strategies can be recognized.
- Universal grouping could be of little efficacy and robustness. Strong emphasis can be given to choosing traits and groups (response vs effect groups) that are relevant to the function of interest (e.g. response to grazing or drought, or effect on soil nutrients). We can determine functional groups in two main ways, focusing on the similarity in response traits or the similarity in effect traits. We can then define all these groups by different approaches (Fig. 3.5): *a priori* based groups (which are data-driven or hypothesis-based) or *a posteriori* defined groups. The careful selection of traits is essential to avoid spurious correlations.
- Trade-offs between traits determine different ecological strategies along axes of correlated and coordinated traits. Historically there have been different ways to define the main evolutionary trade-offs between species, resulting in several schemes (Fig. 3.1) and analyses. Techniques such as multivariate axes can identify the main 'spectra' of trait differentiation between organisms, although they do depend on the choice of traits.
- Undoubtedly, trait differentiation between organisms has created some discontinuity in evolution, for example woodiness in plants, or C3 vs C4 photosynthetic strategies. Hence categorical traits generally reflect discontinuous trait trade-offs between species. Quantitative traits, on the contrary, are often associated to create a more continuous trait trade-off.
- The Gower distance is a useful way to characterize the differences between organisms in terms of multiple types of traits. But care should be used when mixing categorical and quantitative traits, and when many correlated traits covering the same 'functions' are considered.

4 Response Traits and the Filtering Metaphor

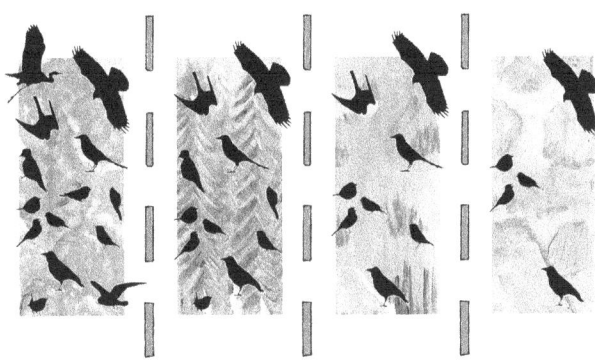

Explaining the distribution of species through their interactions with the environment has been at the heart of ecological research. In fact, the idea that the environment constrains where an organism can exist dates far back into antiquity, including to the times of the great Greek philosopher Aristotle (384–322 BC). We will therefore start off this chapter with a short section giving a historical perspective on examining trait–environment relationships as a scientific endeavour. This is followed by underpinning the importance of the concept of the niche, which is a crucial theoretical concept when studying trait–environment relationships. Several researchers have suggested viewing the constraints that different environmental conditions put on the distribution of species as a series of 'filters' that select which species can live where, and this filtering framework has remained popular up to the present day. Before we introduce this framework, we will describe in more detail what kinds of environmental gradients we usually deal with and how these can be categorized. Towards the end of the chapter we take a slightly more methodological viewpoint and describe the different analyses that are possible when studying functional traits along environmental gradients (further detailed in the corresponding 'R material Ch4'). We then end with some important applications of trait–environment relationships, focusing on so-called species-level analyses.

4.1 From Early Biogeography to Trait–Environment Relationships

With the birth of what we call now biogeography, through works by eighteenth- and nineteenth-century naturalists like Georges-Louis Buffon, Alexander von Humboldt, and Augustine de Candolle and his son Alphonse de Candolle, the question of where a species can exist, and why it is able to do so, was tackled for the first time in a serious scientific manner. These outstanding scholars laid the foundations of natural history by connecting the flora and fauna found in different locations to their physical and chemical environments. At roughly the same time, the idea that *form equals function*, i.e. morphological traits of organisms can reflect adaptation to different ecological conditions, became obvious. This equation was a central building block of early and later manifestations of evolutionary concepts by Lamarck, Wallace and Darwin. Though biogeographical observations played a very important role in the development of evolutionary theory, the ultimate focus of these observations was to contribute to understanding selection in the struggle for existence.

The idea that morphological and physiological requirements are the basis for understanding why certain organisms exist under specific conditions slightly predates the development of evolutionary theory (Schouw 1823; Liebig 1842; Watson 1847–1859). Towards the end of the nineteenth century this idea was pursued further, but by now was incorporating evolutionary aspects, as well as drawing from experimental physiological work. Schimper (1903) and Warming (1909) constitute two seminal works of that era for plants, describing in detail the different factors that affect their distribution and how these factors translate into morphological and physiological adaptations. For the animal kingdom, Semper (1881) proved to be equally important and influential.

Another critical development came about with the notion of the community, or *biocoenosis*, as it was initially termed by Möbius (1877). This concept would later divide ecological research more strictly into the field of autecology, concerned with individuals or populations of the same species and their relation to the environment, and the field of *synecology*, the branch of ecology that studies how interacting groups of different species that co-occur in the same locality relate to their environment while emphasizing the interactions between the different species themselves. The term synecology is virtually out of use today and is mostly synonymous with community ecology.

4.1.1 Trait Variation within Species

Considering the *response of populations* of a single species to environmental gradients, Clausen, Keck and Hiesey, published an influential monograph showing the geographical variability within *Achillea lanulosa* (Clausen et al. 1948). They detailed morphological and physiological adaptations occurring across the species' geographical ranges, and their figure highlighting the different heights and growth forms of *A. lanulosa* populations along an elevational transect through the Sierra Nevada became a prominent textbook example of intraspecific geographical variation (Fig. 4.1). Importantly, they also pioneered what is now known as the 'common garden' experiment, as a way to tease apart genetic and environmental causes for phenotypic variation (see also Chapter 6, Box 6.1) when organisms from different origins develop or grow under the same environmental conditions.

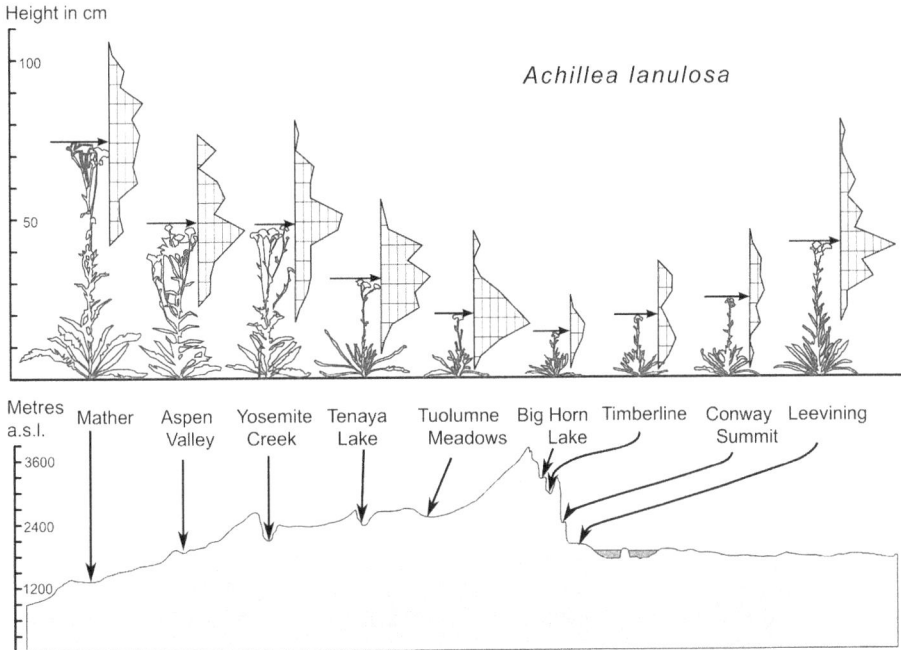

Figure 4.1 Variation in the overall height distribution within populations of *Achillea lanulosa* (now *Achillea millefolium*) across a geographical and altitudinal gradient in the USA. Taken and redrawn from Clausen et al. (1948), with permission from the Carnegie Institution for Science.

Subsequent works studying geographical variation among populations have mostly done so from an evolutionary angle and adopting genetic methods. Examples of more recently conducted studies cover the whole spectrum of living organisms, from growth-rate differentiation driven by temperature in the eukaryote *Neurospora crassa* (Ellison et al. 2011), latitudinal and elevational variation in multiple traits in the plant *Clarkia unguiculate* (Jonas & Geber 1999), latitudinal body-size variation in the fruit fly *Drosophila subobscurato* (Gilchrist et al. 2001), competition-driven variation in beak and body morphology of Darwin's finches (Bowman 1961), and the variation in skin colour of humans as a response to ultraviolet radiation (Jablonski & Chaplin 2010).

The driving question in much of this line of research is whether the phenotypic variation within a species across its geographical range (or parts of its range) is caused by genetic differences among the populations, or by the environmental variation that occurs across these populations. In many cases, there will be a role for both. The important distinction is that genetic differences can lead to local adaptations and eventually form new species. Apart from these genetic effects, nowadays there is also the realization that the performance of individuals can be altered by environmental effects experienced by the parental generations of the individual, giving rise to its own new research field, epigenetics (e.g. Bossdorf et al. 2008). These issues will be further discussed in Chapter 6, focusing on the drivers of phenotypic variability within species.

4.1.2 Trait Variation across Species

Looking at the response of multiple species to the same environmental gradient simultaneously, Robert H. Whittaker was interested in how the abundances of different co-occurring tree species changed along the elevation of the Great Smoky Mountains in the Southern Appalachians (Whittaker 1956). He found continuous changes in abundance (in a round or *bell-shaped pattern*) along a moisture gradient, with peaks and boundaries of the abundance curves of single species mostly not co-occurring with those of other species, but rather scattered across the gradient (Fig. 4.2). Based on his own work and that of several peers who had used similar approaches to study vegetation, he reviewed and formalized this way of studying vegetation, calling it *gradient analysis*. Over time, the statistical methods to analyse vegetation–environment relationships became more sophisticated, leading to the development of numerous so-called ordination methods. The general principle of these methods is to order species along multiple gradients. Ideally, this ordering reflects the environmental gradients that underlie the changes in community composition. An important aspect of ordination is thus to condense the multivariate information presented by multiple samples (e.g. vegetation plots, insect traps) by graphically representing them in a simpler way, mostly in the form of two-dimensional graphs. The statistical tools for the task of ordination in ecology were often forged by vegetation ecologists, but their use is now widespread throughout ecology (Šmilauer & Lepš 2014; Paliy & Shankar 2016).

The last step on our conceptual ladder is to link traits and abundances/occurrences of species with the environmental conditions present along environmental gradients. This

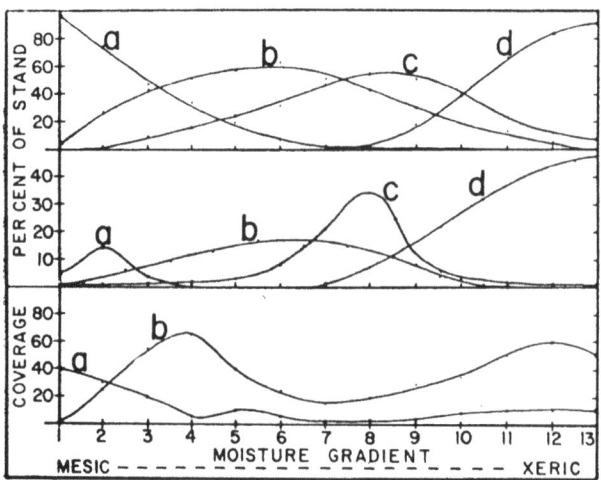

Figure 4.2 Abundance distribution of four species classes (top panel; a = mesic, b = submesic, c = subxeric, d = xeric), four tree species (middle panel; a = *Betula allegheniensis*, b = *Cornus florida*, c = *Quercus prinus*, d = *Pinus virginiana*), and two growth forms (bottom panel; a = herbs, b = shrubs), along a moisture gradient in the Rocky Mountains, USA. Taken from Whittaker (1956), with permission from Wiley. Ecological Monographs © 1956 Wiley.

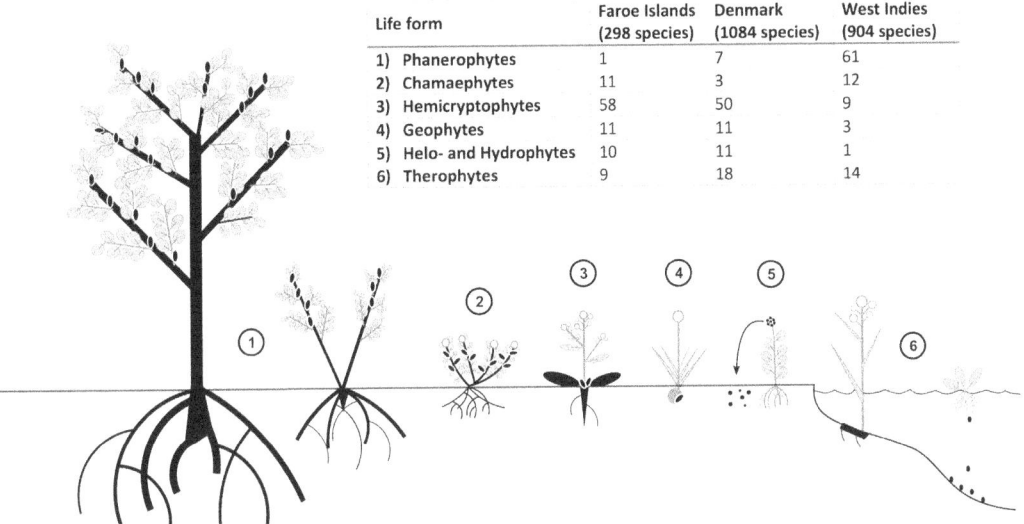

Figure 4.3 Different Raunkiær life forms are shown with the position of the persistent organs (buds or seeds). These are located at different heights in relation to the soil (or water) surface and thus play a role in how plant species are able to survive unfavourable times, especially frost periods. The data in the accompanying table are from Ostenfeld (1908), who showed proportions of these different life forms across three different floristic regions of differing climates. For simplicity, we omitted numbers of Epiphytes and succulent forms and summarized different types of Phanerophytes from Ostenfeld's data.

step merges the preceding work of both Clausen et al. (1948) and Whittaker (1956) with that of the Danish botanist Carl Hansen Ostenfeld, who combined the occurrence of species in three climatically distinct localities and trait information in a simple way to study the association between traits and climate (Ostenfeld 1908). He used the proportion of species belonging to one of Raunkiær's plant life forms in those three climatically distinct localities to demonstrate the importance of this trait (Fig. 4.3). Raunkiær's plant life forms (Raunkiær 1934), a simple way to consider functional groups of plant organisms (see Chapter 3), is based on the position of perennial buds in unfavourable conditions. In particular, if a plant is able to avoid exposing its buds (or seeds) to freezing by keeping them close to or under the soil surface, it increases its chances of coping with winter-cold climates. In consequence, warmer climates (Caribbean islands) host a higher proportion of species that have regenerative buds positioned high above the soil surface, whereas in colder climates with winter frosts (Faeroe Islands and Denmark) more species rely on saving the buds from the cold by having them close to or in the soil.

From the table in Fig. 4.3 we can also conclude that groups of species with similar traits, in this case the described life forms, tend to inhabit similar environmental conditions (which we defined as 'response groups' in Chapter 3). In a similar direction, Grime (1979) computed proportions of different plant strategies (C-S-R strategies; see Chapter 3) in different communities and assessed how this proportion changed along disturbance and stress gradients. Likewise, from our earlier example, Whittaker not only considered the changes in abundance of single species along a moisture gradient; he also pooled species of the same growth form, showing for instance that herbs were more prominent at the mesic end of the gradient, while shrubs were more abundant at xeric sites, which could hint at shrubs being better adapted to using deeper water sources in the soil via their deeper root systems. All these approaches correspond to the definition of *a priori*, hypothesis-based, functional groups (see Fig. 3.5), but below we discuss other ways to understand this type of analysis.

4.2 Environmental Gradients

In the previous section we discussed examples of changes in traits along altitudinal gradients (Fig. 4.1), moisture gradients (Fig. 4.2) and even across countries with different climates (Fig. 4.3). Hence, we have already seen several examples, and probably have at least a vague understanding, of what *environmental gradients* entail in general. But we still miss an exact definition of what constitutes an environmental gradient, and a possible categorization of the different gradients that we usually deal with in ecology. We follow here the definition given by Garnier et al. (2016) by which an environmental gradient is regarded as 'a gradual change in a given biotic or abiotic environmental factor through space or time'. This is a very general definition which encompasses many types of gradients. Most commonly, gradients are categorized in the literature (Austin & Smith 1989; Guisan & Zimmermann 2000) as one of three different types: resource gradients, direct gradients or indirect gradients.

- *Resource gradients* describe any variable that constitutes a source of organic or inorganic matter that is necessary for metabolism by the organism. By this definition, the resource is at least partly consumed by the organism.
- *Direct gradients* are based on variables that have a direct impact on the physiological performance of the organism, but which are not 'used up' or consumed. They characterize the physical and chemical environment experienced by the organism (e.g. soil pH, temperature).
- *Indirect gradients* are essentially approximations of single or multiple direct gradients and/or resource gradients. An example is the use of altitude to approximate a climatic gradient. Indirect gradients are often easier to assess than direct gradients. The risk of using indirect gradients lies in making wrong assumptions about what their underlying direct gradients are. For instance, temperature and precipitation are correlated with altitude, but further unrecognized factors, such as human-made disturbances, could be influential at the same time (e.g. higher grazing pressure by livestock at lower altitudes). Often it is impossible to separate the effects of multiple direct gradients when they are reflected by a complex indirect gradient.

Commonly, any of these gradients are used to define species' environmental preferences, i.e. species optima (see Box 4.1), and then these optima can be related to species

Box 4.1 Environmental preferences and the species niche

For several analyses described in this chapter it is necessary (implicitly or explicitly) to estimate species' environmental preferences along a gradient. Following Fig. 4.2, we can expect that many species have a sort of unimodal-like abundance distribution along a given gradient. The maximum of this distribution is generally referred to as the *optimum* for the species along that gradient, and the width (or breadth) of this distribution generally represents how specialist (narrow distribution) or generalist (wide distribution) a given species is with respect to the studied gradient. Essentially, the shape of the distribution of species along a gradient characterized by an optimum and a breadth characterizes the *species niche* (see also Chapter 1). In general, accounting for the fact that species abundances vary in response to multiple gradients, but also to other species being present, multivariate analyses can be used to define such niches. For example, constrained multivariate analyses (redundancy analysis, RDA, or canonical correspondence analysis, CCA) generally provide species scores along the gradients considered, which can be used as an indication of species optima along these gradients (see also Fig. 3.5). Note that RDA assumes a linear relationship between species abundance and a gradient, which is useful when the extent of the gradient considered is not very broad, and CCA assumes a unimodal-like relationship (i.e. like those depicted in Fig. 4.2; see also Šmilauer & Lepš 2014 for suggestions on multivariate analyses).

As just described, the observed abundance distribution of a species along a gradient reflects both the outcome of the direct effect of a gradient on the species' ability to grow and the effect of the interactions with other species. This makes the observed niche, also called the '*realized niche*', generally different from the *fundamental niche* of a species, which should reflect only the environmental conditions in which a species can sustain and grow its population, without considering biotic interactions. The realized niche can be assumed to be often smaller than the fundamental one, reflecting the effect of competitive, stronger species. However, other ideas that complement the concept of the realized niche have been put forward, where the realized niche of a species is enlarged by coexisting species, e.g. through facilitation (Bruno et al. 2003). One example is provided by plant species that are sensitive to grazing and when growing alone show a reduced performance (small fundamental niche), yet when protected from herbivores by a coexisting spiny plant species will have a larger realized niche. Niche expansion is also conceivable when there is direct interaction with other trophic levels, e.g. in the case of the symbiosis between mycorrhiza and plants (Peay 2016).

The realized niche is sometimes called the 'beta-niche' (Silvertown et al. 2006a), with reference to the fact that it reflects the niche across different environmental conditions. On the contrary, the alpha-niche (Pickett & Bazzaz 1978) generally reflects differences within a local community between the types of resources used by different species. We will focus on this aspect of niche differentiation within a community in Chapter 7.

traits, explicitly or implicitly (Fig. 3.5 and this chapter). Such types of tests, further described in section 4.4, are built on an important assumption, i.e. that the abundance of species along the gradient is the product of the gradient itself. The impact of the biotic interaction of a given species with other species (Chapter 7) is thus expected to be represented by the gradient and is in fact reflected by the abundance of the species. Such biotic interactions include any kind of within- or across-trophic-level interactions, i.e. commensalism, mutualism, predator–prey relationships, parasitism, competition or facilitation. Hence, predicting species abundance only by a given gradient can be limiting, as observational data recorded along gradients are the product of both abiotic and biotic effects, and distinguishing them is often difficult. The discussion on the separation of abiotic and biotic effects on species distributions is important, but we think it largely falls outside the scope of this book. However, we give some important information related to the tests shown in this book, in Box 4.1, sections 4.3 and 4.4 and Chapter 7. What is clear is that observational data along a gradient are the product of both abiotic and biotic effects, and these cannot be often distinguished easily. Some solutions can also be found using experimental approaches, which allow more precise control of the explanatory factors one wants to study. Moreover, initiatives have formed that set up experimental field sites in different parts of a country or even on a global scale, allowing the investigation of whether hypothesized relationships can be found across the different sites or not, and therefore have greater potential to draw more general conclusions (e.g. Nutrient Network, www.nutnet.org; Biodiversity Exploratories, www.biodiversity-exploratories.de).

Finally, it is important to specify over what geographical and temporal *scales* a given gradient varies. Scales can vary from millimetres to kilometres, and from hours to years. The mean annual air temperature can be very similar for two points 10 km apart, whereas soil properties might be distinctively different for two points less than a metre apart. Even at smaller scales, ecological conditions can change in very short spans of space and/or time, leading to different compositions of inhabiting species and communities at taxonomic and functional levels. For instance, the vertical soil profile can select different soil organisms and their traits on spatial scales of a few centimetres, leading to different soil-food-web structures (Berg & Bengtsson 2007), while across a horizontal range of several metres in the same location, food-web structure remains largely the same. This differentiation in scale for different gradients defines the kind of geographical and temporal scales that can be expected to have an effect on trait distributions of species and communities.

4.3 The Environmental Filtering Metaphor

The study of the relationship between traits and environment has often been associated with a metaphor, which is generally referred to as *environmental* or *habitat filtering*. This metaphor is based on the idea that species are 'filtered' from a regional species pool into local communities according to their traits, and it was originally proposed by Keddy (1992b; see also Díaz et al. 1998), though the use of the term filtering goes back at least to MacArthur and Wilson (1967). In other words, the species with traits less adapted to a given condition are filtered out, i.e. the probability that they will be present and/or their abundance will be lower. The filtering is expected to occur based on multiple abiotic and biotic constraints. Within a geographically defined space (i.e. a community or, more generally, an 'assemblage'), a series of *hierarchical filters* determines the traits that confer better performance under these ecological conditions, and, hence, which species are more likely to exist. As an analogy, one can imagine sieving soil in a stack of sieves with meshes of increasingly smaller size, so that larger soil grains like pebbles get caught in the upper sieves, while only very fine soil grains filter all the way down into the sieve at the bottom of the stack. The small grains that finally make it to the last sieve are a subset of all soil particles in the sample.

In an ecological framework, filtering can be caused by a set of environmental factors that select for or against species from the pool of available species, which could potentially disperse to a given site. Species are allowed to 'pass' if they possess the right traits to survive, grow and reproduce. We have a group of species, for example the flora or avifauna of an entire continent, but the local environment filters out those species that are less adapted. The crucial addition to the simple sieve/filter analogy is that the filtering of the species is based on their functional traits, which is analogous to the sizes and shapes of the soil grains, which determine if they can pass through a certain sieve or not. If the set of functional traits of a species allows membership in the habitat or leads to superior competitive ability, it can pass the filter. If not, it will not be part of the community. This is the main idea of environmental filtering. As a set of environmental conditions can define a habitat, the term habitat filtering instead of environmental filtering is also often used.

One possible representation of which filters can be considered and how they are arranged is presented in Fig. 4.4 (see also Zobel 1997; Götzenberger et al. 2012). The different filters are thought to operate on different spatial and temporal scales. In the particular scheme shown, processes on evolutionary timescales lead to the formation of regional species pools. Regional species pools are a subset of the total global pool that have been created through speciation, extinction, and migration of species over evolutionary time (Ricklefs 2004). This regional pool is still defined at the relatively large biogeographical region scale. Dispersal limitation, i.e. the dispersal filter, can still filter out species that occur in one part of a continent or biogeographical region, but that are not able to disperse to other parts, even though climatic conditions are suitable in those parts. Penguins and polar bears, which occur at the South and North poles, respectively, are good examples. But dispersal can also affect the occurrence of species at a smaller scale, e.g. in a meta-community context where a species could fail to reach an isolated patch with favourable conditions because its potential for long-distance dispersal is limited.

Finally, to become a member of the actual local community, it is necessary for the species not only to reach this locality, but to also find suitable environmental conditions, regarding both *abiotic and biotic factors* (Weiher & Keddy 1995). Many similar schemes, like the one presented in Fig. 4.4, exist in the ecological literature, and scholars differ in their opinions about which of the filters are more important, and

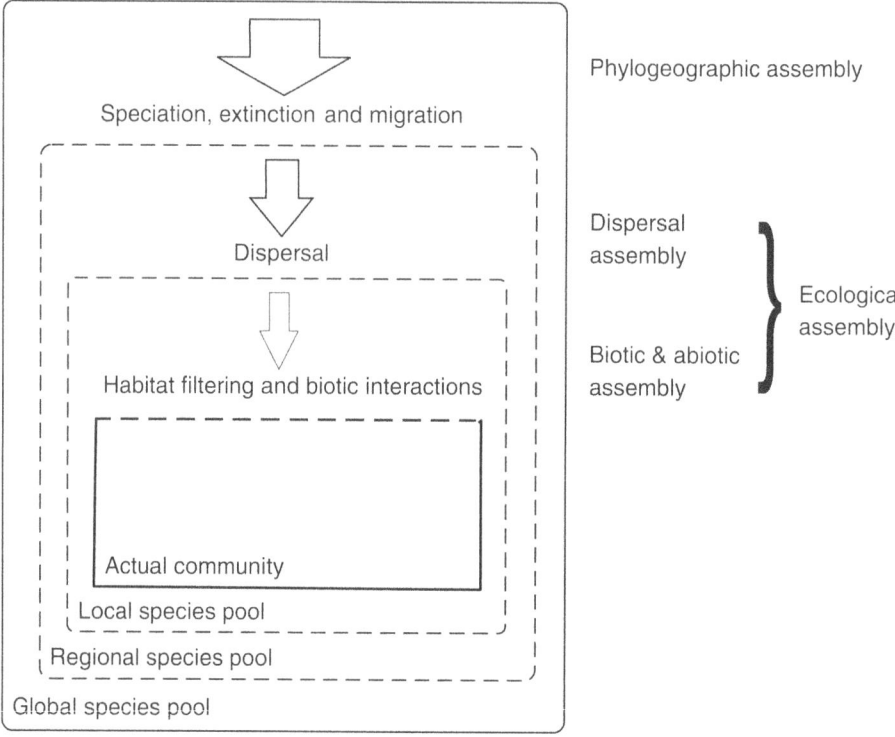

Figure 4.4 Graphical representation of the filtering concept. The different filters are depicted as a hierarchical series of processes that restrict membership in the species pool at the next finer scale, ultimately resulting in the actual observed community. The attributes global, regional and local refer to the idea that the different processes prevail at different scales. It is important to note that despite the supposed hierarchy, the different processes act simultaneously and can also largely overlap. Taken from Götzenberger et al. (2012), with permission from Wiley. © 2011 The Authors. Biological Reviews © 2011 Cambridge Philosophical Society.

under what conditions, and how they should be organized in this hierarchical representation of nested filtering. One thing that is shared among the different representations of the filtering framework is that with the passing of each filter, some species (and subsequently their traits) are excluded, or at least their abundance is modified, and this happens progressively as we decrease the spatial scale at which we observe the assemblages (de Bello et al. 2013b).

At the local level, the different filters represent the different environmental conditions that prevail where a species exists. Therefore, these filters include, and are sometimes separated into, abiotic and biotic filters (Mayfield & Levine 2010), although their distinction is often not easy or straightforward (see also Chapter 7). The abiotic filters represent all those abiotic conditions that we have described as direct or indirect gradients in section 4.2. Biotic filters include interactions within trophic levels (competition, facilitation etc.) and across trophic levels (pollination, mycorrhizal symbiosis, parasitism etc.). These interactions can restrict the occurrence of a species in a locality in which it could be otherwise exist, because the species gets, for instance, outcompeted by other species or falls prey to a predator.

While the filtering scheme implies a sort of hierarchy where higher-level processes act first and lower-level processes are only effective after this more initial filtering, it is important to note that this hierarchical structure is a conceptual construct and not necessarily a meaningful representation of how the assembly of ecological communities actually takes place (Kraft et al. 2015). Indeed, the filtering metaphor is more of a conceptual tool to communicate concepts among ecologists, rather than a perfect description of reality. Alternatively, processes can be bottom-up as well as top-down. Most importantly, very often it is virtually impossible to precisely differentiate the effect of different filters using observational data only (Mayfield & Levine 2010). We will elaborate on these ideas in Chapter 7, which provides an overview of how ideas about community assembly change over spatial and temporal scale with an emphasis on the more recent developments in coexistence theory.

4.4 From Species to Communities and Back

Combining species abundances, their traits, and environmental conditions, is the basis for what we understand today as studies of trait variation along environmental gradients. The task of combining and analysing these data is the subject of vivid debates and the methodology is still developing (Kleyer et al. 2012; Peres-Neto et al. 2017; Zelený 2018). Examples of how to face this challenge are provided in 'R material Ch4'. We here provide some basic arguments that need to be considered when performing these analyses. The first important distinction when assessing how traits change along gradients is that the trait–environment relationship can be dealt with at different levels, as we have partially seen in the previous sections. We envision three main types of analyses: (1) at the population level, i.e. within a single species, (2) at the species level across multiple species, and (3) at the community level, i.e. in assemblages of coexisting species. These different types of analyses are schematically summarized in Fig. 4.5.

At the '*within-species* level', each individual, or population, of a species usually represents a single data point. Most often, we will likely average individual measurements for a certain trait for a particular species at the population level, so that the different populations along an environmental gradient are each represented by a mean trait value (but see also Chapter 6 on the matter of intraspecific trait variability). For example, we could average plant height of measured individuals at the level of different populations across the altitudinal gradient in Fig. 4.1. At an even finer level, we could use each individual measured and its position on the gradient, to test for significant changes across the gradient.

For the second type of analysis ('*across-species* level', or simply '*species level*'; Kleyer et al. 2012) each species is represented by its optimum along a given gradient and its value for one or more traits. Each observational unit in this analysis is a species. In the simplest of cases we can define this optimal environment as the one where the abundance of a species is the highest. We can also use multivariate analyses to define

Response Traits and the Filtering Metaphor

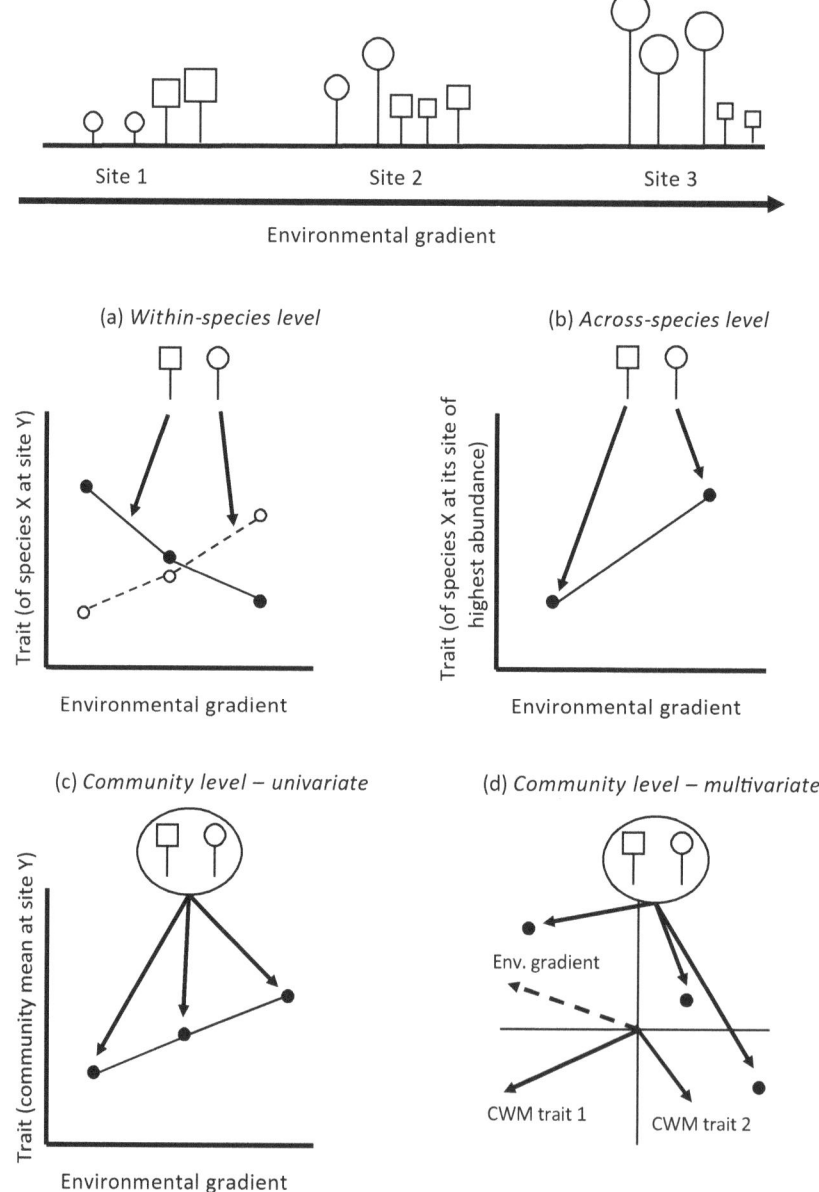

Figure 4.5 Trait–environment relationships at different levels. Different ways to analyse the data are possible, and are depicted under (a), (b), (c) and (d). Note that the x-axis, i.e. the environmental gradient under consideration, is always the same. In each of the different graphs, we project different ways of testing the relationship between traits and the environment. At the top, a simplified depiction of two coexisting species that occur along an environmental gradient is shown. At the within-species level (a), each point on the graph symbolizes the mean trait value over the individuals sampled from a population at a location along the gradient. Note that we can

species optimal conditions (Box 4.1). It is important to note that in many cases ecologists are interested in the reverse relationship between the environmental gradient and traits, such that the trait value is the predictor (i.e. on the x-axis), whereas the position along the environmental gradient (i.e. the species' niche) is the response (i.e. on the y-axis).

To take traits to the '*community level*', there is one more methodological step that we need to take, which is to combine species trait information into a single community-based trait measure, such as the trait average of species in a community. With this simple approach, all we need to do is to take the trait values of the species that co-occur in a given locality and calculate their average. It is worth pointing out that the study by Ostenfeld (see Fig. 4.3) made use of exactly this measure; because the trait (life form) is a categorical one, the average trait at the community level is simply the proportion of species that are assigned to a given life form. In Chapter 5 we will see in more detail how this measure can be calculated also taking into account species abundances, via a frequently used index of community trait composition called community weighted mean (CWM).

Whereas the within-species-level approach is rather straightforward, there are some more complex possibilities for statistical analyses regarding the across-species and community-level analyses that also somewhat blur the distinction between these two levels of analysis. For example, the so-called *fourth-corner* approach provides a solution that can be seen as a combination of both species- and community-level approaches (Dray & Legendre 2008). Other tests can combine population-level and species-level analyses (Jamil et al. 2013). An interesting approach in this sense is explained in Chapter 6. Importantly, many of the methods considered nowadays leave the realm of univariate statistics (i.e. having a single response variable, as is the case for Fig. 4.5a–c) and analyse trait–environment relationships in a multivariate fashion (as in Fig. 4.5d). In less statistical terms: we try to explain the abundance or occurrence of several species simultaneously (i.e. the community composition) by their position along environmental gradients and the traits that these species possess, acknowledging that their traits are – or should be – the factors that determine species' ecological performance, and therefore the abundance of these species.

Caption for Figure 4.5 (cont.) depict such a relationship for multiple species on the same graph, as done here for only two species for simplicity. In the across-species-level approach (b), each point on the graph is the mean trait value of that species on the y-axis, and the 'optimum' position of the species on the gradient along the x-axis (for example where the species is most abundant). The relationship depicted in (b) can often be flipped (swapping x- and y-axes), following the idea that a species' traits define its position along the environmental gradient. Finally, on the community-level graph (c), each point is the average of trait values of species co-occurring at a location along the gradient. We can also combine these community-level measures for different traits and different gradients in a single analysis, as depicted in (d), where each site's community average is projected into a multivariate space that is constrained by the environmental gradient(s).

There are several statistical methods that aim at establishing a trait–abundance relationship, or trait–environment relationship in the above manner. They all face the challenge of combining the three different matrices (species composition, also called species × community matrix in Chapter 5; species × traits matrix, as introduced in Chapter 3; and community × environmental conditions), and they achieve this task in different ways. We cannot present them here in detail, but a good overview and comparison can be found in Kleyer at al. (2012), and we also introduce some typical analyses in 'R material Ch4'. For community-level approaches we provide more material in Chapters 5 and 6.

We can see many examples of trait–environment analyses in the recent literature. In fact, there has been a boom in such studies since the early 2000s. While, as we mentioned, the trait–environment relationship can be tackled with different analyses, we think that the specific comparison of species- vs community-level analyses is helpful to understand what the analyses reflect biologically. One early trait-based study, which highlighted the distinction between species- vs community-level analyses, was Ackerly et al. (2002). The authors showed a weak relationship between leaf traits (specific leaf area, leaf size) and species environmental preference along an insolation gradient (species-level analysis; Fig. 4.6a–b). However, when traits and species composition were combined into community means (either weighted or not by abundance; see Chapter 5) these relationships became quite strong (community-level analysis; Fig. 4.6c–d). It is important to realize that analyses at different organizational levels can give different insights into questions surrounding how traits vary along environmental gradients. Explanations of these differences are partly mathematical in their nature (Ackerly et al. 2002; Hawkins et al. 2017; Peres-Neto et al. 2017), and to infer a biologically meaningful interpretation of a trait–environment pattern needs careful consideration of the relationships that are being hypothesized in a given study (see Zelený 2018). In Box 5.1 we provide some deeper discussion about this topic. In summary, we think that the two types of analyses provide an answer to different types of questions, mainly if environmental filtering applies to most species in the community; in this case, we would expect to see a relationship at the species level, or at least for one or a few dominant species, i.e. expressed by community means.

4.5 Relating Functional Traits to Fitness

One other typical example of species-level analysis is in examining the relationship between traits and fitness, which is one of the essential premises of the trait-based approach (see Fig. 1.1; Violle et al. 2007). In practice, relating traits to fitness can be faced as summarized above for the species-level analyses, where a trait serves as an explanatory variable to predict species fitness components. However, the strength of the relationship between traits and fitness remains little tested, mostly because data on entire life cycles of species are needed, though recent efforts have begun compiling such data (Salguero-Gómez et al. 2015, 2016). Adler et al. (2014) used such data for plants in combination with functional trait data, showing that traits like seed mass, wood

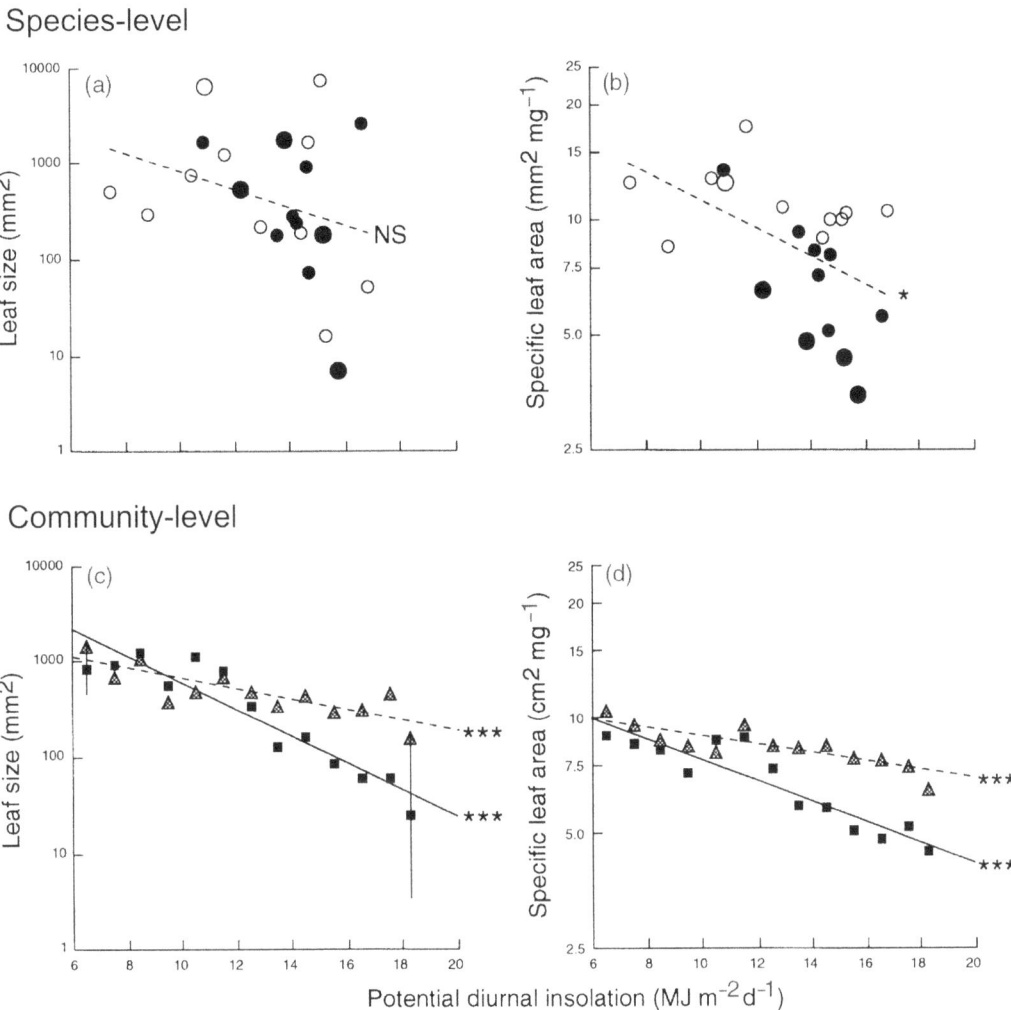

Figure 4.6 The relationship between potential diurnal insolation and two traits – leaf size and specific leaf area – at the across-species level (a and b), and at the community-mean level (c and d). At the species level (i.e. each point representing species' preferred insolation on the x-axis and the species' trait average on the y-axis), traits and environment are only weakly related or unrelated. At the community level (i.e. each point representing the mean trait value across species within a plot) the relationships are stronger. The solid and broken lines in c and d depict relationships with community-mean values that are either weighted, or not, respectively, by the abundance of the species. As such, the graphs here represent the types of analyses depicted in Fig. 4.5, panels b and c, i.e. species-level and univariate community-level analyses. The horizontal lines in panel c indicate trait variability (as standard deviation) around the trait mean, indicating an increase in functional diversity along the gradient (see Chapter 5). Taken from Ackerly et al. (2002), with permission from Springer Nature. Copyright © 2002, Springer Nature.

density and specific leaf area are indeed related to the fitness components survival, fecundity and growth. However, different traits were associated with different fitness components, and much of the variation in fitness components remained unexplained. This difficulty in finding consistent links between species fitness and functional traits is most likely also one of the main reasons why we often see that environmental gradients only poorly predict the functional traits at the species level, whereas the link between the environment and traits is more evident at the community level, especially when we mostly address the dominant species (e.g. Ackerly et al. 2002).

More recently, Pistón et al. (2019) proposed that rather than single traits, the combination and interaction of traits can better explain fitness of species. The main idea is that certain trait combinations can lead to multiple functional strategies resulting in similar fitness. When addressing plant fitness, the whole multidimensional functional space has to be taken into account, not just single, one-dimensional aspects. For this reason, methods based on regression trees can be used to relate traits to fitness components or even species' environmental preferences (de Bello et al. 2005; Kleyer et al. 2012; see 'R material Ch4').

Another potential reason why evidence of strong trait–fitness relationships is scarce in the literature is the overriding environmental effects on fitness (Adler et al. 2014). It is possible that the trait interactions that cause alternative trait combinations to have similar fitness are different in different environmental conditions (Laughlin & Messier 2015; Pistón et al. 2019). Despite well-known effects of both trait interactions and environmental changes on species fitness (Dwyer & Laughlin 2017), trait-based studies traditionally focus on single-trait–fitness relations and do not consider habitat-suitability limitations on demographic rates.

4.6 Traits and Species-Distribution Models

One application of the variation of species along environmental gradients is the modelling of species distributions on the basis of their niche, so-called *species-distribution modelling* (SDM, among a whole bouquet of other names for the same general approach). There has been an enormous development and much scientific success during the last two decades on this topic. This concept opens the door to answering important questions, such as where species are going to be distributed in the future under scenarios of global change, or how much and to where an invasive alien species is likely to spread. These models try to predict the occurrence of a species in a location based on multiple parameters that define the (realized) niche (Box 4.1) of the species along environmental gradients. The predicted occurrence map of a species can then be compared to the actual observed species distribution (Fig. 4.7). A contemporary and extensive treatment of SDM, including its application in the R software framework, can be found in Guisan et al. (2017). Though the SDM approach is based on species' responses to environmental gradients and does not care so much about what traits are responsible for this response, there are several aspects of SDM for which traits are helpful and relevant, which we will briefly overview here.

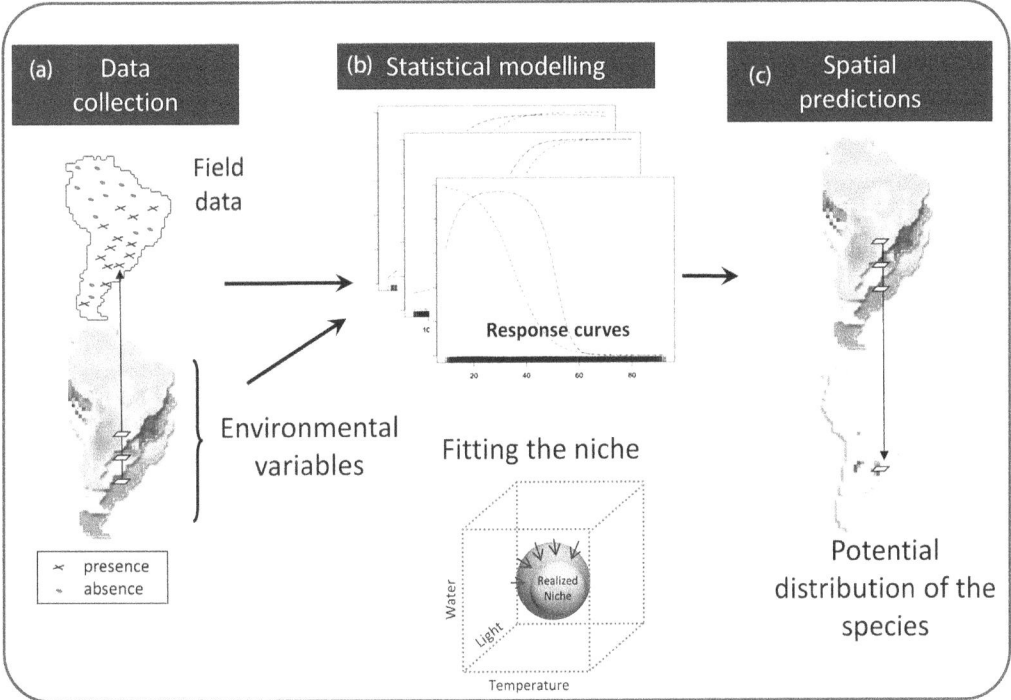

Figure 4.7 The general workflow for modelling the distribution of species. Species observations and environmental data from the observation localities are used as input in statistical models. These then predict where a species can potentially occur, given the geographical distribution of the environmental predictors used in the models. Taken from Guisan et al. (2017), with permission from Cambridge University Press. © Antoine Guisan, Wilfried Thuiller, and Niklaus E. Zimmermann 2017.

It is more and more recognized that predictions of future distributions of species solely based on abiotic responses need to be complemented by *biotic interactions* to predict more precisely the occurrence of species (Schweiger et al. 2008; Berg & Ellers 2010; Wisz et al. 2013). For instance, range extensions of many plants are often lower than those of their specialized herbivores, which puts constraints on the range extension of these herbivores. If these biotic responses include heterotrophic interactions, the information about whether necessary prey, symbionts etc., are locally available will obviously determine if a species occurs in a given location, independent of its dispersal ability and physiological limits, which are set by, e.g., climate. The same is valid for biotic interactions within trophic groups; for example, if we know that certain species exclude each other due to competition, we can refine our prediction of the occurrence of a species by the occurrence of other, potentially competitively superior, species.

Regarding the abiotic variables that determine the occurrence of species in terms of their niche, it has recently been recognized that traits can be used to define the response of species to abiotic gradients, and therefore can assist and improve the prediction of species distributions. For instance, Pollock et al. (2012) showed that the inclusion of traits of the

L-H-S scheme (i.e. specific leaf area, plant height, seed mass; Chapter 3), including their interactions, in modelling the occurrence of multiple *Eucalyptus* species clearly demonstrates that traits modify the responses of species to environmental gradients.

Similarly, Shipley et al. (2017) demonstrated that traits can be used to predict habitat affinities of species and the potential occurrence of species at a given site. The maximum entropy (*Maxent*) approach of Shipley et al. (2006) predicts the abundances of species as a function of the environment and both the community trait means as well as the single-species trait values. A second approach, called '*trait space*', is basically an extension of the Maxent approach that allows incorporation of the entire trait-distribution structure (not only the mean), and thus accounts for intraspecific trait variability (Laughlin et al. 2012). In this way, the trait space model of Laughlin et al. (2012) could be used to model more precisely how co-occurring species would interact with each other, given that intraspecific trait variability is available or can be reasonably well estimated. Thus, species functional traits have great potential to be incorporated into SDM, making this approach less correlative and more process-orientated. This is also the case for studies that link predicted habitat suitability to functional traits like individual growth rate (e.g. in Grass Carp; Wittmann et al. 2016).

Besides the purpose of SDM to directly predict the geographical distribution of species, the predictions gained by the models can be used to predict species richness by 'stacking' species distribution models of multiple species. In the same way, such stacked SDM can be used to predict other aspects of biodiversity, including trait-based functional diversity and mean trait values of assemblages. The latter is especially interesting and promising for predicting the functioning of ecosystems under scenarios of future climate and land-use change, as attempted in a number of case studies (Buisson et al. 2013; Grigulis et al. 2013; Albouy et al. 2014; Thuiller et al. 2014).

Take-Home Messages

- Different environmental conditions 'filter out' individuals with less suited functional traits. As a result, we can observe trait changes within- and across-species along environmental gradients.
- There are different types of environmental gradients. When associating trait changes to a given gradient it is important to establish, as precisely and in as standardized a way as possible, which environmental conditions vary along the studied gradient.
- It is possible to run different tests relating traits to environmental gradients, i.e. at the within-species level (multiple individuals of a species measured along a gradient), at the species level (multiple species with different environmental preferences, i.e. niches) and at the community level (aggregating trait values for a group of species coexisting at a given spatial scale).
- Species level analyses are particularly important for relating traits to species fitness, hence defining which traits are more functional, which in turn allows parameterization of species distribution models.

5 Community Metrics

In the last decade, ecologists have proposed a plethora of methods to characterize the functional trait structure of biological communities. These measures offer different solutions to summarize the trait information of the species in a community as a single number. Most of these efforts have focused on 'functional diversity' (e.g. Villéger et al. 2008; Pavoine & Bonsall 2011; Carmona et al. 2016), and the resulting 'diversity of indices' can be overwhelming, both mathematically and conceptually. In this chapter we intend to navigate ecologists through this labyrinth of approaches, outlining their similarities, limitations and practical applications. Since a comprehensive exploration of all indices is well beyond the scope of a single book chapter, we will focus on those that are currently most often applied and, when possible, on the use of existing tools in R to compute them (see 'R material Ch5' for details).

While R tools generally provide results with relatively little effort, it is essential for users, even those without a strong mathematical background, to understand the basics of how each index is computed. This allows users to develop a deeper understanding of the ecological relevance of the indices' values. As we noted in Chapter 3, it is essential to consider which data are required, in what format, and what the effects of data transformation and missing trait data may be when using functional diversity indices. This chapter explores these considerations in detail for the simplified case where information on intraspecific trait variability is not available. In Chapter 6 we will further deal with ways to introduce intraspecific trait variability data, when available, using the existing tools (note though, that in the R material Ch3 we also provided tools to compute trait dissimilarity accounting for intraspecific trait variability).

5.1 Community Functional Trait Structure

A biological community is understood in this book as an assemblage of species coexisting (or co-occurring) at a given time and place (Chapter 1). Biological communities are commonly characterized using data from sampling units (such as vegetation plots, pitfall traps etc.), including the identity, and often the abundance, of co-occurring species.

For ease of interpretation, the functional trait information of a biological community is often summarized into a limited number of indices. One family of approaches summarizes the number of species within, or abundance of, certain 'functional groups' or 'functional types' (see Chapter 3). For example, we may choose to document the relative abundance of nitrogen-fixing plant species in a meadow, or how many carnivorous fish species are in a pond. These sorts of indicators attempt to summarize how *dominant*, or rare, a given *type* of species is. In terms of a quantitative trait, like plant height or body size, this approach would correspond to estimating the most frequent value expected in a collection of organisms in a locality. For example, in a classroom of students, it could correspond to the average height of students in that classroom (Fig. 5.1).

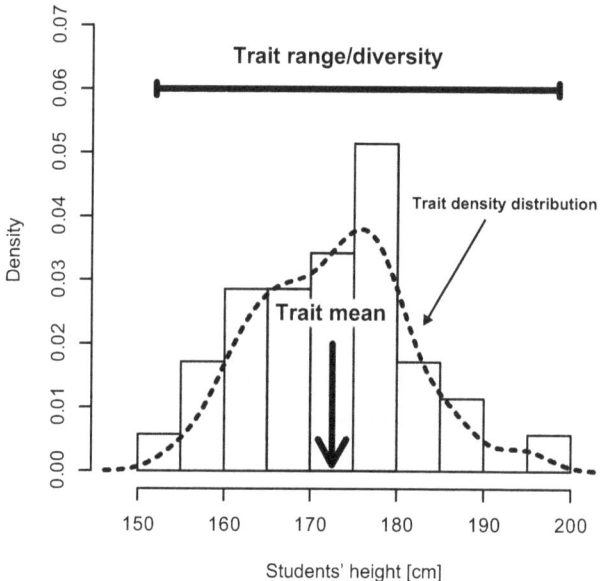

Figure 5.1 A graphical representation of the two most intuitive parameters to characterize the functional trait structure of a community, i.e. (1) the trait mean and (2) trait range. In this example the 'community' represents the height distribution of a classroom of 35 students. The 'trait density distribution' approximating the trait distribution in the communities (dotted line) can then be measured by many parameters, like variance, mean dissimilarity etc., further described in this chapter.

Another type of index quantifies the variability of traits across organisms, i.e. how much organisms differ between them. It is now 60 years since George Evelyn Hutchinson asked, 'why are there so many *kinds* of species?', when observing water bugs in the Santa Rosalia pond (Hutchinson 1959; see also Chapter 7). Similarly, by observing a coral reef one could easily conclude that a great variety of forms coexist in a given habitat. Accordingly, a family of indices intends to quantify how many kinds of species there are in a study unit. The extent of such phenotypic differences between organisms can be defined as functional trait diversity, or more simply, *functional diversity*. In the same way that the 'average' can be quantified in different ways (like the mean and the median), functional diversity can be derived from metrics like the range, the standard deviation, the variance etc. In the case of height across a group of students, each of these metrics represents a component of the phenotypic diversity among the students for the trait considered (Fig. 5.1).

In mathematical terms, the 'shape' of a set of values, such as the probability density distribution of the students' heights shown in Fig. 5.1 (which we call 'trait density distribution'), can be characterized by various indices. Each of these indices is called a 'moment' in mathematical terms. The so-called 'first moment' of a distribution of values corresponds to their mean. The 'second moment' is the variance, the 'third moment' is the skewness, and the 'fourth moment' is the kurtosis. Similarly, in functional ecology, these four 'moments' can be useful in characterizing the distribution of traits in a community, or in a region, or on whatever other scale of interest (Le Bagousse-Pinguet et al. 2017). While all four moments are interesting in their own particular way, the first two already provide a considerable amount of information and we will focus on them in this chapter.

Before introducing the variety of indices to characterize the distributions of traits in a community, it is important to mention that there has been a great degree of inconsistency in terminology in the literature – unfortunately also throughout some previous publications by the authors of this book. We argue that any index that attempts to summarize the 'composition' of a community in terms of the traits of the constituent species can be referred to as an index of *community functional trait structure* (Garnier et al. 2015), also called *trait distribution* (Carmona et al. 2016). Often, also, authors refer to a *functional trait space*, because most of the indices can be represented in multivariate trait space (see later in this chapter) to represent the extent and distribution of traits in a particular study unit (e.g. population, community, region etc.). These terms can be employed, basically, as synonyms.

The functional structure of communities cannot be assessed by a single measure but rather needs a multi-index approach (Díaz et al. 2007). We have outlined above two main families of indices, either characterizing (1) the dominant traits in a community, or (2) the extent of functional differences between organisms (which we defined above as 'functional diversity'). The general consensus is that these two types of indices (which correspond to the first two 'moments' describing a trait distribution) constitute the most typical toolbox for functional ecologists. While this basic toolbox can be further enriched, they have been proposed as a general framework to tackle a great variety of ecological questions (Ricotta & Moretti 2011).

Some authors have referred to both these types of indices (referring to dominant traits and trait differences) collectively as functional diversity (Díaz et al. 2007). Here, to remove this semantic ambiguity, 'functional diversity' will be used only in the narrower sense, i.e. to describe the extent of trait variability between organisms (Fig. 5.1), as mentioned above. Therefore, in this book we formally exclude all measures referring to community trait means from the definition of functional diversity. On the contrary, we will use 'functional structure' and 'trait distribution', as synonyms to describe all the possible indices summarizing the traits of species in a community (i.e. both dominant traits and trait differences).

5.2   Community Weighted Mean (CWM)

5.2.1   Computing CWM

A simple yet powerful index to describe the functional structure of communities is the so-called community weighted mean, or CWM, which was conceptually introduced in Chapter 4. The index has sometimes also been called the 'community aggregated trait' or 'community functional parameter' (Violle et al. 2007). The index corresponds, for each trait (either quantitative or qualitative), to the average trait values in a community weighted by the species' relative abundances. In other words, the most abundant species have a bigger 'weight' on the average. The formula can be summarized as:

$$CWM = \sum_{i=1}^{N} p_i x_i \qquad \text{Eqn. (5.1)}$$

where N is the number of species found in a given community, p_i is the proportional abundance of species i in that community (and goes from 0 to 1), and x_i is the trait value of species i.

In Fig. 5.2 we show, in practical terms, how CWM is computed for two different types of traits. The figure shows that two types of objects are needed for computing CWM (and all other indices described in this chapter as well): (1) a 'species × community' matrix, i.e. an object including species composition across a number of sampling units defining the different communities, and (2) a 'species × traits' matrix, summarizing trait information for a considered set of species (which was already introduced in Chapter 3). In the 'species × community' matrix, users might have some quantitative data of species abundance (e.g. number of individuals) or might have only presence/absence data (i.e. only zeros and ones).

The 'species × traits' data matrix usually includes a single value per species, often corresponding to the average across multiple measurements. In this matrix, categorical traits with two levels are generally converted into a series of binary variables (note that if categories follow an order, e.g. height classes, then categorical traits can be expressed as a single ordinal variable; see Chapter 3). This issue will be further discussed below and in this chapter's accompanying R material (some R functions already automatically convert categorical traits into binary codes).

5.2 Community Weighted Mean (CWM)

INPUT DATA

'Species × community' matrix

	comm. 1	comm. 2	comm. 3
species 1	10	49	0
species 2	10	0	6
species 3	10	5	0
species 4	10	0	4
species 5	10	5	0
species 6	0	0	2
species 7	0	11	8
Sum	50	70	20

Species relative abundances (p_i)

	comm. 1	comm. 2	comm. 3	comm. 3.NA
species 1	0.20	0.70	0.00	0.00
species 2	0.20	0.00	0.30	0.33
species 3	0.20	0.07	0.00	0.00
species 4	0.20	0.00	0.20	0.22
species 5	0.20	0.07	0.00	0.00
species 6	0.00	0.00	0.10	0.00
species 7	0.00	0.16	0.40	0.44
Sum	1	1	1	1

'Species × traits' matrix (x_i)

	Body size	Carnivory
species 1	10	1
species 2	20	1
species 3	30	0
species 4	40	1
species 5	50	0
species 6	NA	1
species 7	70	0

$x_i * p_i$ for body size trait

	comm. 1	comm. 2	comm. 3.NA
sp1	2.0	7.0	0.0
sp2	4.0	0.0	6.7
sp3	6.0	2.1	0.0
sp4	8.0	0.0	8.9
sp5	10.0	3.6	0.0
sp6	0.0	0.0	0.0
sp7	0.0	11.0	31.1
Sum	30.0	23.7	46.7

$x_i * p_i$ for carnivory trait

	comm. 1	comm. 2	comm. 3
sp1	0.20	0.70	0.00
sp2	0.20	0.00	0.30
sp3	0.00	0.00	0.00
sp4	0.20	0.00	0.20
sp5	0.00	0.00	0.00
sp6	0.00	0.00	0.10
sp7	0.00	0.00	0.00
Sum	0.6	0.7	0.6

OUTPUT: CWM for body size ; CWM for carnivory

Figure 5.2 Input and output data in the calculation of the community weighted mean (CWM) for one quantitative and one qualitative trait.

In the example of Fig. 5.2, with a total of seven species found across three communities, we consider two traits, body size (say in cm) and carnivory (yes/no), for some fish species. The first step is computing the relative abundance of species in the three communities based on the input data of the species × community matrix. For example, species 1 ('sp1') has 20 individuals in community 1, where the total number of individuals in that community is 50. The same species has 49 individuals in community 2, where the total number of individuals is 70. So its relative abundance would be 10/50 = 0.2 (i.e. 20%) and 49/70 = 0.7 (i.e. 70%) in communities 1 and 2 respectively. Species 1 is absent from community 3, so its p_i there is 0. Obviously the sum of p_i in a community must be equal to 1, i.e. 100%, since p_i reflects the proportional abundance. Doing this sum can be a good way to check if you are computing things correctly, both when doing calculations by hand (which we recommend doing at least the first time you do this, to make sure you understand the process) and with R. The second step in the calculations is computing the product of x_i and p_i, i.e. the multiplication for each species' relative abundance and its trait value in each plot. So the computation is done for each trait separately. Finally, the CWM is simply the sum of all these products ($\Sigma x_i * p_i$) in each community.

The CWM value for quantitative traits can be understood as the mean value which would be more likely to be found if we drew one individual randomly from the community. In community 1, the CWM for body size is 30. In this case, all five species present in that community (species 1 to species 5) have the same abundance (10 individuals each) and the CWM in this case corresponds exactly to a simple mean of the traits, i.e. $(10 + 20 + 30 + 40 + 50)/5 = 30$, as all species have the same 'weight'. By randomly selecting, for example, 20 individuals out of the 50 in community 1, the mean value would likely approximate 30. It is important to note that this case (species have the same weight) also reflects the case in which we do not have data on species' abundances. In this case, the relative abundance of species, p_i, will be $1/N$ (i.e. 1 divided by the number of species in that community). This corresponds exactly to the case of users having presence/absence data only (i.e. 0 and 1 in the species × community matrix).

The case of community 3, for body size, is also important, because it includes species 6 with no trait information available. Missing values are an important issue in functional ecology (Chapter 2) and can sometimes be replaced by phylogenetically informed estimations (Chapter 8). In the most simple, and common, scenario, the species for which trait information is missing is completely discarded from the calculations, for all communities where the species is present. In doing this, a new relative abundance for the 'remaining' species is computed (see 'comm.3.NA' in Fig. 5.2). There were two individuals of species 6 in community 3, accounting for 10% of the total number of individuals. However, because body size information is not available for this species, the CWM will be computed as if this species was actually absent, i.e. it has an abundance equal to zero. If we 'remove' species 6 from community 3, the new total number of individuals would be 18, instead of 20. Then, the relative abundance of, for example, species 2 in community 3 would go from 0.3 (6/20) to 0.33 (6/18) and so on. This is important because it means that each time we are missing trait information for a given species, we actually remove that species from the analysis. The impact of doing this will be proportional to how abundant, in relative terms, the species 'removed' is (Pakeman & Quested 2007; Majeková et al. 2016a). In the case of community 3, the effect is probably relatively small, because the species represents 'only' 10% of the community, although it could still be biologically very relevant. If, for example, the trait value of species 6 would be 60, then the 'real' CWM would be 52.7, instead of the computed value of 46.7, with a difference of ~11%. At the same time, the CWM value of community 3 would still be bigger than for communities 1 and 2, so it is likely that biological conclusions about ranking (Kazakou et al. 2014) would not be altered considerably, in this case at least. We will discuss in other sections (Chapters 8 and 11) the problems and solutions concerning missing trait values.

We can now move our attention to the case of qualitative traits, also called categorical traits, such as carnivory, in Fig. 5.2. The way CWM is computed with this sort of trait is exactly the same as for quantitative traits, once the trait has been transformed (implicitly or explicitly) into some quantitative scale (see further below). The only difference is with the interpretation of the index. The CWM in this case represents the percentage of a given category of the trait in a community. For example, CWM for carnivory in community 2 is equal to 0.7, indicating that 70% of the individuals in the community are carnivorous and, hence, 30% are not. CWM for qualitative traits also reflects, as in the case of quantitative traits, the expected trait value (carnivory yes vs no) when picking random individuals from that community. But the interpretation of CWM for qualitative traits as a percentage is probably the most straightforward. Otherwise, the difference between community 1 and community 2, both having CWM for carnivory equal to 0.6, is that in the first (with all species having the same abundance) it can be interpreted as the percentage of *species*, being carnivorous, while in community 3 it can be interpreted only as the percentage of *individuals* being carnivorous, but not as the percentage of *species*, as species have different weights.

Note that we have made all exemplary calculations in this section based on individuals as a measure of the abundance of species, which makes it a bit easier to explain the implications of the computations. However, as we discuss in the next section, we can use any measure of species abundance to calculate CWM (and in fact abundance-weighted functional diversity measures; see section 5.3), such as biomass, estimated cover in a vegetation plot etc.

5.2.2 Considerations on Species Abundance

As outlined above, in calculating indices we standardize species abundances relative (i.e. *proportional*) to the total abundance of the community (p_i). This critically affects both CWM and many other indices discussed below (Majeková et al. 2016a). Relative abundance could reflect, for example, the number of individuals of species i proportional to the total number of individuals in the community (see Fig. 5.2). However, the number of individuals is often difficult to measure, as individuals from clonal species are not easily distinguishable. Fortunately, abundance can be expressed in many other ways, for example in terms of biomass (biomass of one species over the total community biomass), cover (cover of the species over the sum of all species cover) or frequency (e.g. number of subplots, or contacts in a point quadrat method, over the total of all subplots/touches in the community).

Here we show why the CWM can be a practical solution when the traits have not been measured for each individual (Chapter 11). Imagine now a simple community composed of three species (Fig. 5.3). The first species in the example in Fig. 5.3 has five individuals with sizes 20, 25, 30, 35, 40 (i.e. mean = 30 cm). The second species has three individuals with sizes 35, 40, 45 (mean = 40 cm) and the third one has only one individual of 80 cm. If we consider all individuals independently, the mean will be 38.9, i.e. (20 + 25 + 30 + 35 + 40 + 35 + 40 + 45 + 80)/9. If we can consider a plain species average, where all the species have the same weight, the average is computed as (20 + 25 + 30 + 35 + 40)/5 = 30 and (35 + 40 + 45)/3 = 40 for species 1 and 2 respectively, while species 3 has only the value 80. The average of these three values is (30 + 40 + 80)/3 = 50. These two averages (38.9 and 50) are quite different and this is because the one individual of species 3 has a disproportionately large effect in the second approach.

Most importantly, notice that the first 'average' approach can also be computed by Eqn 5.1 making a weighted average of species means, with the weight of each species

Community Metrics

	Sp1	Sp2	Sp3	MEAN
$CWM = \sum_{i=1}^{N} p_i x_i$				
Size of different individuals (cm)	20 25 30 35 40	35 40 45	80	38.9
Mean per species (x_i)	30	40	80	50
				SUM
Species relative abundance (p_i)	5/9 = 0.56	3/9 = 0.33	1/9 = 0.11	1
$x_i * p_i$	30*0.55 = 16.7	40*0.33 = 13.3	80*0.11 = 8.9	38.9

Community Weighted Mean (CWM)

Figure 5.3 Understanding the CWM and the meaning of species relative abundance. In the example, CWM is computed for a community with three species (Sp1, with 5 individuals, Sp2, with 3, Sp3 with 1) and one trait.

proportional to the number of individuals per species, i.e. 5/9 for species 1, 3/9 for species 2, and 1/9 for species 3. This corresponds to the CWM. As such, the mean of species (30, 40 and 80 for the three species respectively) can be weighted by their relative abundance, i.e. 30 * 5/9 = 16.7, 40 * 3/9 = 13.3 and 80 * 1/9 = 8.9. The sum of these three values, i.e. 16.7 + 13.3 + 8.9, which is actually what the formula above shows, is actually 38.9!

Most of the time it will be impossible to measure traits for each individual (Chapter 11) so that having a measure of species abundance and some average trait value for each species would be enough to answer several ecological questions. Obviously the way the abundance is estimated will affect the results. Imagine we want to compute CWM of plant height in a forest and we express the abundance of species in terms of number of individuals (assuming that individuals are easily distinguishable), both for trees and herbaceous species. Most likely, the number of individuals of herbaceous species will outnumber the amount of tree individuals. As a result, herbaceous species will have a bigger impact (greater p_i) on the CWM, which will be very similar to the height of herbaceous species. Imagine now that we express abundance in terms of species cover. Most likely, in this case, tree species will have a greater p_i, and the CWM will be closer to that of the trees. Finally, if we use biomass to estimate abundance, then trees will have a much bigger abundance and the CWM will very much approximate the values of tree species. Obviously, none of these alternatives (individuals vs cover vs biomass) is better *a priori*, or closer to reality. Rather, each approach considers a different type of species as dominant (e.g. the species with a greater abundance) and thus focuses on a different part of the community structure (i.e. either trees or herbs). In some cases, you might even prefer to compute CWM only on trees or only on herbaceous species to avoid such problems and focus more directly on a given portion of the community structure. To do this, you can split the species × community matrix into two matrices and compute the CWM for each of them.

A similar issue is found when deciding if species abundance data need some transformation. The input data 'species × community' matrix can be transformed before computing species' relative abundances. A transformation that is frequently applied to data based on number of individuals or biomass is the $\log(x + 1)$ transformation (adding 1 to account for zeros in the species × community matrices). For example, in community 2 in Fig 5.2, species 1 has 49 individuals, thus $\log(49 + 1) = 3.91$, while species 7 has 11, so $\log(11 + 1) = 2.48$. It is clear that the abundance of species 1 after the transformation decreases much more than that of species 7; the relationship between the two species is $49/11 = 4.45$ without log and $3.91/2.48 = 1.57$ with log, the second being a much smaller difference. Log transformation, thus, changes the perspective of abundance and dominance, making communities less dominated by a few species.

By decreasing dominance, log transformation most often increases evenness in species abundances. Hence, CWM based on relative abundances computed after log transformation will be much more dependent on species that were originally not so dominant. The effect of transformation will be bigger, in general terms, the more unevenly the abundance (without transformation) is distributed among species (Majeková et al. 2016a). Consequently, the effect of data transformation will be bigger when abundance is estimated as number of individuals or biomass, because these methods detect very few dominant and many rare species. Discussing whether species × community data need transformation and what type of transformation is needed is outside the scope of this chapter (but we invite you to read Šmilauer & Lepš 2014). In general, the most important message to keep in mind is that transforming species abundance data gives much more importance to the signal of subordinate species (Cingolani et al. 2007; de Bello et al. 2007). This can be used to our advantage; for instance, computing different CWMs based on different abundances of species enables us to see if changes in CWM along environmental gradients are due to dominant or subordinate species (e.g. Kichenin et al. 2013).

Box 5.1 Is CWM a trustable index?

Indeed, just like any index of community structure, CWM is also not an absolutely perfect index. It presents issues that need to be kept carefully in mind. One important issue when using CWM is to assess changes in trait values along gradients. Why so? In this box we provide a view on recent discussions on the use of CWM to assess trait changes along gradients.

The main potential 'problem' is a rather straightforward property of the index, i.e. that one single species can cause important changes in CWM, as long as it is a rather dominant or frequent species in some plots and less abundant in others. If the abundance of such a species changes along a gradient, and it possesses certain traits, though not related directly to the gradient, it might appear that the trait value of the

Box 5.1 (*cont.*)

entire community responds to that gradient when it is calculated, when in fact the trait should not be considered as a response trait. Peres-Neto et al. (2017) have shown this, in a statistical sense, while (a) discussing that this phenomenon can lead to an overestimation of significant relationships between CWM and environmental gradients, and (b) arguing that their alternative analytical fourth-corner approach is superior. Zelený (2018) offers a less dismissive solution, where the appropriateness of CWM is dependent on the ecological question being asked. We here summarize this discussion, starting with a simple example to illustrate the issue.

Imagine two plant communities, one fertilized and one not, and that we are interested in assessing changes in traits along the fertilization gradient (to ultimately understand what happens to community structure and its relationship with productivity and litter decomposition, for example). Now imagine an extreme (and possibly rather rare) case in which the two communities have exactly the same species – say 10 species, the same 10 taxa in both communities. Everything important stays the same in the two communities except for one thing – one species is more abundant in one of the two communities. Imagine that this species is *Taraxacum officinalis*, the dandelion, and that it trebles its abundance in the fertilized community. Obviously any trait that *Taraxacum* has will appear differently in the CWMs of the two communities, no matter if it actually represents a response to fertilization or not. For example, CWM of flower colour will change towards a more yellow-dominated community and, possibly, specific leaf area (SLA) of the community will be higher. If we have several plots under this exact same scenario, these traits will both result as statically 'responding' to changes in fertilization. However, the yellow flower trait is, logically, not the real response trait that causes *Taraxacum* to be more dominant in fertilized conditions. Maybe the real response trait is that this species has a greater SLA than any other species, and this causes *Taraxacum* to assimilate resources faster than other species. Of course we cannot trust the results on flower colour, so we have a problem! Both tests for flower colour and leaf type will give a significant result, but maybe only one is the real response trait (or even worse, neither of them is and we did not measure any of the actual response traits). It is true that in this case we cannot blindly accept the results, as already pointed out by Šmilauer and Lepš (2014). Luckily, as biologists we are trained not to accept all results as meaningful.

There are, of course, less extreme cases in which it is less clear if changes in CWMs can be considered, or not, as a true reflection of trait filtering (Chapter 4). Peres-Neto and colleagues (2017) suggested altogether that other methods, like an improved version of the so-called RLQ, should be preferred compared to CWM (see also Chapter 4). Zelený (2018) further suggests that randomizing species identity could be applied to compare CWM to a null expectation to test how strong the relationship is between CWM and the environment. As Zelený nicely summarizes, it is very important to specify what the null hypothesis being asked is. With the null model he suggested, similarly to Peres-Neto et al., we can interpret such null expectations as that filtering is NOT a widespread phenomenon across species along

Box 5.1 (cont.)

a gradient (remember the null hypothesis is that nothing is happening – in this case that there is no filtering operating on most of the species). Hence, if the observed relationship between CWM and environment is stronger than the null model, we can interpret this as a pronounced turnover in species identities along the gradient and an association of this turnover with trait differences. This is a fine ecological hypothesis, but the problem is that it is just one possible way that trait filtering can operate (Chapter 4). In an extreme case, filtering can operate, for example, only on one or very few species, changing the abundance of these species along a gradient (see the dandelion example above). Filtering can even operate because of changes in trait values within species (intraspecific trait variability; Chapter 6).

So, let's go back to the dandelion example. If we apply the null model suggested by Zelený, the relationship between SLA and fertility will likely be similar between the observed and randomized data. Similarly, the data won't likely show a significant relationship between traits and environment using the approach by Peres-Neto and colleagues. So we can indeed conclude that filtering based on that particular trait does not operate on the *majority* of species. But can we really dismiss the fact that maybe one species, or a few species, and not the majority, actually respond to the gradient because of some response trait? Maybe filtering does not operate on all species but only selects the few winners. And if this is the case, using a null model to test the null hypothesis that filtering operates on the majority of species would then miss this pattern.

From an effect trait point of view (see Chapters 3 and 9), the changes in flower type and leaf type along a gradient are also important, in terms of the functioning of ecosystems, irrespective of whether they are caused by one species or many. If one species with high SLA increases, e.g. between the communities along the fertility gradient mentioned above, it is possible that productivity will increase (Garnier et al. 2016) and litter decomposition will be accelerated because of greater litter decomposability (Cornelissen et al. 1999). Even the increase in yellow flower types can affect pollination. So a simple test of a significant change in CWM along a gradient can still be very much indicative of important changes in the community structure, irrespective of whether this change is due to a few or many species.

Maybe, together with the solution proposed by Zelený and Peres-Neto and colleagues, it is useful to compare results using CWM with and without abundance (Ackerly et al. 2002; Cingolani et al. 2007). Greater significance when accounting for species abundance reinforces the notion that filtering occurs mostly through a replacement of dominant species. This implies a different null expectation than the hypothesis implicitly used in the randomizations, i.e. filtering acting on 'all' species. Maybe not all species are filtered by the main gradient under study (which is essentially the hypothesis tested with the null model by Zelený), but still, there are some important (and predictable) changes. Surely, from an ecosystem-functioning point of view, the fertilized and unfertilized communities might have different processes and associated services. It could be important to know if the effect of this

> **Box 5.1** (*cont.*)
>
> is due to (one or more) dominant species (Chapter 9), but likely there will be an effect that one species with specific traits becomes abundant, at least based on the expectation of the mass-ratio hypothesis (Grime 1998).
>
> We conclude that the CWM is a very simple, yet powerful, quick-to-compute and rather easy-to-interpret indicator, and should not be discarded without a second thought. Combining results using abundance or not, or even considering log-transformed abundance, can be done as a further test to assess trends of dominant and not-so-dominant species. We can also conclude that the null model suggested by Zelený (2018) is a simple and powerful way to check if filtering occurs on the majority of species or not.

5.3 Functional Diversity Indices

After having seen computations and issues with CWM we can now move to functional diversity. To compute functional diversity, it is often necessary to quantify the trait differences between each pair of species in an assemblage. Potential tools, and pitfalls, in such calculations are outlined in Chapter 3. In the present chapter we assume that readers are familiar with methods for estimating species' pairwise dissimilarities, and thus we move directly to describing some of the most commonly used functional diversity measures (summarized visually in Fig. 5.4). Indices of functional diversity, as for CWM, can be commonly computed from two matrices of data, the species × community matrix and the species × trait matrix (see above). For most of the indices a trait dissimilarity matrix is computed before the computation of functional diversity (Fig. 3.2, Chapter 3). Some exceptions exist, of course, mostly for indices computed on single traits (as opposed to multivariate indices). We will introduce the main indices of functional diversity following several criteria, basically starting from the simplest ones, and trying to group them into 'families'.

Before treating these indices, it is important to stress that one of the first measures of functional diversity that have been applied in the literature is the number of functional groups present in a community (Díaz et al. 2001). While this approach is apparently simple, Chapter 3 shows that the definition of functional groups can often involve a degree of subjectivity. In cases where we can assume such grouping of species is reasonable, then the number (or evenness) of functional groups could indeed constitute a useful index of functional diversity.

5.3.1 Range and Convex Hull

The *trait range* in a community is probably one of the simplest indices. In the example in Fig. 5.1 it would be simply the difference between the tallest and the shortest

5.3 Functional Diversity Indices

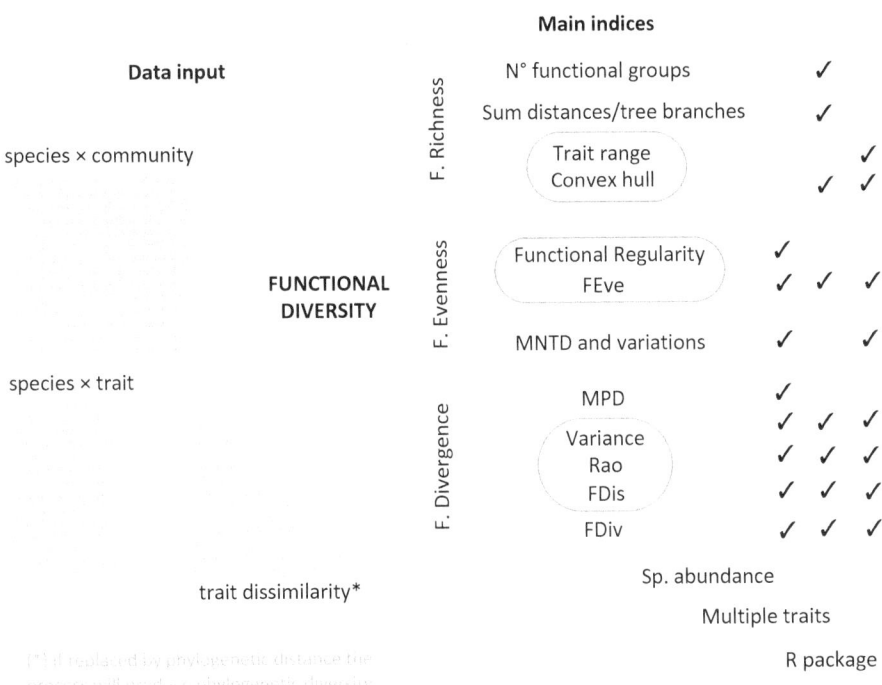

Figure 5.4 Main functional diversity indices, their proposed categorization, and information on whether or not (i) species abundance is considered, (ii) they can consider multiple traits at once, and (iii) R functions are well included, at present, in some R package. Circles indicate indices that are strongly mathematically linked. Not all indices need necessarily computing a trait dissimilarity, see text, although in practice is the most common approach.

heights: ~50 cm. Such a range would say that for this trait there is quite some variability in the heights of the students (probably exceeding, by far, the variability observed during our periodic lectures!).

Trait range can be adapted to account for multiple traits, and be expressed as a multivariate volume including all species in a community. This can be visually represented for two (or more) traits as the geometric shape that includes the trait values of the species in the target community (Fig. 5.5). The bigger the volume, the bigger the functional diversity is. In mathematical terms, such a shape is called a *convex hull*; this is an approach proposed by Cornwell et al. (2006) and called *FRic* (short for functional richness) in the R function dbFD (Laliberté & Legendre 2010). In practice, the index is the minimum volume that can include all species in a community. The convex hull approach has been further refined and adapted to other cases for example, as we will discuss in Chapter 6, to define trait hypervolumes (Blonder et al. 2014; Carmona et al. 2016). In practice, most often these approaches do not consider how densely populated the 'functional space' is, but they consider 'only' how big the portion of this space is realized in the community. With these examples in mind you can now understand why the functional structure of the communities is often understood in terms of 'functional

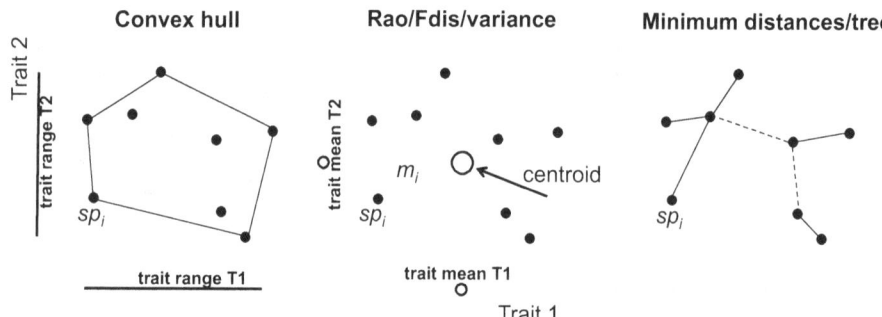

Figure 5.5 Visualization of three of the approaches to quantify functional diversity discussed in this chapter: convex hull, multivariate mean dissimilarity and uniformity. Each point in trait space is a species (sp_i) with two trait values (T1 and T2). The position of a point on the axes reflects the value of the species for both traits. The black points define the species in a target community, and the grey points define the species from other communities. Notice that the computation can be done for more than two traits and that the two (or more) traits in the figure can be replaced by multivariate axes (for example, treating a dissimilarity matrix with a PCoA as suggested by Villéger et al. 2008).

space'. It should also be noted that, for single traits, the convex hull corresponds exactly to the trait range (Fig. 5.5). Similarly to the trait range, there is a good chance that the convex hull will increase with the number of species considered, so that standardization is required if we want to remove such an effect, for example through null models (Cornwell et al. 2006). Similarly, log transformation or square-root transformation of traits is needed to avoid problems with potential increasing variance of trait values with larger trait mean (see below).

5.3.2 Sum of Distances

Another family of rather intuitive indices of functional diversity is those concerning the sum of trait dissimilarities in a community. Let's now go back to Chapter 3 and specifically to Fig. 3.2. Let's then consider community 3 in Fig. 5.2, composed of species 2, 4, 6 and 7. When we consider the dissimilarity of the three traits in Fig. 3.2 we obtained these values: $d_{2,4} = 0.44$, $d_{2,6} = 0$, $d_{2,7} = 0.94$, $d_{4,6} = 0.5$, $d_{4,7} = 0.83$, $d_{6,7} = 1$. The sum of these dissimilarities is 3.72 and can represent a measure of functional diversity. Different authors have proposed indices that reflect this information. For example, Walker et al. (1999) proposed a simple sum, as we just illustrated, called functional attribute diversity (*FAD*). Actually, in their original formulation, all the distances are summed twice, because the two sides of the symmetric dissimilarity matrix are considered (see Chapter 3). Hence in our example FAD = 3.72*2.

Similarly, Petchey and Gaston (2002) proposed creating a functional dendrogram based on species dissimilarities. This is the sort of tree in which the distances on the tree reflect functional distances, which we have already described in Chapter 3 as a tool to define functional groups. This can be done using a variety of clustering routines (see example in the R material accompanying Chapter 3). Functional diversity can then be estimated as the total *dendrogram branch lengths* connecting community members together. Note that, as mentioned in Chapter 3, the type of dissimilarity considered and the dendrogram produced can considerably affect the results. Some solutions based on producing a sort of 'consensus' tree have been proposed (Mouchet et al. 2008). Furthermore, to account for possible effects of groups of correlated traits, Cadotte et al. (2009) proposed to use non-metric multidimensional scaling (NMDS) to create, from the original distance matrix, a new distance matrix that accounts for correlated traits (we will go back to this type of multivariate analysis later on). The clustering and branch length computation can then be performed using this new distance matrix.

It is important to note that none of the approaches described above includes species abundances (although some modifications have been developed to do this). Most importantly, the functional diversity computed with both approaches is strongly related, linearly, to species richness (Pavoine et al. 2013). Hence, the results will be highly correlated with the number of species. For this reason, these indices should only be used, in our view, accompanied by a standardization after using null models (see Chapter 7) to remove the inherent variation due to species richness. Otherwise, these indices will not give much more information than species richness itself. In general, while these options to sum dissimilarities are calculated somewhat differently and could have some slightly different properties (Petchey & Gaston 2006), we think they are overall fairly similar in their meaning and properties and, most importantly, in their strong positive relationship with species richness.

5.3.3 Variance

Another very simple and familiar index would be *variance*, which basically summarizes the (squared sum of the) distances of each species from the community weighted mean, and can be expressed as:

$$Variance = \sum_{i=1}^{N} p_i(x_i - CWM)^2 \qquad \text{Eqn. (5.2)}$$

As mentioned above, the variance is the second 'moment' of a trait distribution. The main difference between the range and the variance is that the former considers only the trait values of the two species with lowest and highest values, while variance considers all species, and therefore it is less strongly affected by outliers. The second important difference is that the trait range cannot consider species abundance, while variance can (by using p_i; see above). Of course, users can neglect potential information on inequality in species abundances, thus giving a weight of $1/N$ to all species (N = number of species). In this case, variance would represent, in practice, a plain average of squared distances of species from the CMW. Note here that users could also use $1/(N - 1)$, instead of $1/N$, when assuming that not all species (or individuals; see Chapter 6) have been sampled. The R function *var* actually uses this alternative. Finally, another difference between trait range and variance is that former should be more positively correlated with the number of species.

Trait range and variance, in principle, can be computed only using one trait (but see below). These two indices are the only ones shown in Fig. 5.4 that do not require the computation of trait dissimilarity. Finally, these indices are designed for quantitative traits, but the results will not be unitless if we do not standardize them, (because they depend on the unit on which traits have been measured). Hence, some type of standardization is needed to compare results between traits. For example, traits could be standardized *before* computing the indices (dividing each value of a particular trait by its maximum value or using 'scale' in R). Alternatively, the values of trait range or trait variance can be standardized after their computation (e.g. dividing each trait range, or each variance, of a trait by its maximum value in a dataset).

It is important to note that for some traits, trait variability is expected to increase with the mean. For example, if we compare different species based on size-related traits (plant height, seed mass, body size etc.) we should be aware that the difference between species could increase for higher trait values. As such, in a community composed of bigger species, the observed variability will appear larger compared to a community with smaller species, although this might not necessarily reflect biological patterns. Imagine a plant species (A) which is 20 cm tall, and another (B) which is 40 cm tall. Their difference is only 20 cm, but B is double the height of A, and therefore it might more easily outcompete the shorter one when they compete for light. Imagine now a species (C) which is 1 m, and another species (D) which is 1.20 m. The difference C–D is again 20 cm, but this 20 cm might be biologically less relevant than the 20 cm difference between A and B, which are probably relatively more dissimilar than C and D regarding how successfully they compete for light. Because size-related traits often increase exponentially, a 20% increase in a community with trait mean of 1 m would be 20 cm, but in a community with a trait mean of 30 cm this would be 6 cm. While, obviously, $20 > 6$, they both reflect, to a similar extent the proportional increase. The risk of not accounting for this proportional scaling can be understood by an example with CWM: imagine you have a species in a community with an extreme seed mass value; this species is going to disproportionately affect the CWM of seed mass, particularly if the species has a considerable relative abundance. To take such effects of species with disproportionally large trait values into account, traits are often transformed (generally using *log transformations*) before computing indices. Many traits, such as the L-H-S ones (Chapter 3), for example, often need such log transformation (Westoby 1998). It is therefore good practice to check, with simple histograms, if traits need some transformation to improve normality and take into account non-linear increases (for example with seed mass, leaf area, body mass). A further control could be done by testing whether the potential increase, or decrease, of trait range and variance with the trait mean is weaker than in random expectations. We treat this issue in 'R material Ch5'. Notice that various other indices discussed below are, directly or indirectly, also connected to trait range and variance. Similar care is also required in these cases.

5.3.4 Mean Dissimilarity

An important family of indices measures functional diversity (implicitly or explicitly) as the mean dissimilarity between species. These indices take into account species' relative abundances (with one exception). Let's consider again community 3 in Fig. 5.2 (with dissimilarities being $d_{2,4} = 0.44$, $d_{2,6} = 0$, $d_{2,7} = 0.94$, $d_{4,6} = 0.5$, $d_{4,7} = 0.83$, $d_{6,7} = 1$, following Fig. 3.2). Although, maybe surprisingly to some, there are different ways to compute the mean of these dissimilarities.

The first decision is whether to consider the dissimilarity between each species and itself, for example $d_{2,2} = 0$, $d_{4,4} = 0$. In other words, the decision is whether or not to consider the diagonal in the dissimilarity matrix (see Fig. 3.2). If we exclude the diagonal, the index produced is called Mean Pairwise Dissimilarity (*MPD*; Weiher & Keddy 1995). This considers only one of the triangles in the dissimilarity matrix, without the diagonal: the example above would be $(0.44 + 0 + 0.94 + 0.5 + 0.83 + 1)/6 = 0.62$. Note that in the case where all species would have the same abundance, the weight of each dissimilarity is 1/6, as there are six pairwise combinations. More generally, in the case that each species has the same weight, the weight of each dissimilarity between pairs of species is $1/(N*(N-1)*2)$ (see Ricotta et al. 2016). It should also be noted that, in the case of species having the same weight, MPD = FAD/N(N − 1). We will show below how MPD can also be computed accounting for species abundance.

In another approach, Schmera et al. (2009) proposed using the sum of the distances (as in FAD) divided by the number of functional units (*MFAD*) as another approach to compute mean dissimilarity. This requires some discrete division of species into functional units, which might be practically challenging with quantitative traits and does not take into account species abundances. MFAD is also positively correlated with the number of species. It should be noted that, in the case of species having the same weight, then MPD = FAD/N(N − 1).

If we do not exclude the diagonal in the dissimilarity matrix, hence consider the dissimilarity between one species and itself, then two related indices can be computed. These are Rao Quadratic Entropy (hereafter '*Rao*'; Botta-Dukat 2005), which is a multivariate form of variance, and Functional Dispersion (*FDis*; Laliberté & Legendre 2010). As we show below, Rao and FDis have inherently the same mathematical basis as variance (Pavoine & Bonsall 2011), and results obtained with them will be very much correlated. Authors have suggested several justifications for preferring one of these indices over the other, but, in our view, no strong biological or mathematical reason seems to generally support this choice. We show the similarity between Rao, variance and FDis in the following formulas. Let's first introduce the Rao index, following de Bello et al. (2016):

$$Rao = \sum_{i=1}^{N} \sum_{j=1}^{N} p_i p_j d_{ij} = 2\sum_{i>j}^{N} p_i p_j d_{ij} \qquad \text{Eqn. (5.3)}$$

In practice, Rao is an average of the dissimilarity between each pair of species i and j (d_{ij}) in a community, weighted by the abundances of both species (by using p_i and p_j).

Note that this formally includes the dissimilarity within species (when $i = j$). This is apparently omitted on the right side of the formula, because it equals zero for each species, but we show below that the effect of the diagonal is present when the weight of each species pair is computed.

It should be noted that Rao is a generalization of the Simpson index of diversity. As such, if $d_{ij} = 1$ (i.e. dissimilarity between all species pairs is equal to 1, or, in other words, all species are maximally distinct), Rao is equal to Simpson (expressed as 1 minus dominance with dominance being $\sum_i^N p^2$). This is a nice property because it shows that the maximum value of Rao equals Simpson when all the species are functionally unique (any $d_{ij} = 1$); hence, this allows comparison between taxonomical and functional diversity within the same conceptual and mathematical framework. Note, that for this property to be true, trait dissimilarity should be constrained between 0 and 1, as we suggested in Chapter 3 (Botta-Dukat 2005; Pavoine et al. 2009). Although other approaches are possible, they can limit the potential comparison with species diversity.

Since the Simpson index represents the maximum value that the Rao index is theoretically able to attain, if we replace $d_{ij} = 1$ in the Rao equation to express Simpson diversity, it is clear that Simpson can be expressed as $2\sum_{i>j}^N p_i p_j$. When expressed in this way, the formulas give an interesting relationship with MPD, which can be expressed as:

$$MPD = \frac{1}{\sum_{i>j}^N p_i p_j} \sum_{i>j}^N p_i p_j d_{ij} = \frac{Rao}{Simpson} \qquad \text{Eqn. (5.4)}$$

This formula for MPD (derived from Swenson 2014a) shows that MPD and Rao provide a different way to deal with the species' contribution to average trait dissimilarity (the former disregarding the diagonal in the dissimilarity matrix, the latter taking it into account). This leads to a different relationship with species diversity and a different use in community ecology studies (de Bello et al. 2016). In practice, MPD and Rao will be often correlated at high species richness values, but at low species diversity they will not be! Also, while Rao diversity tends to be positively correlated with species richness when species richness is low (note that above species richness values of about 10 species, this correlation fades), MPD is not correlated linearly with species richness. It is also important to note that in the case of communities with no species or one species only, species richness equals 0 or 1 respectively. Here, different algorithms might approximate this issue in different ways, but we understand that they correspond to a case with no functional diversity, hence both MPD and Rao can be considered equal to 0.

Let's now show why variance and Rao are related, following Champely and Chessel (2002). Let's assume, for now, that we have only one quantitative trait, so that the dissimilarity is based only on that trait. Then variance can have different forms:

$$\sum_{i=1}^N p_i(x_i - CWM)^2 = \sum_{i=1}^N \sum_{j=1}^N p_i p_j \frac{d_{ij}^2}{2} = \frac{1}{2}\sum_{i=1}^N \sum_{j=1}^N p_i p_j d_{ij}^2 = \sum_{i=1}^N p_i m_i^2 \qquad \text{Eqn. (5.5)}$$

These formulas show that variance is tightly related to Rao, i.e. it is a special case of Rao when the dissimilarity between species is squared and divided by two (compare with Eqn 5.3). This similitude also suggests a new type of formulation where variance is expressed in terms of dissimilarity between species, via Rao. Expressing variance in terms of pairwise dissimilarity allows us to account for any type of dissimilarity based on single and multiple traits (or phylogeny; see Chapter 8). The last part of the formula in Eqn 5.5, derived from Pavoine and Bonsall (2011), shows that Rao and variance can also be expressed in another way, i.e. in the distance of a species i (i.e. m_i) from a (multivariate) average of a community (called the 'centroid' by Laliberté & Legendre 2010; Fig. 5.5). As variance is the average distance of the species from the CWM, parameters can also be expressed in the same way: for single traits $m_i = x_i - CWM$, and for two traits it could be expressed as $m_i = \sqrt{((x_i - CWM)_{trait1}{}^2 + (x_i - CWM)_{trait2}{}^2)}$, i.e. the Euclidean distance. Then Pavoine and Bonsall (2011) further showed that:

$$FDis = \sum_{i=1}^{N} p_i m_i \qquad \text{Eqn. (5.6)}$$

If we compare this formula with the ones for Rao shown above, we clearly see that FDis corresponds, for single traits, to the squared root of variance, i.e. SD of trait values. We can then conclude that the index has the same mathematical basis as the Rao index. Not surprisingly, Laliberté and Legendre (2010) showed that the correlation of FDis with Rao was extremely high ($R > 0.96$). Note that the correlation in that test was not perfect; we think this was simply because (1) the relationship between Rao and FDis is not expected to be linear based on Eqn 5.5 and 5.6, and (2) in their test Rao was computed directly from dissimilarity values, while FDis was computed based on new dissimilarities obtained after transforming the original dissimilarities into multivariate axes, thus creating slight noise, which affected the results. See 'R material Ch5' for more information about this. In practice, and especially if the calculations of trait dissimilarity are done in the same way, using FDis and Rao should obtain very similar results.

The similarity between Rao and FDis, and the relationship between Rao and CWM above, both show that these indices of functional diversity represent how distant species are from a 'centre of gravity' in a functional space represented by the CWM, i.e. the centroid in Fig. 5.5. Another similar index in this family of mean dissimilarity indices is 'functional divergence', or *FDiv*, as proposed by Villéger et al. (2008). The index firstly computes the mean distance of species from the CWM, then it computes the distance of these species from this mean dissimilarity. The index was designed to be independent of FRic, although in some cases (more frequently than expected by chance) we detected a slight positive correlation, which is statistically significant – see 'R material Ch5'. In practice, FDiv will be independent of the number of species and thus considerably resembles MPD in terms of properties. In general, this index has been used less frequently than the others. Another, possibly related, index is Trait Onion Peeling (*TOP*), proposed by Fontana et al. (2015), which combines information on species' positions from the centre of the distribution with convex hull volumes. However, this index has been used mostly for cases where trait information for multiple individuals of a species is available within each community, a case which we discuss in the next chapters.

5.3.5 Regularity

Another type of index tries to quantify how regularly species are distributed in functional trait space. Imagine one community with five species (species 1, 2, 3, 4 and 5) with trait values of 10, 21, 30, 40 and 50 cm, respectively. The range is 40, the variance is ~245. The species are almost 'equidistant' in the sense that the most similar pairs of species are, in all cases, ~10 cm apart. What do we mean by that? If we focus on species 1, the most similar species is species 2, with $d_{1,2} = 11$. The most similar species to a target species can be called a 'neighbour' species. As such, for species 1 the minimum distance between to any other species is 11. It is 9 between species 2 and 3, and it is 10 for species 4 and 5 (or species 3). Now imagine another case, also with five species, but trait values 10, 30, 30, 45, 50. The range and variance will be very similar to the previous example (40 and 245 respectively). But now there are two pairs of species with very similar neighbours (i.e. some of the neighbour species are functionally very similar). Two species are identical (trait value = 30) and two are also quite similar (45 and 50). We can thus conclude that the traits in the first community are more uniformly distributed.

Regularity in trait values has been theoretically attributed to niche differentiation and the response of species to competitive exclusion, i.e. limiting similarity (MacArthur & Levins 1967). Obviously, a community with many functionally similar neighbours implies that there are some pairs of species that occupy the same functional space and might therefore compete more intensively. However, while the ecological concept underlying these indices is rather clear, there is no consensus on how to express this concept with existing indices, and not all existing approaches are well understood.

One reason that creates confusion with these sorts of indices is that they need to account for the regularity of spacing between species in terms of trait values, but also for the evenness in the distribution of abundance across species. Accordingly, the indices can decrease due both to a lack of regularity in traits, but also to the dominance of some species, making their interpretation rather complicated, at least in our view. This might partially explain the relatively lower number of studies in the literature employing these types of indices. Also, we have reason to believe that these indices are either dependent on the number of species, or not always independent from the trait range of the communities (and therefore from the convex hull). We show this below and in 'R material Ch5'. For these reasons, and also following the work by Botta-Dukat and Czucz (2016), we believe that these indices should be used carefully, even more so than other ones.

Possibly the first index of this type that was proposed is the functional regularity index (*FRO*), which can be used for single traits (Mouillot et al. 2005). FRO takes a value of 1 when the distances between all nearest-neighbour species pairs are identical and when all species have the same abundance. Conversely, FRO will approach 0 when some species are tightly packed in terms of trait similarity, and with a high proportion of abundance concentrated within a narrow part of the functional trait space. The index was expanded for usage with multiple traits by Villéger and colleagues (2008), who proposed an index called *FEve*. The approach works by employing a minimum spanning tree that connects points in a multi-trait approach to estimate the nearest neighbour species (Fig. 5.5). The index FEve, for single traits, provides very similar results to FRO, although sometimes multiple potential minimum spanning trees exist for a given set of points. At the same time, in various simulations we have run (see 'R material Ch5'), we detected that FEve, and hence FRO, can be negatively correlated with trait range. In other words, the bigger the range, the lower the chance that species will be unevenly distributed, which is somehow counter-intuitive to us. This feature is particularly evident when simulating cases with a fixed number of species and varying range. As a matter of fact, Villéger et al. (2008) showed some indication of a negative relationship between FRic and FEve ($p = 0.1$). This relationship could be stronger when using simulations in which independent species richness and traits range scenarios are considered.

The last indices in this group are based on mean neighbour taxon distance, MNTD, which was initially developed for studies of phylogenetic community structure (see Chapter 8) where the part 'neighbour taxon' in MNTD refers to pairs of species that are most phylogenetically related to each other, i.e. have the smallest phylogenetic dissimilarity (Webb et al. 2002). As suggested in the example at the beginning of this section, the index does not consider all the dissimilarities between all pairs of species in a community, but just computes a mean dissimilarity between closer species (the neighbours). In the example above (with five species in two different communities), this would be the following: for the first community above (traits being 10, 21, 30, 40, 50), MNTD would be $d_{1,2} = 11$, $d_{2,3} = 9$, $d_{3,2} = 9$, $d_{4,5} = 10$, $d_{5,4} = 10$. The mean of these five values is 9.8. For the second community (traits being 10, 30, 30, 45, 50) it would be $d_{1,2} = 20$, $d_{2,3} = 0$, $d_{3,2} = 0$, $d_{4,5} = 5$, $d_{5,4} = 5$. The mean is 6. Instead of the mean, some authors have also used the SD of the neighbour taxon distance, i.e. SDNTD (Stubbs & Wilson 2004; Kraft & Ackerly 2010). Both MNTD and SDNTD covary negatively with the number of species (more species meaning less regularity) and, therefore, null models need to be used to remove this effect. MNTD and SDNTD also tend to vary with trait range, so a version of SDNTD divided by trait range has been proposed to take this effect into account (Stubbs & Wilson 2004; Kraft & Ackerly 2010). Finally, other modern indices consider different aspects of the regularity and evenness in trait space, like, for example, the Trait Even Distribution (TED) of Fontana et al. (2015) and functional evenness proposed by Carmona et al. (2016). These indices, as we will discuss in the next chapters, need some estimation of traits for multiple individuals within a species, or some indirect estimation of these values (e.g. by assuming a certain standard deviation around the available mean trait value). For this reason we do not describe them here in detail, but in principle they work like the other indices described so far, trying to estimate the regularity in distances between species (TED) and the uniformity of trait distribution in a community (Carmona et al. 2016) with respect to a hypothetical case of perfect evenness/regularity.

5.4 Functional Diversity Components

As we saw in section 5.4, functional diversity can be measured in many different ways. Such a wealth of indices reflects the fact that functional diversity can be quantified by different and, ideally, independent components. How can we 'classify' these different indices? Various authors have proposed that functional diversity can be decomposed into three primary components: *functional richness, evenness and divergence* (Mason et al. 2005; Villéger et al. 2008; Carmona et al. 2016). This distinction is based on a similar idea of trying to group the variety of indices of species diversity into different groups, reflecting various components of the structure of biological communities (for example, species richness, evenness and dominance). There is no definitive consensus on how the variety of indices presented in Fig. 5.4 should be classified within this scheme, but we anyway summarize this view for teaching purposes.

It is generally assumed that *functional richness* (defined as the amount of functional space occupied by the organisms in an ecological unit) does not account for species abundance. As such, in Fig. 5.4 we included in this family of indices all those that do not account for species abundance, despite the fact that they have varying relationships with species richness (sections 5.3.1 and 5.3.2), and despite the fact that all other indices can be computed giving the same weight to all species (thus discarding potential species abundance data).

Functional evenness, i.e. the regularity in the distribution of the abundance in trait space of the organisms that comprise an ecological unit, should be estimated by the indices described in section 5.3.4. At the same time, as we saw, the existing indices that are based only on mean trait values need to be used with special care because they might, not negligibly, depend on trait range and number of species.

Functional divergence, i.e. the degree to which the abundance in trait space of the organisms that comprise an ecological unit is distributed toward the extremes of its functional volume, can be then quantified by all indices linked to the weighted mean dissimilarity between species, whatever the specific form used is (see section 5.3.3). Although Villéger et al. (2008) recommend that this index should be independent of trait range/volume, many also include in this family indices that tend to increase with the range (Rao, FDis, variance).

We do not claim that the classification of the indices, as summarized in Fig. 5.4, is the best way possible (if any perfect classification exists), but hopefully this helps users to get a general idea of the main groups of indices. This in turn should help to interpret the indices that are eventually used for study within a general conceptual framework. In the next chapter we will see how this framework can be expanded while using trait probability approaches. We also remind readers that other indices reflecting other moments of trait distribution, such as kurtosis and skewness, can be also considered (Le Bagousse-Pinguet et al. 2017), acknowledging that there are different views on how to group indices of functional diversity.

In recent years, a number of other interesting indices describing other aspects of functional diversity have been proposed and considered. A great deal of attention has been devoted to functional redundancy in the literature (Laliberté et al. 2010; Ricotta

5.4 Functional Diversity Components

et al. 2016). *Functional redundancy* should reflect how stable a community's functional structure is to the potential loss of species. A community with high functional redundancy should be minimally impacted by the loss of one or more species, because there are several species with similar traits. Low functional redundancy should thus imply lower buffering capacities of communities towards potential environmental changes. How to estimate functional redundancy though is not completely clear in the literature. To the best of our knowledge, Walker et al. (1999) were the first to attempt quantifying functional redundancy. They assessed the dissimilarity between dominant and rare species before and after increased disturbance of vegetation. Possibly the first formal index of functional redundancy was proposed by de Bello et al. (2007), taking advantage of the fact that the Simpson index reflects the maximum dissimilarity in the case that all species in a community are functionally unique (see above). Accordingly, they proposed an index expressed as Simpson diversity minus Rao diversity. In a similar fashion, Ricotta et al. (2016) proposed an index of redundancy measured as (Simpson − Rao)/Simpson. If we apply the formulas shown in section 5.5.3 this index can be summarized as:

$$Redundancy = 1 - MPD \qquad \text{Eqn. (5.7)}$$

Alternatively, Laliberté et al. (2010) proposed using the idea that redundancy can be measured as the functional dissimilarity between those species with similar effect on ecosystem processes, an approach that can be applied in different configurations. Imagine a community with several species that can fix nitrogen. Imagine now that these species have different traits (besides nitrogen fixing). It is then possible that these species will respond differently to the environment, because of their functional differences, giving a sort of insurance that the maintenance of the considered ecosystem function (nitrogen fixing) will be conserved at least by some species after some environmental change. To apply the approach, users can determine, for example, the average number of species per functional group (here defined as functional effect group (see Chapter 3), i.e. defining functional group in terms of their effects on ecosystem properties) or the average trait dissimilarity within functional groups. One potential drawback of this approach is that it can include some subjective decisions, such as the number of groups considered and the classification of species into groups. Through experience, we think that the index depends very strongly on the number of species, which otherwise reflects the common ecological expectation that more rich communities are more redundant. In this sense, we think that it is reasonable to expect that greater species richness should provide greater redundancy, since the concept of redundancy inherently depends on species richness. Conversely, too tight a relationship between redundancy and species richness might mean that the redundancy index will not provide much more information than species richness itself, overlooking cases with differences in species richness but comparable redundancy.

Another group of indices intends to quantify the *rarity* or originality of a given species or set of species (Violle et al. 2017). In practice, it can be essential to understand how 'special' species are, in terms of their traits, in a given dataset. Functionally rare species can be particularly important in this sense, owing to very original combinations of traits (Mouillot et al. 2013a). Most often, estimating rarity or originality includes a measure of how similar each species is, on average, to all other species, or how different the nearest neighbour is (see section 5.3.4). In practice, these indices resemble forms of MPD and MNTD. They are, however, measured only for single species, and not the whole community, so for each species we can estimate what the average distance from all other species in a community or a region is, or how close the closest species (neighbour) is. While redundancy gives us an average estimation of the effect of losing one species, these measures consider the uniqueness of species that could be lost. In this sense, these approaches have been proposed (Sasaki et al. 2014; Carmona et al. 2017b) to define extinction risk scenarios connected to how functionally unique (i.e. 'rare') each species is compared to other species in a community or region.

5.5 Partitioning Functional Diversity

Up to now in this chapter, we have analysed indices that compute functional diversity within a given study unit, for example the diversity between species *within* a given community, or *within* a given region. However, ecologists are also interested in quantifying the dissimilarity *between* two (or more) study units. In other words, ecologists are interested in quantifying *beta diversity*. The field of quantification of beta diversity has a long history, often populated by very polarized points of view, both in terminology and mathematics, which can indeed result in confusion and indecision regarding the concepts and indices one should use. Here we provide some basic tools to assess beta diversity using functional traits, compare it with taxonomic (i.e. species) beta diversity and, potentially, also with phylogenetic beta diversity. We do this without the intention of solving any debate about the partitioning of diversity, but just to provide some practical solutions.

Estimating beta diversity implies, in general terms, partitioning diversity across different scales, for example partitioning total diversity in a region into diversity within vs between communities. Following a general understanding of such partitioning, mostly derived from studies on taxonomic diversity, the diversity within a community is called alpha diversity (α), the dissimilarity between communities is called beta diversity (β) and the 'overall' regional diversity is called gamma diversity (γ). Beta diversity is also often referred to as 'turnover', as it is defined by the species or traits that are not shared between two communities. In practice it is very important to define, in each study, what a community is and what a 'region' is. Imagine a collection of plots within a big meadow, or a collection of pitfall traps in a forest patch. The total diversity contained within the ensemble of samples could be considered gamma diversity. Imagine another case, where the samples are all from the same habitat but are from different locations. Again, gamma diversity is the diversity of the entire set of samples, irrespective of them being from different locations. In both these cases, though, beta diversity represents the dissimilarity within one habitat type. Further, communities could be different samples along an altitudinal gradient, with all samples together considered the gamma diversity. In this case, beta diversity would represent the dissimilarity across habitats. Moreover, to confuse users a bit more, some authors consider a 'community' to be a habitat (for example, dry meadows). As such, we think that the most crucial step is to properly define what alpha, beta and gamma diversity represent in each particular study, while there is no strict limitation regarding the spatial scale on which these diversities can be calculated. It is, for instance, absolutely legitimate to calculate gamma diversity at the level of an entire continent, and then decompose it into alpha and beta diversity within and between assemblages at a largely defined 'community' level, e.g. 10 by 10 km grid cells.

Once the scale at which alpha, beta and gamma diversity are defined is clear, it is then possible to estimate each of these components. In general, it is assumed that the partitioning of diversity can be additive or multiplicative (with the resulting never-ending debate definitely being outside the scope of this book; see Jost (2007) and related works). The additive approach implies that gamma is equal to the sum of beta and the average alpha ($\gamma = \alpha + \beta$). In the multiplicative approach, gamma is the multiplication of beta by the mean alpha ($\gamma = \alpha*\beta$). In the case of species richness, imagine two communities with 10 and 15 species, respectively, and five of those species are in both communities (shared species). The total number of species will be 20 ($\gamma = 20$), and the mean alpha diversity is $\alpha = (10 + 15)/2 = 12.5$. In the additive approach $\beta = 20 - 12.5 = 7.5$, while in the multiplicative approach $\beta = 20/12.5 = 1.6$, meaning that γ diversity is 1.6 times α. The criticisms towards the additive approach can be generally solved by summarizing β as a proportion of γ, i.e. $\beta_{prop} = 7.5*100/20 = 37.5$ implying that 37.5% of the diversity is found between communities. If γ diversity is 1.6 times α, then it means that α represents $1/1.6 = 0.625$, which is exactly $1 - 0.375$. Accordingly, we see no major problems in expressing β using the additive approach in proportional terms, as discussed in de Bello et al. (2010b) and by several other authors.

Regardless of the choice between the multiplicative or additive approach, partitioning of diversity could be attempted, in principle, with several of the indices described in section 5.3. For example, all indices classified as 'functional richness' in Fig. 5.4 can be computed for any α and γ, and thus β can be derived either using the multiplicative or additive approach. In this same vein, Chiu and Chao (2014) also proposed a way to use FAD to partition functional or phylogenetic diversity. It is less clear if indices of functional evenness can be used for such a type of partitioning, particularly because they could decrease with the number of species (like MNTD), thus β can be negative, which is biologically meaningless in our view.

In general, all indices related to trait variance and MPD (section 5.3.4) could be used for partitioning diversity, although some specific recommendations should be followed. It is known that variance can be partitioned at different hierarchical levels (de Bello et al. 2011; Violle et al. 2012), which can be very useful and intuitive for functional diversity partitioning. However, variance cannot be computed for taxonomic diversity. Therefore, beta diversity cannot be computed with the same formulas for taxonomic and functional diversity, which can result in the reduced comparability of results between these different aspects of diversity. Another intuitive approach would be to use the mean dissimilarity approach, i.e. MPD, computing mean dissimilarity between species *within* a community (α) and the mean dissimilarity between species *between* communities (β). Unfortunately, using the MPD approach α + β ≠ γ, as γ can be lower than α (resulting in negative β).

There is a growing consensus that Rao can be used for several of the goals of partitioning diversity across scales. It is generally possible to avoid obtaining negative beta values and use Rao as an index that can be directly scaled and compared to taxonomic diversity and even phylogenetic diversity (de Bello et al. 2010b). This, unfortunately, requires some new specific guidelines and some tricks. The first one is related to the following question: in which way should the relative abundance of species be calculated to compute γ (de Bello et al. 2010b)? This is generally where students and researchers (including various published papers) make some mistakes. In most cases, the relative abundance can be safely computed at the scale of γ diversity, i.e. as the mean of p_i across all plots (Fig. 5.6). This is (only apparently) contrary to what we have learnt above, when we treated γ simply as a plot and computed relative abundance accordingly. In the example in Fig. 5.6 with a region with five species, with abundances 1, 1, 2, 2, 1 (from species 1 to species 5 respectively), the 'mistake' would be to divide each value by 7, i.e. their sum, resulting in values of 0.14, 0.14, 0.28, 0.28, 0.14. This approach can produce negative beta values. The approach we recommend instead is to compute p_i for each plot and then make an average across the plots. For example, species 5 is present only in community 3 and its $p = 0.33$ in that plot. So, the mean relative abundance across the three plots composing the 'region' would be $0.33/3 = 0.11$, and not 0.14 as we have shown above. Following this approach, β diversity will not be negative! As demonstrated in de Bello et al. (2010b), it is not actually a mistake to compute the relative abundances as 0.14, 0.14, 0.28, 0.28, 0.14, but users of this approach have to correct the way that the mean α in the region is computed because each plot will need a different weight, otherwise they can easily get negative β values. Again, in most cases, computing the relative abundance for γ as in Fig. 5.6 (i.e. mean of p_i across plots) is recommended as the best approach. See de Bello et al. (2010b) for more discussion on this topic.

The second important adjustment with the partitioning of diversity is that several authors recommend transforming the values of the Rao index, and other indices as well, into equivalent numbers (de Bello et al. 2010b; Chiu and Chao 2014). For example, the Simpson taxonomic diversity index (and, therefore, its general form Rao) can be expressed as $1 - D$ (with D = dominance, i.e. $\sum_i^N p^2$) or as $1/D$. The first approach scales the values between 0 and 1. The second approach would give a value which corresponds to species richness, in the case that all species have the same abundance. Species richness also presents the upper limit of possible values, and any deviation from perfect equality in abundances results in a decrease in diversity, to the degree that species have such unequal abundances that the value reaches its lower limit of 1 (in fact, this lower limit is reached when there is only one species, all other species have 'zero abundance'). Rao can also be expressed with 1 as the lower limit (meaning there is only one species, or all species are functionally identical), and the upper limit is the number of species (when all species are functionally unique and have the same abundance). In the example in Fig. 5.6, as also shown in 'R material Ch 5', community 1 has a total of two species but they are identical regarding their life form (i.e. both are grass).

5.5 Partitioning Functional Diversity

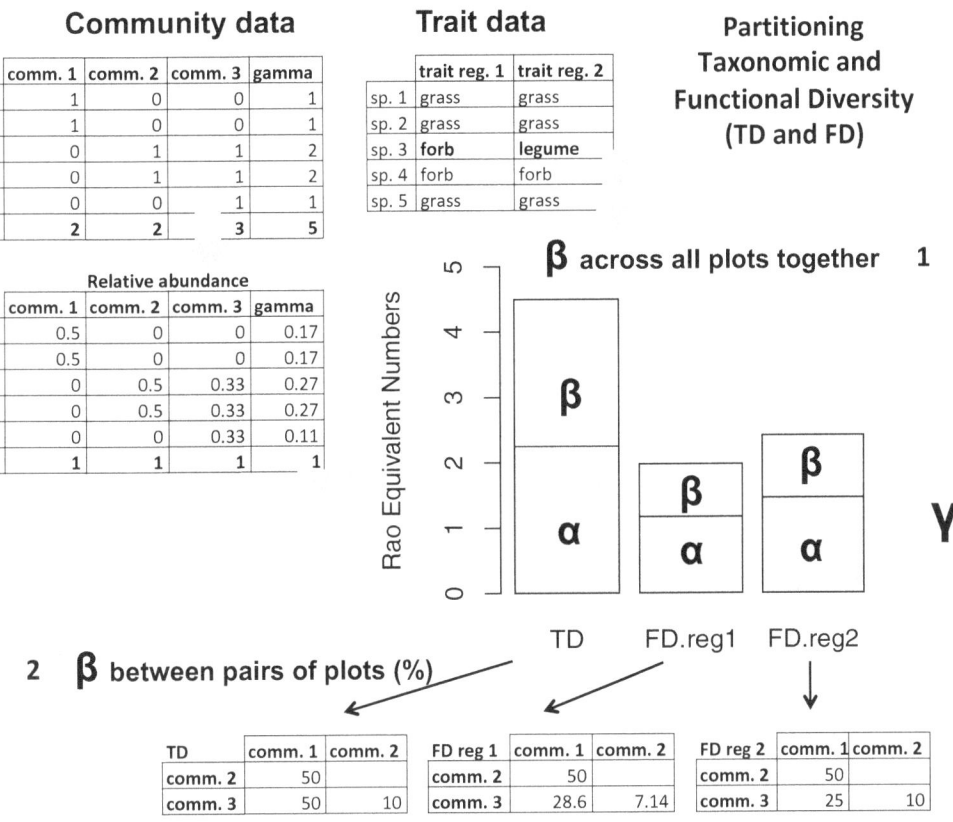

Figure 5.6 Partitioning taxonomic and functional diversity into alpha, beta and gamma diversity (α, β, γ). β can be computed like (1) an overall value for a given dataset (or set of plots), or (2) between each pair of plots in the dataset. In the example with partitioning-based Rao diversity (expressed in equivalent numbers), there are three communities, each with two species and different traits. Trait data reflect two cases of mimicking two possible regions with exactly the same species composition but with species having different traits. See text for more details.

Therefore, Rao expressed in equivalent numbers will be 1. Now consider the case of community 3, in which all species have a unique life form (i.e. in the case of 'region 2', one species is a grass, one is a forb and one is a legume, and we assume that the dissimilarity between these three types of species is always 1). Rao in this case will be equal to the number of species. How do we express Rao values in equivalent numbers? This is simply obtained, by:

$$Rao_{eq.numb} = \frac{1}{1 - Rao} \qquad \text{Eqn. (5.8)}$$

with Rao expressed in this formula in the 'classic' way we learned, as in section 5.3.4. For example, in community 1, with the two species being equal, Rao = 0, so that $Rao_{eq.numb} = 1$; in community 3, where we have one grass, one forb and one legume, Rao = 0.67, thus $Rao_{eq.numb} = 3$.

The approach using equivalent numbers to partition diversity has been specifically suggested to avoid underestimating beta diversity. For example, communities 1 and 2 in the examples in Fig. 5.6 are both taxonomically and functionally completely different (they do not share species or types of traits: in the first community there are only grasses, and in the second there are no grasses). Therefore, both taxonomical and functional beta diversity should be maximized. If we express beta diversity between these two plots, i.e. $\beta_{prop} = (\gamma - \alpha)*100/\gamma$, then a value of 50% would correspond to the case that a set of samples does not have any common information. This represents, in other words, a complete turnover and the maximum beta diversity between a pair of plots (each plot contains 50% of the total diversity). The only way to obtain this result is by expressing diversity in equivalent numbers (Jost 2007). For Simpson diversity, communities 1 and 2 in Fig. 5.6 would have a value of 0.5 each (α), and the gamma (γ for the combination of communities 1 and 2) would be 0.75, so that β in the additive form would be 0.25, and thus $\beta_{prop} = (0.75 - 0.5)*100/0.75 = 33\%$. This would be misleading, because from Fig. 5.6 it is clear that both communities 1 and 2 contain half of the total diversity, so the value should be 50%. However, expressing values in equivalent numbers solves this issue. In equivalent numbers, Simpson for both communities is 2 (α), and for the two communities combined is 4 (γ), so that $\beta_{prop} = 50\%$. Note that in this formulation, 50% corresponds to the maximum beta diversity for a pair of plots, but the maximum value for β_{prop} depends on the number of plots. So, for comparing β_{prop} values between datasets with different numbers of plots, a normalization is needed:

$$\beta_{prop.norm} = \beta_{prop}/(1 - 1/n) \qquad \text{Eqn. (5.9)}$$

where n is the number of plots in a dataset: see de Bello et al. (2010b).

A simple way to understand why both Simpson and Rao underestimate beta diversity (when not using equivalent numbers) is that they have an upper limit, so α and β cannot increase indefinitely. Note that the use of equivalent numbers for the partitioning of diversity can be applied not only with the Rao and Simpson indices; it can be applied for other indices as well, such as Shannon diversity. As mentioned above, Chiu and Chao (2014) proposed a way to use equivalent numbers with the FAD index. We would like to remind readers, however, that the FAD approach depends very much on species richness, which sometimes results in patterns likely to be counter-intuitive to many ecologists. For example, with the approach proposed by Chiu and Chao (2014), communities 1 and 3 in Fig. 5.6 do not share any common species, so that the taxonomic beta diversity should be maximized between them. This will also result, a bit surprisingly to us, in a complete turnover in terms of beta functional diversity (maximum beta diversity). This could look suspicious to many ecologists, as the two communities do share some traits (species 5 is a grass, like the two species in community 1), so that in reality the functional beta diversity should not be maximized. We think (and believe that most field biologists would agree) that the two communities share some functional information, hence the beta diversity between communities 1 and 3 should not be at the maximum (see Botta-Dukat 2018 for more discussion on this topic). Although we understand that subtle mathematical explanations can support almost any approach, we think that estimations of beta based on the FAD index should be used with care.

Finally, we want to stress that by using the Rao approach for partitioning diversity, it is possible to compare results across taxonomic, functional and phylogenetic diversity. This comparison is possible if the maximum distance between species is set to 1, as we have recommended before. This opens up exciting options to test changes in these three components of diversity within the same mathematical framework. Of course the change, say in functional diversity, cannot be completely independent of the change in taxonomic diversity, i.e. we cannot obtain a higher functional replacement than taxonomic replacement, so that β_{prop} for functional diversity will always be less than or equal to β_{prop} for taxonomic diversity. At the same time, this also comes with a consequence. As we showed in Chapter 3, with species dissimilarity expressed with Gower distance, the dissimilarity between many species will be small, particularly when using quantitative traits. As such, β_{prop} with the Gower distance will be, in practice, always quite small because of the small dissimilarity values. This problem can be minimized by using other approaches to estimate species dissimilarity that include information on intraspecific variability, which we will discuss in the next chapter. In practice, this dependence of the partitioning of diversity on the way in which dissimilarity is computed is important, but mostly overlooked (but see Pavoine et al. 2016).

5.6 R Tools to Compute Functional Diversity

In the accompanying R material, we show how to calculate most of the indices discussed here. We would like to stress that there could be quite some confusion when using the R functions for some of the indices described in this chapter, particularly Rao and MPD. We claim that none of the mainstream functions produce the expected results based on the formulas above. One existing misconception, in our view, is that Rao diversity is an abundance-weighted form of MPD, a 'myth' that we hope we have dismantled by showing the relationship between MPD, Rao and Simpson above (de Bello et al. 2016a; Ricotta et al. 2016). As a matter of fact, the function *mpd* (in the library *picante*) when considering species abundance (argument *abundance·weighted = TRUE*) surprisingly produces the values expected from the Rao index, because it considers the diagonal in the dissimilarity matrix (but it does not consider the diagonal when *abundance·weighted = FALSE*). In our view this is a reductive approach, disregarding the fact that a weighted version of MPD exists and it has important biological relevance (Ricotta et al. 2016). Similarly, the function *dbFD*, in the package *FD*, uses the function *divc* from the package *ade4* to compute Rao. This function actually computes the variance form of Rao $\left(\sum_{i=1}^{N}\sum_{j=1}^{N} p_i p_j \frac{d_{ij}^2}{2}\right)$ (i.e. the distances are squared and divided by two). Hence, the 'Rao' index produced this way is not exactly a generalization of the Simpson index of diversity (even when the argument *scale.RaoQ = TRUE*). Note also that, while the majority of indices in the *dbFD* function uses a dissimilarity produced using PCoA, the form of Rao offered is computed directly from the Gower dissimilarity (without using a PCoA first). In terms of diversity partitioning, one of the handy existing functions is the *disc* function in *ade4*, as it provides a dissimilarity between pairs of plots (such as in option 2 in Fig. 5.6). However, as with the function *divc*, the function *disc* actually computes variance when attempting to compute Rao, which can be quite useful for some applications, but again does not result in a Rao value that is a direct generalization of the Simpson index of diversity. We offer an integrated tool to partition taxonomic, functional and phylogenetic diversity (the function *Rao*), together with other tools, in the accompanying 'R material Ch5'.

Take-Home Messages

- We can characterize the functional trait structure of communities using various indices, mainly community weighted mean (CWM) and functional diversity (FD), among other 'moments' of community trait distribution.
- CWM is a powerful and simple index which represents the traits of the most dominant species in a community. By definition, it can be affected by variation in the composition of very few species, so that it needs to be interpreted carefully.
- FD expresses the extent of trait differences between organisms. There are different indices of FD, which can complicate its use and interpretation. Sometimes the different indices are summarized into 'families' of indices, i.e. functional richness, functional evenness and functional divergence. Since there is no perfect index, it is important to have a general idea of what each index represents and how much they are related to each other and to species richness.
- Among the different indices, the Rao quadratic entropy, at present, is apparently the best-suited index for the partitioning of functional diversity into within- and between-community components. This is because of its relationship with multivariate trait variance and because it is possible to scale it against species diversity.
- Care should be used when computing functional diversity with some of the existing R functions, to avoid obtaining unwanted results.

6 Intraspecific Trait Variability

Species are the fundamental unit in ecology and evolutionary biology. In functional ecology, researchers have typically treated single species as static entities in space and time, characterizing each species using just *fixed mean trait values*. This would mean, in the context of the previous chapter for example, that a species gets the same mean value in all plots within a dataset irrespective of where it is living (Chapter 5). The *mean field theory* (MacArthur & Levins 1967) suggests that mean trait values capture the majority of the dynamics of species, with the explicit assumption that interspecific trait variation is larger than intraspecific trait variation, which is actually often the case (Westoby et al. 2002; Siefert et al. 2015). However, even in the seminal paper by MacArthur and Levins (1967) authors recognized the importance of expressing how variable a trait is around such a mean (Fig. 6.1). As we will discuss in this chapter, many studies have shown significant differences between conspecific individuals for a great variety of functional traits.

Intraspecific trait variability (ITV hereafter), or *within-species trait variability*, can be defined as the set of trait values expressed by a set of individuals within a species (Albert et al. 2011). We can measure ITV in a group of individuals from a single population within a site, or in multiple populations across the landscape or geographical range in which this particular species occurs, allowing within-species trait variability to be expressed on different scales (see section 6.3 for details). Intraspecific trait variability is not negligible and contributes significantly to overall trait variation (Albert et al. 2010; Violle et al. 2012; Siefert et al. 2015), as well as affecting ecological processes such as population growth and biomass production (Ellers et al. 2011), community structure and ecosystem processes (Bolnick et al. 2011). Moreover, ITV is essential for assessing eco-evolutionary dynamics, as it allows for rapid adaptation of species to

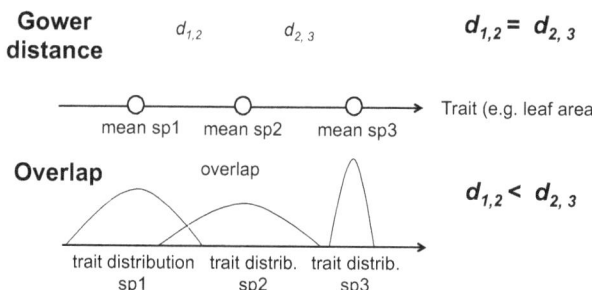

Figure 6.1 Trait differences between species can be expressed in terms of the differences in species' mean trait value (for example, using the Gower distance, expressing the distance between the mean of different species; Chapters 3 and 5); in the figure, each circle is the mean of one species, with species 2 equally distant from species 1 and 3, i.e. $d_{1,2} = d_{1,3}$ where $d_{1,2}$ is the dissimilarity between species 1 and 2). Alternatively, it is possible to account for the variability within species, intraspecific trait variability. This can be done, for example, using the distributions of trait values within species (see also Fig. 5.1), represented by a curve for each species in the figure. The overlap between these curves representing trait distributions can be used to estimate how similar or how dissimilar the species are (less overlap, greater dissimilarity). At the end of this chapter we show how the overlap in trait distributions can be used to measure dissimilarity between species (Fig. 6.7).

environmental stress, which may then feedback as selection pressure to species (Schoener 2011; Turcotte et al. 2011). Altogether this means that individual variation, even if lower than interspecific differences, results in different ecological strategies, and that the mean trait value does not take the role of intraspecific trait variability into account (Violle et al. 2012).

In light of this, a central issue in functional ecology is whether we can use only fixed trait values for single species or if we must also include, and how, trait variability within individuals. This chapter attempts to provide some answers to this question. To do so we first describe the source, role and importance of within-species trait variation, and finally give key examples of how to assess and deal with ITV once the data have been collected (for details see 'R material Ch6'). For further discussion of different sampling schemes, i.e. to decide how and when best to account for ITV, see Chapter 11.

6.1 The Source of Intraspecific Trait Variability

There are two major sources of ITV, namely the different *genotypes* present in a population and the *plasticity* in response to environmental heterogeneity experienced by an individual (and in some cases its parents). The local environment in which most species live can be relatively heterogeneous, and each individual can be subjected to (slightly) different environmental conditions throughout its lifetime. As a result, populations are composed of variable *phenotypes* with different trait values, i.e. individuals have different values for certain traits. The variation in phenotypic expression is even larger when we compare individuals between populations, for instance populations across an environmental gradient (Fig. 4.1). The sources of phenotypic variation can be manifold, but species typically respond to environmental variation through two mechanisms, which may also interact with each other: (i) genetic variation and (ii) acclimation or phenotypic plasticity (Geber & Griffen 2003).

ITV due to *genetic variation* is the phenotypic variability between individual genotypes resulting from various evolutionary processes such as gene recombination, mutation, genetic drift, selection and immigration (Blanquart et al. 2013). These processes generate individuals within a population that differ in their genetic information, and thus likely their phenotype. Phenotypes with traits that best fit the local environmental conditions have a higher chance to survive and, subsequently, have a higher reproductive success, contributing more to the gene pool of the next generation. Their offspring in turn generally inherit the phenotype of their parent(s) through parental genes. Offspring are then better adjusted and have a higher fitness under the local environmental conditions. We thus see that, because of genetic variations, ITV occurs across different environmental conditions (either microhabitat differences within a location or across locations), selecting for and against different genotypes. In other words, genetic variation between populations is often due to adaptation, but genetic variation within a population may not be. Ecological processes, such as climate change, habitat fragmentation by land-use change or founder effects, i.e. the founding of a new population by a few individuals, may decrease genetic variability, and thus ITV, which in turn may increase the risk of extinction (Spielman et al. 2004) and impair ecosystem processes (Hughes et al. 2008).

The second important source of ITV is *phenotypic plasticity*, which is the production of alternative phenotypes by a given genotype under various environmental conditions (Schlichting & Levin 1986; DeWitt & Scheiner 2004). In other words, it is the dependency of a trait value, for a single individual, or several individuals with the exact same genotypes (clones), on local environmental conditions experienced by this individual during its development (Fig. 6.2). For example, genotypes of the plant *Polygonum lapathifolium* grown under low light levels have lower biomass but larger leaves compared to genotypes grown under high light levels (Sultan 2000). In *Tetrix* ground-hopper species, adult body colour is strongly determined by the substrate colour experienced during nymphal development (Hochkirch et al. 2008), while in the springtail *Orchesella cincta*, temperature-induced egg size plasticity occurs after temperature stress, with subsequent consequences for fitness (Liefting et al. 2010). Similarly, biotic cues such as the presence of a specific predator, strongly affect prey behaviour and body morphology, which in turn affects predation risk. This is seen in the tail of Pacific treefrog (*Pseudacris regilla*) tadpoles in the presence of diving beetles (i.e. an enlarged tail is formed to lure predators away from the head) or fish predators (i.e. a shallow tail is formed for speed to escape predation; Benard 2006). In these examples each individual experiences a slightly different environment during its life cycle which may thus cause phenotypic plasticity leading to ITV. Plasticity,

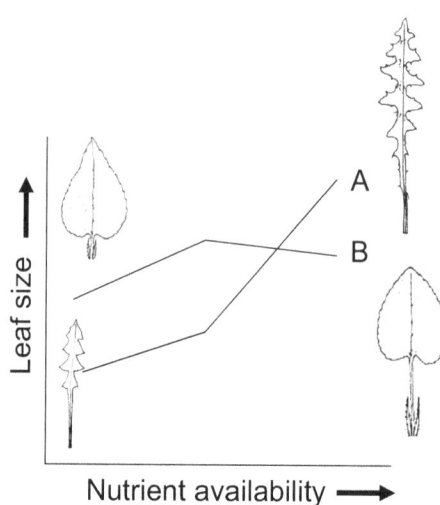

Figure 6.2 Phenotypic plasticity and reaction norm. In this hypothetical case, seeds of one parent plant (hence with similar genetic basis) are grown on either nutrient-poor or nutrient-rich conditions. Leaf size of species B is not a very plastic trait when plants are grown under different nutrient conditions, while species A shows a strong difference in leaf size with a change in soil fertility, hence leaf size is very plastic for this species. The slopes of the lines represent the 'reaction norm'. Note that the reaction norms cross with a change in soil fertility. This indicates that if leaf size might have an effect on competitive ability, the outcome of competition between species A–B differs depending on nutrient conditions.

strictly speaking, is the phenotypic variation *within* an individual, not in a population, so that if we sample multiple individuals in a community, the resulting ITV cannot be called plasticity. However, in the literature a population-derived trait variability is often referred to as phenotypic plasticity.

The actual change in the value of a trait of a single genotype across a range of environments is called a *reaction norm* or norm of reaction (Fig. 6.2; Sultan 2000). It describes how each genotype within a single species, or genotypes of multiple species, phenotypically respond to varying environments. This can be used to understand intra- and interspecific sensitivity and response to environmental change. There can be different reaction norms for every genotype, phenotypic trait and environmental variable. A shallow reaction norm (i.e. the slope) indicates that the value of a particular trait at one extreme of an environmental variable does not change much when this individual is placed at the other extreme. On the contrary, species with a steep slope for a particular trait show a strong increase or decrease in the trait value when placed at opposite environmental extremes. We can compare the response of different species to a selection of environments by their reaction norm slopes (Le Lann et al. 2014). If the reaction norms of two species cross, this means that the first species performs better at one extreme, while the second species is more successful at the other extreme. It is important to realize that for every genotype within a species, and for every environmental variable, a different reaction norm can emerge (Sultan 2000). Similarly, for many functional traits there are reactions norms across individuals within and between populations.

The range of phenotypes that can be produced within an individual plant or animal is not only determined by genetic effects (V_G, variation due to genotype) and by systematic environmental effects (V_E, variation caused by the environment in which a genotype develops) but also by their interaction ($V_{G \times E}$). This *genotype–environment interaction* term indicates that the environmental effect is different for some genotypes relative to others (DeWitt & Scheiner 2004; Sgrò et al. 2016). An example of this $V_{G \times E}$ interaction is tree height and diameter increment in aspen hybrid clones (*Populus tremula* × *Populus tremuloides*). Clone height and stem growth are significantly higher in forests compared to agricultural land, i.e. there is a strong environmental effect, while clone growth differs within a site, i.e. there is a genetic effect. However, the ranking of clones also changes significantly across sites, indicating a strong genotype–environment interaction affects growth (Yu & Pulkkinen 2003).

Finally, an increasing body of literature shows that organisms can not only alter their own phenotype in response to environmental conditions, but also affect the phenotypes of their offspring, an example of *trans-generational plasticity* (Sultan 2000; Zizzari & Ellers 2014). These effects are expected to operate often via *epigenetic changes*, where epigenetic alterations lead to differential gene expression without causing irreversible genetic mutation (Bossdorf et al. 2008). Yet, these relatively flexible epigenetic patterns can persist through mitotic and meiotic division and transmitted to the offspring, thus creating an epigenotype pre-adapted for the same environmental conditions that were experienced by its parent. Offspring development, structure and morphology can be influenced by the parents' environment. This allows the offspring to maintain critical functions, even if this function is reduced due to the experience of the parent. For example, *Persicaria maculosa* seedlings of nutrient-deprived parent plants can increase their allocation to roots compared with seedlings of genetically identical plants grown with ample nutrients, and seedlings of light-deprived plants can increase shoot growth relative to root growth compared to seedlings of genetically identical plants grown at high light levels (Sultan 1996). Passing a *parental stress-adapted phenotype* to offspring does not enrich the offspring plastic potential itself. Instead it can make it pre-adapted to the environmental stresses they are most likely to encounter, the same as those experienced during the parental generation (Herman & Sultan 2016). Seedlings without this trans-generational information will reach similar phenotypic traits through their own individual plasticity, but with some time delay, which can be costly in extreme conditions. Such stress pre-adaptation can be especially important for invasive plants or plants occupying new niches within a few generations. This field of research is expanding rapidly as tools that account for such parental transgenerational effects are developed. As genetic analyses remain expensive, techniques such as DNA de-methylation can be useful to remove the 'memory' of epigenetic changes received from the conditions in which parents grew (Puy et al. 2018 for more information).

6.2 The Importance of Intraspecific Trait Variability

Patterns and processes in community ecology are influenced by four major groups of factors, namely speciation, drift, selection and dispersal (Vellend 2010; see Chapter 7). We argued above that ITV can result in different ecological strategies and that the mean trait value approach can underestimate this process, for example in environmental filtering as described in Chapter 4 (Fig. 6.3). Therefore, ITV plays an important role in adaptation of individuals and species to environmental stresses, affecting speciation, species density, distribution, invasion, interactions between species, community composition and ecosystem processes (Violle et al. 2012). The range of within-species trait variability strongly depends on the spatial scale of the study (Violle et al. 2012). At each scale, different factors may generate within-species trait variability, or the strength of the factor may differ across space.

On the spatial scale of a *community*, individuals in a population may differ in trait values due to spatial heterogeneity in abiotic conditions or because they have different biotic interactions with organisms on the same or another trophic level. For instance, small-scale variation in light, soil moisture, air temperature, prey availability or predator presence leads to local adaptation and selection of individuals with trait values that match these conditions (i.e. the actual trait values maximize their fitness), leading to variation in traits within the population (Jung et al. 2010). Of course, neutral processes like ecological drift can also cause different phenotypes within and across populations. When we scale-up to the *landscape* level, we can observe populations across different environmental conditions and with different levels of connectivity, which decreases as the distance between populations increases. Environmental variation, particularly in abiotic conditions, usually increases with the distance between populations, and results in the overall trait variation within the meta-population being larger than it is within a single population. At the landscape scale, genetic exchange decreases with the distance between populations, reinforcing processes that lead to local genetic selection. We can

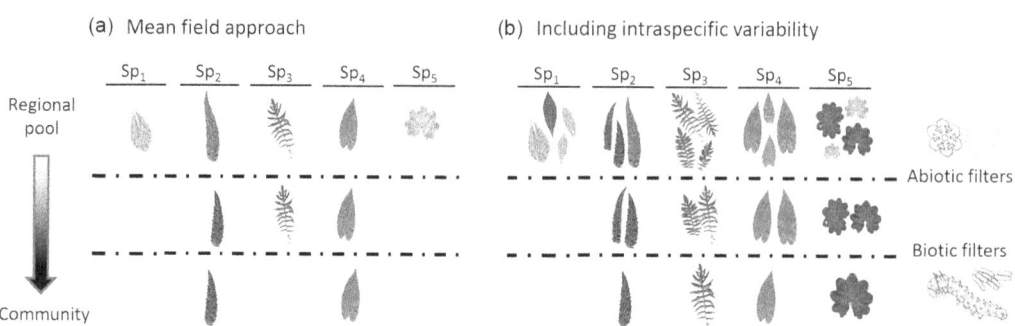

Figure 6.3 Community assembly theory (a) under the classical approach where only mean trait values are considered for species of the regional pool; and (b) incorporating intraspecific variability. Each leaf shape represents a species. Dashed lines represent abiotic and biotic filters (see also Chapters 4 and 7). Figure created by Javier Puy.

thus conclude that the full extent of ITV is found across the entire geographic range of a species. Geological and climatic gradients, in combination with local adaptation and genetic selection of individuals, often produce strong trait variation or shifts in trait values. For example, in the springtail *Folsomia quadrioculata*, hatchling size significantly increases with latitude (Sengupta et al. 2016). Similarly, many European freshwater fish show large-scale intraspecific variability in several life history traits (Blanck & Lamouroux 2007), and variability in leaf and root traits tends to be relatively higher at coarser spatial scales (Liu et al. 2010). These differences in ITV have an important impact on biodiversity, which we will see later in this chapter.

This means that ITV emerges at different levels of biological organization, i.e. from the individual to the ecosystem, and subsequently that ITV and spatial scale interact (see end of this chapter). Below we present some examples of studies that show the importance of ITV across different scales, i.e. for individuals, populations, communities and finally ecosystem processes.

6.2.1 Speciation

The formation of new species is the evolutionary process by which populations of a given species evolve to become two, or more, new discrete species. Sometimes the process is more subtle and subspecies are created. Traditionally, the general paradigm was that populations undergoing speciation should be spatially isolated by a barrier preventing gene flow, for instance by oceans, mountain ridges or living on separate islands, for a lengthy (geological) period of time (*allopatric* speciation). Each population, composed of individuals with variable phenotypes, will encounter different environmental conditions and selection pressures, and will eventually evolve by adaptation to new niches, which progresses until populations are no longer capable of gene exchange if/when they come back into contact. Later it was argued that populations that are partially separated in zones or peripheral populations might also encounter new niches, leading to reduced gene flow and eventual speciation. Although ITV plays a role in these forms of speciation, it is of particular importance in *sympatric* speciation (Via 2001; Bolnick & Fitzpatrick 2007). In this type of speciation, two or more species descend from a single ancestor, all of which live in the same geographic, often local, area. There is no geographic isolation preventing gene flow (unlike allopatric speciation). As ecological factors often appear to drive sympatric speciation, particularly food or microhabitat conditions, this process is also referred to as *ecological* speciation.

The initial step of sympatric speciation is genetic polymorphism in a specific phenotypic trait or a set of traits across individuals in a population. As a result, individuals become ecologically separated in dissimilar niches, leading to trait divergence and partial or complete reproductive isolation within the population, and eventually to speciation. Classic examples are found in host race formation in invertebrate herbivores, such as the apple maggot fly *Rhagoletis ponomella* (Bush 1969) or the pea aphid *Acyrthosiphon pisum* (Via 1999), among others. In these species a host-shift has occurred, and each host race predominantly feeds on its own plant species, which enhances adaptation to that particular plant and increases reproductive isolation between host races. Host races often differ in traits such as body size, colour or life history traits. Sympatric speciation through trait polymorphism is also found in other animals and plants (Bolnick & Fitzpatrick 2007). In vertebrates, sympatric speciation has occurred through, for example, polymorphism in call frequency in the large-eared horseshoe bat *Rhinolophus philippinensis* (Kingston & Rossiter 2004), mating call differences due to beak size in the Darwin Finch *Geospiza fortis* (Huber et al. 2007), or phenotypic colour variation, and acoustic reproductive communication in male African cichlids (Barluenga et al. 2006).

6.2.2 Population Size and Genetic Diversity

Genotypes usually differ in phenotypic expression. Intraspecific *genetic diversity*, or the number of genotypes in a (sub)population, can have a large effect on population size, as well as on a community and ecosystem. An increase in the number of genotypes has been shown to enhance plant primary productivity, with potential cascading effects on green-web species, such as herbivores and their associated predators (Hughes et al. 2008; Kotowska et al. 2010). The positive effect of genetic diversity on population processes can be due to genotype additivity, determined by the phenotypic properties of the mixed genotypes, as well as non-additive interactions between genotypes, such as inhibition, interference or facilitation of one genotype by another. These negative or positive genotype–genotype interactions are believed to be dependent on the level of functional trait dissimilarity between genotypes. If genotypes have similar phenotypes (i.e. homogeneous phenotypes due to inbreeding or cloning) then inhibition will likely occur, while if genotypes differ strongly in phenotypic expression, facilitation will be dominant in a population (Heemsbergen et al. 2004). The less phenotypes resemble each other in functional trait values, the smaller the chance that they will strongly compete for basal resources. On the other hand, it is also possible that a genotype for traits associated with greater competitive abilities might outcompete the other genotypes. The former has been observed for the springtail *Orchesella cincta*, using inbred iso-female lines to create nearly homogeneous genotypes. In a microcosm experiment with one or a mixture of two, four and eight genotypes, obtained from different iso-female lines, an increase in the number of genotypes strongly enhances population size and biomass production of this springtail (Ellers et al. 2011). The degree of phenotypic differences in life history traits associated with population growth rate in the inbred line, i.e. egg size, egg development time and juvenile growth rate, determines the magnitude of the genotypic number effect. This may be explained by micro-niche differentiation between genotypes. Other examples of ecological success of inter-individual variation in genotypes are known, for example, for bacteria, vascular plants, crustaceans, insects and fish (Forsman & Wennersten 2016).

6.2.3 Adaptation

The human impact on ecosystems worldwide, via climate change, pollution and alteration of land-use types and habitat fragmentation, has resulted in modified habitat quality for many species. This means that for species to persist, especially if they have low dispersal ability, adaptation to these new conditions is crucial for their survival (Berg et al. 2010). It has been shown that *population fitness* increases with increased ITV for many groups of organisms. More variable populations perform better in general, and have been shown to be less vulnerable to changes in habitat quality (Forsman & Wennersten 2016). The positive effect of phenotypic variability on population persistence via adaptation of species to stress also affects the extinction probability of species. For instance, in mammals and birds, especially passerines and parrots, interpopulation variability in adult body mass, time to sexual maturity and litter size reduces extinction vulnerability, indicated by red list data (Gonzales-Suarez & Revilla 2013). The importance of ITV for species survival even holds on a geological timescale, as it may reduce the global extinction risk of species. For instance, high colour polymorphism in species and genera of benthic Ostracodes (Trachyleberididae) is positively associated with greater geological longevity (Liow 2007), and similar findings have been reported for amphibians, reptiles and snakes (Forsman & Hagman 2009; Pizzatto & Dubey 2012). The effect of colour variation on species performance is explained by the association of colour polymorphism to other functionally important ecological traits.

A high phenotypic variability might also be an *adaptation to biotic factors*. For instance, plant secondary metabolites are commonly produced by plants in response to the presence of grazers, reducing leaf damage. Intraspecific variation in secondary metabolites may promote switching of the herbivore between individual plants, as has been shown, for instance, in the bushwillow *Combretum fragrans*, which is fed upon by the herbivore caterpillars of the lappet moth *Chrysopsyche imparilis* (Mody et al. 2007). Herbivore switching shortens their feeding bouts on plant individuals, while individual plants did not experience a marked reduction in leaf tissue. Similarly, interpopulation variation in condensed tannins in cottonwood, *Populus* sp., affects food tree selection in beavers, which in turn greatly affects the fitness of that particular tree (Moore et al. 2014). As populations adapt to local environmental conditions, ITV contributes significantly to the strong latitudinal trait variation and shifts in trait values across the macro-scale.

6.2.4 Distribution

The distribution of species is strongly influenced by interspecific variation both in genes and phenotypes (reviewed in Forsman & Wennersten 2016). This may be due to several mechanisms, which may operate simultaneously, and affect the ecological success of species. First, as we have shown above, ITV contributes to local adaptation resulting in dampened population fluctuations, reduced vulnerability to environmental variability, and lowered population extinction risk. Second, ITV has a positive effect on the ability of species to colonize and establish in new areas. Experimental studies have shown, for a variety of organisms and experimental conditions, that founder phenotypic diversity (as well as genetic diversity) increases establishment success in new areas (Forsman 2014). Many studies have shown large *distribution ranges* for species with high ITV compared to species with low ITV. For instance, polyphagous noctuid moth species have a larger range distribution than oligo- or monophaguous species (Franzen & Betzholtz 2012), and in fishes and amphibians polyphenic clades have a broader range size compared to non-polyphenic clades (Pfennig & McGee 2010).

Some species show variation in a specific functional trait resulting in a broader range size. For example, species may show intraspecific variation in reproduction mode, with both sexually and asexually reproducing individuals within a single species (i.e. facultative parthenogenesis). Asexual reproduction or parthenogenesis is often linked to polyploidy (gene duplication), and is observed in both animals and plants. Asexual-dominated populations regularly have a much broader distribution than their sexual conspecifics, and female-based polypoid species are often overrepresented at range margins, a phenomenon that is known as geographical parthenogenesis (Cosendai et al. 2013). For instance, a recent study on harvestman shows that for *Leiobunum manubriatum*, a species with diploid and tetraploid individuals, asexual reproduction is associated with *range expansion* towards more marginal habitat in the north of their range (Burns et al. 2018). Polyploidy not only influences the mode of reproduction but also other traits such as, among others, body size (Lavania et al. 2012). The presence of gene duplication in populations may thus affect the variability in a suite of traits that in turn influence species adaptation, colonization and survival.

6.2.5 Invasion Predictability

We explained above how adaptation allows species to alter their phenotypic traits to maximize fitness after a human-induced environmental change. However, human activity also creates new suitable habitat for many plants and animals, locally, regionally and even globally. The ability of species to invade this newly formed, disjunct habitat depends largely on three factors: dispersal power (Walther et al. 2002; Berg et al. 2010; Schloss et al. 2012), adaptation to the same habitat conditions and competitive ability (Shurin 2000; Seabloom et al. 2003). Species can reach new localities if they can overcome the distance between their current range and new habitat location, but only if they have the right set of dispersal traits. Upon arrival, species must be adapted, or adapt to, to those habitat conditions to be able to compete with species already present. Two different theories predict that the chances of establishment and invasion depend upon having traits sufficiently dissimilar to avoid competitive-exclusion by niche partitioning, or sufficiently similar to withstand the prevailing biotic and abiotic conditions (Loiola et al. 2018). There are considerable interspecific differences in dispersal and competitive ability between interacting species, and this dissimilarity shapes community composition under environmental change (Cadotte et al. 2006; Livingston et al. 2012).

Species can track shifts in climate and move across environmental gradients via dispersal (Walther et al. 2002; Parmesan & Yohe 2003). Dispersal ability not only differs between species, but there is also ITV in dispersal ability, which has proved to be important for recent changes in species ranges. Along newly established range margins, some species exhibit a higher frequency of dispersal vs non-dispersal phenotypes, or an increase in the number of phenotypes with a higher vs lower dispersal ability. Roesel's bush cricket *Metrioptera roeselii*, for instance, has either dispersing or non-dispersing individuals (Thomas et al. 2001). Two morphs exist, a short-winged morph that cannot fly and a long-winged morph that can. This later, macropterous form is particularly abundant at recently established range margins, but is completely absent in older populations far from the range margin, where the brachypterous, flightless form resides. In another bush cricket, the long-winged cone head *Conocephalus discolor*, two dispersive phenotypes are described, a long-winged and an extra-long-winged form. The frequency of this highly *dispersive morphotype* increases in recently established populations (Thomas et al. 2001). Similar examples can be found in plants, such as in the invasive South African ragwort *Scenecio ineguidens*, where individuals at the edge of expansions have seeds with a larger pappus and plume load, resulting in lower seed terminal velocity and, correspondingly, a higher dispersal distance (Monty & Mahy 2010). Some other life history traits undergo rapid phenotypic change at expanding range margins; these include body size, growth rate, feeding rate, coloration, reproductive ability and age to maturity (Chuang & Peterson 2016).

6.2.6 Community Assembly

One of the central problems in community ecology is understanding how communities are built, i.e. how a local community is assembled from the regional species pool (Fig. 6.3; see also Chapter 7). Can we construct general rules, so-called ecological assembly rules, that predict the arrival and persistence of species in a community? These rules are believed to act as 'filters' between the regional species pool and the community, and 'filter out' species according to their traits, as we discuss in several chapters of this book (see Chapters 4 and 7). ITV plays an important role in community composition and assembly theory as it determines which species can potentially pass these filters, i.e. have the right traits that enable them to become established in the community. When species are described only by their mean trait value, we assume that they can enter the community if they possess the right dispersal, tolerance and competitive traits or if their mean traits match the external and internal conditions. However, we have seen that ITV can increase the ecological niche of species via trait variation and this might mean that, at least partially, some individuals with trait values different from the species mean could establish in the community, or that they can, to a certain extent, plastically change their traits, to match the filters acting on the community. In other words, although a species should be, in principle, absent based on its mean trait values, individuals with higher or lower traits values than the mean might pass the filter and occur in the community. In other words, accounting for ITV in assembly processes results in a higher community *species diversity*, as ITV allows some individuals of species that are excluded based on a mean trait value to pass the environmental filter.

6.2.7 Trait-Mediated Species Interactions

Species not only adapt their traits to changes in their abiotic environment but also to the 'presence' of other species, for instance to *neighbours, prey, predators and parasites*,

Figure 6.4 Trait-mediated predator–prey interactions. The development of the tail of tadpoles of Pacific chorus frogs depends on the predator present. When exposed to info-chemicals of diving beetles the tail is enlarged to lure the predator away from the head. When exposed to info-chemicals of fish the tail becomes shallow to increase swimming speed and escape predation. Figure created by Janine Mariën.

with implications for populations, communities and even ecosystems. For instance, in some plant species growing in semi-arid shrublands, specific leaf area and plant height of individuals is, among other things, determined by the trait value of a neighbouring individual, as well as the packing density of these neighbours. These neighbour effects on trait values are often even stronger than precipitation regimes (Le Bagousse-Pinguet et al. 2015). There are interesting examples, especially from the aquatic environment, where prey species exposed to the smell of their predators (i.e. so-called info-chemicals) respond by modifying their own traits. These modifications are often predator-specific and differ between prey species. A classic example is tadpole tail development (Fig. 6.4). When tadpoles of the Pacific chorus frog are exposed to the smell of diving beetles, the tail is enlarged to lure the predator away from the head. However, when exposed to bluegill sunfish, *Lepomis macrochirus*, info-chemicals, the tail becomes shallow to increase swimming speed and escape predation (Benard 2006). Hence, trait modification is information-sensitive. Trait-mediated interactions can also influence community composition when species react differently to predator info-chemicals. Tadpoles of green frogs, *Lithobates clamitans*, and bullfrogs, *Lithobates catesbeianus*, differ in foraging behaviour when exposed to the smell of dragonfly larvae. The feeding rate of the green frog does not change, but tadpoles of the bullfrog significantly reduce their feeding rate (Relyea & Yurewickz 2002).

Another example is the effect of spring temperature changes, and the subsequent *phenological reaction* of the pedunculate oak, *Quercus robur*, winter moth, *Operophtera brumata*, and great tit, *Parus major*. Spring temperatures in European temperate regions have significantly increased over the last couple of decades, resulting in temporal mismatches between species. In response to an advancing springtime, oak bud burst has moved 14 days earlier over the last 40 years (Visser & Holleman 2001). The caterpillars of the winter moth feed on young oak leaves, because these have a rather low phenolic content at the time of egg hatching. However, winter moth hatching has advanced only 10 days, resulting in caterpillars starving from lower food quality. Therefore, phenotypic plasticity of growth of the caterpillar cannot maintain this developmental correlation with timing of bud burst. Moreover, the egg-laying dates of great tits have not advanced enough to keep track with biomass peaks of the winter moth, which the young birds depend upon for growth (Visser et al. 1998). Hence, the correlation between environmental cues used by oaks, moths and tits has been disrupted, resulting in modified species interactions.

6.2.8 Ecosystem Processes

We have given many examples above of how ITV helps species respond to their fluctuating environment, increasing their fitness in a variable milieu, and interact with other organisms. However, species traits (and particularly effect traits; Chapter 1) also have an effect on ecosystem processes (de Bello et al. 2010a; Díaz et al. 2013). Therefore, ITV can influence ecosystem processes, such as litter decomposition, nutrient mineralization, pollination or primary productivity (Chapter 9). Nevertheless, although many studies have addressed the importance of genetic and phenotypic diversity for species performance, relatively few studies have focused on how *ITV influences ecosystem processes* (Whitham et al. 2006). Studies on litter decomposition have shown a significant intraspecific variability in phosphorus and lignin content in senescent leaves of alder, *Alnus glutinosa*, obtained from different geographic regions (Lecerf & Chauvet 2008). The variance in litter decomposition rates in this single stream is explained by the intraspecific variability of these two chemical compounds. Interestingly, the observed ITV for phosphorus and lignin in alder was as large as interspecific differences in litter chemistry. Similar results have been found for nutrient release below cottonwood trees, *Populus*, in a common garden experiment, where variance in tannin input from foliage below individual trees explained spatial heterogeneity in soil net N mineralization (Schweitzer et al. 2004). In populations of spurge laurel, *Daphne laureola*, two breeding systems co-occur, i.e. female and hermaphrodite individual plants. Pollination success is higher for hermaphrodite individuals than it is for female plants in populations with a low number of females, with this effect increasing with elevation (Alonso 2005). Similar effects have also been shown to operate through flowering phenology (reviewed in Elzinga et al. 2007).

As discussed in Chapter 1, functional traits with an impact on fitness may interact with traits of species that determine ecosystem processes. The sometime close relationship between response and effect traits within a species, or between interacting species, may improve our ability to predict how environmental stress will affect ecosystem processes. The question that remains is how ITV in response traits interacts with ITV in effect traits, and what are the consequences of this coupling for ecosystems.

Box 6.1 Selection vs phenotypic diversity experiments

ITV may play a role in local environmental adaptation via genetic variation and/or phenotypic plasticity (see above). As a result, individuals collected along the environmental gradient may differ in their trait values, either because of different genotype or because of trait plasticity. How to tease them apart without, or together with, genetic analyses? In so-called common garden experiments, individuals collected from different localities are brought together under the same environmental conditions (Fig. 6.5). This set-up allows us to determine the relative importance of the genetic component of life history variation, and the intensity and direction of natural selection on life history traits.

To illustrate how one could test the relative importance of these two factors, we will use an example of geographic variation in traits of Lesser Antillean *Anolis* lizards (Thorpe et al. 2005). Individuals from one site are split into two groups; one group will be placed in a cage in a common garden, the other group will be placed in a cage at the native plot. The common garden is usually set up in a large patch in the middle of the environmental gradient, often the preferred habitat. The animals are kept under *ad libitum* food conditions. This procedure is done for all populations across the gradient. Populations are placed in a randomized block design, with each

Figure 6.5 A common garden experiment. Individuals collected from different localities are brought together under the same environmental conditions to test if differences in trait values between populations are due to genetic variation or due to phenotypic plasticity in response to different environments in which the species are normally found. Figure created by Janine Mariën.

> **Box 6.1** (*cont.*)
>
> block containing all populations. Traits are then measured after a sufficiently long period has elapsed, enough to allow the traits of interest, either in the adults or the F1 generation, to react to the common garden conditions. The trait values will be compared (1) between individuals from different populations placed in the common garden, and (2) between individuals reared in the native plot and individuals of the same population reared in the common garden (multivariate analysis of variance with fixed factors: original locality ID, novel locality ID, level of herbivory and elevation). Under the hypothesis of *genetic variation*, we expect strong inter-locality differences in trait values between different populations in the common garden which reflect differences in native plots, but insignificant differences between native-plot reared individuals and common-garden reared individuals with the same origin. Under the hypothesis of *phenotypic plasticity*, we expect that individuals of populations placed in the common garden will converge towards similar trait values, i.e. weak inter-locality differences in trait values among populations, but there will be significant differences in trait values between common-garden reared individuals and native-plot reared individuals of the same origin. It goes without saying that in most cases both factors will play a role, but in general it is possible to identify the most dominant factors explaining ITV.

6.3 Assessing Intraspecific Trait Variability

There are different approaches to assess and quantify ITV, and to evaluate the impact of ITV on communities and ecosystems. The experimental and analytical tools to quantify ITV and its ecological effects depend greatly on the ecological question. Below we describe four major approaches, and their technicalities, used to assess ITV and its effect at different ecological scales. The section is split according to the different types of analysis, although analyses can be combined and partially overlap.

6.3.1 Quantifying Intra- vs Interspecific Trait Variability

An important methodological question is how to quantify the level of trait variability within species and how this compares to variability between species. One way to compare ITV across several species is to compute a *coefficient of variation*; see Albert et al. (2011, 2012). CV is the standard deviation divided by the mean, is unitless and is therefore useful for comparing across species, and across traits. One potential limitation of using CV is that it does not allow for a comparison of trait variability within species against trait variability between species. This contrast is important for assessing patterns of community filtering and species coexistence, and simply to understand if interspecific differences in trait values are actually relevant compared to

ITV (Siefert et al. 2015). Following de Bello et al. (2011) and Violle et al. (2012), the goal of determining the relative importance of ITV vs interspecific variability can be quantified by an approach based on the *decomposition of sum of squares*. In this approach the total community trait variance is composed of between-species trait variance plus the (average) within-species variance across all species (see de Bello et al. 2011 for the detailed formulas). Imagine a given community with three species and four individuals each and the following trait values (say body size):

species A: 4, 5, 6, 7 cm
species B: 6, 7, 8, 9 cm
species C: 1, 2, 5, 6 cm

The corresponding sum of squares is obtained as it is in an analysis of variance (ANOVA), taking all of the trait values for the 12 individuals in our community as the response variable and species identity (with three levels) as the predictor (see 'R material Ch6'). In this example a sum of squares due to species (between-factors variability) equals 32 and the residuals (within-factor variability) equals 27. This means that between-species and within-species variability are almost the same, i.e. between-species is $32/(27 + 32) = 54\%$ and within-species, which we call here '*wITV*', is $27/(27 + 32) = 46\%$. However, notice that in this particular case, species would significantly ($P = 0.029$) differ in body size, indicating that interspecific trait variability is larger than ITV, and that differences in trait values between species are greater than expected based on a null model randomizing the identity of the species.

When the above approach was used to analyse variability in plant communities worldwide, the extent of ITV depended strongly on the trait under consideration, but ITV usually contributed less to trait variation than interspecific variability (Siefert et al. 2015). The average *wITV* across all traits in a community was around 25% (Fig. 6.6). In some cases, though, ITV is almost comparable to interspecific variability, as is also found by de Bello et al. (2011) studying two Czech meadows. The results of Siefert et al. (2015) confirm earlier expectations that trait differences between species are generally stronger than within species, but also indicate that ITV is a non-negligible component of community trait variability.

Violle et al. (2012) further suggest that when *wITV* is high, most of the trait variability occurs within species and species will have highly overlapping trait distributions, and vice versa with low *wITV*. If the overlap in trait distribution decreases with an increase in species richness, i.e. a greater functional differentiation between species, this pattern could support the existence of limiting similarity by competitive exclusion as a factor driving community composition (MacArthur & Levins 1967). On the contrary, if the overlap in trait distributions increases with species richness then one would conclude that equalizing mechanisms control community assembly (Le Bagousse-Pinguet et al. 2014). The absence of a relationship between *wITV* and species richness supports neutral assembly rules.

Notice that *wITV* so far can only be used for single traits. However, the index can be expanded based on a form of Rao quadratic entropy which actually expresses trait variance (for details see de Bello et al. 2011; Chapter 5), corresponding to Permanova analyses, with the function 'adonis' available in R in the *vegan* package. Thus, we can use multiple traits measured across a variety of individuals by computing the functional dissimilarity between individuals based on these multiple traits (see Chapter 3), and then repeating the method described for *wITV*.

Another important point to consider when calculating *wITV* is the number of individuals that have been sampled for each species. As we will see in Chapter 11, it is most unlikely that all individuals within a community will be sampled. One can either sample a fixed number of individuals for each species or, for example, sample more individuals for the more abundant species. As it is not easy to collect rare species, the individuals measured can be selected less randomly, and thus there is less room for choosing well-developed, healthy individuals. One consequence is that ITV of rare species could be 'artificially' larger than between species (de Bello et al. 2011) because suboptimal individuals are selected.

When using ANOVA to analyse the relative importance of ITV versus between-species trait variability, *wITV* is computed as a weighted average of the ITV across all species in that community (with the relative number, biomass, frequency, cover etc. of individuals determining the weight of each species (see Chapter 5) in the decomposition of sum of squares). Care should be taken for size-related traits, as variance is expected to be linearly dependent on the mean, leading to higher variance for larger species. This is not a result of higher ITV but instead due to scaling of measurement units. Log transformation is an appropriate solution to this as it results in the independence of the mean and variance, as opposed to the use of the coefficient of variation, which precludes additivity of within and between FD (de Bello et al. 2011).

6.3.2 Intraspecific Trait Adjustments vs Species Turnover

Along environmental gradients trait values can change due to intraspecific adjustments of species or due to species replacement, also called turnover (Cornwell & Ackerly 2009; Siefert et al. 2015). The specific tool to analyse the relative importance of these two causal mechanisms was originally proposed by Lepš et al. (2011), and is based on the different computations of community weighted mean (CWM; see Chapter 5).

Imagine two plots each with a different level of soil fertility, with this affecting plant height. These two plots can be in the same meadow or in two separate meadows. The plots share at least some species, and these species can change their height depending on the level of soil fertility. The changes in CWM plant height between plots can be caused by two mechanisms: different species with dissimilar heights dominate in the two plots, or the same species dominate in both plots but their heights differ in the two plots. If plant height is completely invariable within species (an unrealistic assumption), the changes in CWM plant height between plots can only be caused by species replacement, for instance when taller species replace smaller species under the more fertile conditions. This example also includes changes in species relative abundance, reflecting a specific case where species composition changes without replacement. In most cases (as described in Chapter 11) trait values for the same species across different plots are often not available. As such, this type of analysis is based on what we call CWM_{fixed}, because it uses fixed trait values within species, irrespective of where they grow.

Sometimes trait values for a single species are available across environmental gradients. So, if there are two plots with different fertility levels, individuals of the same species in the fertilized plot will grow bigger. Hence, if we use different values of height measured in different plots, the change in CWM plant height between the two plots can also be due to ITV. Most of the time the change in CWM plant height is caused by a combination of species replacement and within-species trait adjustments. Lepš et al. (2011) suggest that it is possible to compute CWM_{fixed} as mentioned above and $CWM_{specific}$, which corresponds to computing CWM using the plant height measured in a specific plot (for example, plant height in the fertilized plot). The salient concept here is that any difference between $CWM_{specific}$ and CWM_{fixed} plant height can only be due to intraspecific trait adjustment across plots (e.g. individuals are taller in fertile plots). Hence, the effect of intraspecific trait adjustment is equal to $CWM_{specific} - CWM_{fixed}$. Following the R material in this chapter, CWMs can be computed for each of the two plots, one using the plant height measured in that plot and one using a fixed trait value for the average plant height of a species across all plots. The two CWMs for each plot will be basically treated as 'repeated-measures' of the same plot, so statistical models similar to repeated-measures ANOVA can be used to test the contributions of species turnover and ITV to trait replacement along an environmental gradient. This process is explained in more detail in 'R material Ch6' and in Lepš et al. (2011), and allows us to compute how much in ITV is important along environmental gradients, using indices such as the '*aITV*' used by Siefert et al. (2015; Fig. 6.6). The *aITV* is basically the ratio of the effect of interspecific trait adjustments to the total changes in trait values along a selected gradient.

The approach explained above is now implemented in various R scripts (see 'R material Ch6'), which usually also compute a so-called covariation between species turnover and ITV effects. This covariation is the extent of a *joint* effect between turnover and ITV. In the example of plant height and soil fertility, fertilization would 'select' for taller species and make individuals of the same species also grow taller. Hence, filtering across species due to fertilization, and by individuals within species, has the same effect on CWM height, i.e. increasing the average size of plants in fertile plots. A positive covariation in inter- and intraspecific plant height would thus occur in this case. However, a negative covariation indicates that plots with fertile soil 'select' for taller species, while individuals of some species along the fertility gradient grow shorter under fertile conditions. This reflects a sort of buffering of CWM plant height, with species turnover and within-species adjustments somewhat compensating each other.

If traits have not been measured in all plots, although not ideal, it is still possible to apply this framework. For instance, in an experimental design with five unfertilized and five fertilized plots, species are not sampled in all plots within a treatment, but at least in one of the treatment plots (or very few individuals are measured within each plot; see Carmona et al. 2015a). This assumes (but this can be further tested of course) no considerable intraspecific variation in plant height between replicates of the same treatment, or at least not as big as ITV between treatments (see Chapter 11 for various

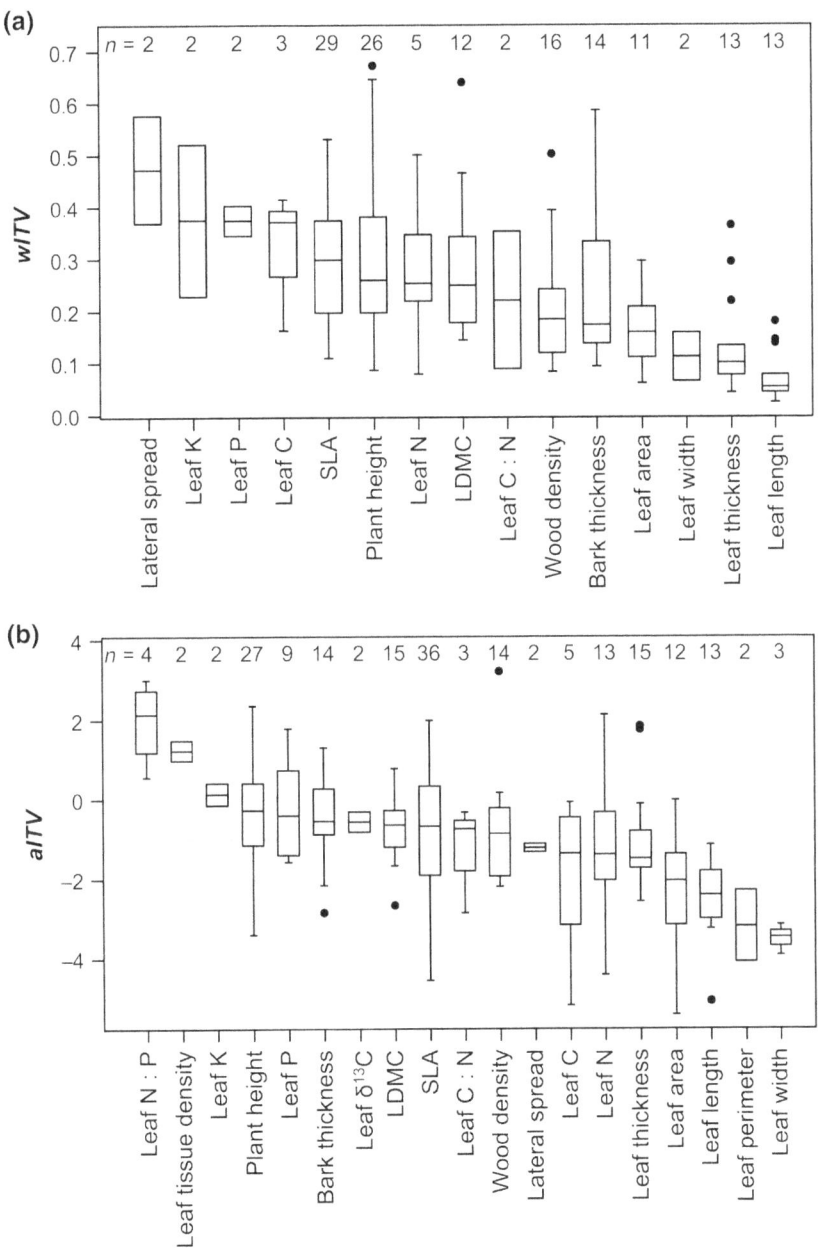

Figure 6.6 Boxplots showing relative magnitude of ITV (a) within communities (wITV) and (b) among communities (aITV). Panel (a) corresponds to the method described in section 6.3.1; panel (b) corresponds to the method described in section 6.3.2. The number of observations (studies) per trait is indicated above each box (total number of studies: 33 for wITV; 37 for aITV). Solid horizontal line indicates overall mean value across all traits. Dashed horizontal line indicates equal magnitude of intraspecific and interspecific trait variation (wITV = 0.5; aITV = 0). Values above dashed line indicate larger intraspecific than interspecific variation and vice versa. Taken from Siefert et al. (2015), with permission from Wiley. © 2015 John Wiley & Sons Ltd/CNRS.

sampling schemes similar to this). At present, the R functions that include both CWM_{fixed} and $CWM_{specific}$ allow for a variety of experimental designs, both with and without replication, for example unreplicated plots along a variety of gradients. In the absence of replicates, it is possible to use the approach proposed by Carmona et al. (2015a), i.e. sampling at least a few individuals of each species in each plot where it occurs.

6.3.3 Trait Variability across Ecological Scales

Trait values differ across a hierarchy of spatial scales; for instance, inter- and intraspecific variability in plant height can vary within and across communities and landscapes etc. In some cases we might be interested in decomposing trait variability even further beyond within-species variability (e.g. within and between branches of a tree). There is a vast number of spatial scales at which traits can vary, starting from the *within-individual scale*. For example, specific leaf area (SLA) can differ significantly within a single tree (Hulshof et al. 2013). Obviously, not all the leaves within a tree have the same SLA values – there is some degree of variability within branches due to age, light conditions etc. (Pérez-Harguindeguy et al. 2013). Therefore, a substantial proportion of trait variability could potentially be explained by differences between light and shade branches within a single tree. Depending on the aim of the study, we may want to look at differences between conspecifics within a population, or between populations of a single species in the landscape, or perhaps at dissimilarity in traits between different species in a community, or to analyse trait differences associated with different environmental conditions, and so on.

Variation in trait values can be assessed across hierarchical spatial scales by means of *variance decomposition* analyses. Messier et al. (2010) decomposed variability in leaf traits by different nested scales. The smallest scale are leaves measured within (nested in) a single branch. The next level would be leaf traits in branches under differing light conditions which are nested within a single tree. Single trees belong to a given population of species that are nested within a community in a given plot with relatively uniform conditions. Finally, trees in a plot belong to a landscape or environmental gradient, which encompasses several plots. Similarly, Carmona et al. (2015a) decompose trait variance on three nested scales: among individuals within species within quadrats, among species within quadrats, and among quadrats. In both cases the authors assign traits (leaf traits and plant height) to appropriate scales, and then fit a mixed effects model that uses the nested scales as a random factor. This kind of approach allows the partitioning of trait variance across the different scales. This can be done in R using the '*lme*' function from the '*nlme*' package, followed by the '*varcomp*' function from the '*ape*' package. Bootstrapping can be performed to obtain confidence intervals for the proportion of variability found at each level (see 'R material Ch6').

6.3.4 Including Intraspecific Variability into Functional Diversity

Using traits to explore species' functional differences and community assembly often relies on the calculation of trait dissimilarity between species. Despite the potential importance of ITV, most methods focus on interspecific trait differences, considering only a single mean trait value for each species. However, as we have seen in this chapter, this assumes no ITV or that ITV is much smaller than interspecific differences, which is not always the case.

In order to estimate functional dissimilarities between species, while also fully considering ITV, *all individuals* within a community should be sampled (Cianciaruso et al. 2009). Unfortunately, measuring all individuals, even in a single community, is virtually impossible (see Chapter 11). A powerful solution is to sample just a representative *selection of individuals* of each species and compute the dissimilarity between species in terms of the overlap between trait distributions (MacArthur & Levins 1967). The rationale behind this method is that conspecifics have some degree of variability in their trait values, with some trait values being more frequent than others (Carmona et al. 2016). This can be represented by means of a species trait distribution (Mouillot et al. 2005), as shown in Figs. 6.1 and 6.7.

First, consider a functional trait, let's say body length (or SLA in Fig. 6.7). The distribution of body length values for each species can be computed, in a simplified way, by using the mean and SD of trait values within species, and building a normal distribution with those parameters, or more elegantly using kernel density estimation. More details on the rationale and maths involved can be found in Mouillot et al. (2015) and Geange et al. (2011), among others. Importantly, the body length distribution of each species is a *probability density function*, which means that its total area is 1, i.e. it integrates to 1. This property is especially interesting because it can be used to estimate the dissimilarity in body length between pairs of species by simply

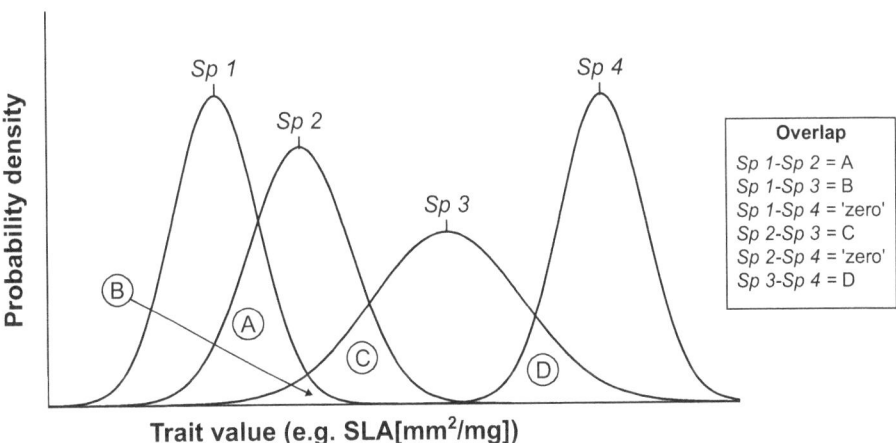

Figure 6.7 Species overlap (O) based on trait values distribution within species, using a probability density curve for each species. The area below each curve is always equal to 1 by definition (100% probability), so that 1-O for each pair of species reflects the dissimilarity between them. Note that with roughly constant variability, the more the means differ, the smaller the overlap between species. Taken from Lepš et al. (2006), with permission from the Czech Botanical Society. © Leps et al.

estimating what proportion of their respective probability density functions *overlap* (Fig. 6.7). The dissimilarity in body length between species i and j is then given by $1 - \text{overlap}_{ij}$. Since we can compare the level of overlap in body size for all pairs of species in the community, and the dissimilarity will be scaled between 0 and 1, all of the analytical methods for estimating functional diversity based on dissimilarities described in Chapter 5 can be used, for single or multiple traits. Interestingly, using trait dissimilarities based on ITV seems to give more biologically meaningful results than trait dissimilarities based on species means, particularly on a local scale (de Bello et al. 2013a).

However, species differences are, by definition, multidimensional. Thus, correctly estimating the overlap between coexisting species requires us to consider multiple traits. In recent years, different methods have been published that account for ITV and multiple traits simultaneously. These include *nicheRover* (Swanson et al. 2015), *dynamic range boxes* (Junker et al. 2016) and *hypervolumes* (Blonder et al. 2014; Blonder 2016). In this line, Carmona et al. (2016, 2019) developed the notion of Trait Probability Density (*TPD*), which combines the ideas of considering probabilistic trait distributions and multiple traits. TPDs are mathematical functions that represent the distribution of probabilities of observing each possible trait value in a functional space whose coordinates are the considered traits. The main advantage of TPD functions is that they can be built for any spatial scale. While previous uses of probability density functions have only considered the differences between pairs of species, the TPD framework allows combining the TPD functions of the species that are present in a community (TPD_S, with or without considering the species relative abundances) to build the TPD of the community (TPD_C; Fig. 6.8). Then, TPD_C functions can be aggregated in a similar way to estimate TPD functions at any spatial scale (regions, biogeographical realms, continents or the whole world). This allows the estimation of dissimilarities not only between pairs of species, but also between any of these scales, i.e. functional beta diversity (Chapter 5). Further, since TPD functions are fundamentally probability density functions (i.e. they integrate to 1), one can also estimate the dissimilarity between TPD functions at different scales. For example, by estimating the dissimilarity between the TPD_S function of a species in a community and the TPD_C of the community, you can say how unique this species is in the community (i.e. how rare the traits of the species are within the community; Carmona et al. 2017a).

However, the TPD framework is not only limited to estimation of trait dissimilarities; it can also be used to estimate practically any aspect of functional diversity at *any spatial scale*, while considering ITV and multiple traits. Among the functional diversity indices we presented in Chapter 5, Carmona et al. (2016, 2019) show how to estimate functional richness, evenness and divergence, functional beta diversity (and its decomposition into nestedness and turnover components, which is beyond the scope of this book, but can be seen in Villéger et al. 2013; Carmona et al. 2016) and functional redundancy. On top of this, it is possible to estimate all indices based on a matrix of dissimilarities between species (Rao, MPD, FDis etc.) while considering ITV. Readers interested in the specific concepts and formulas to estimate all these indices can consult

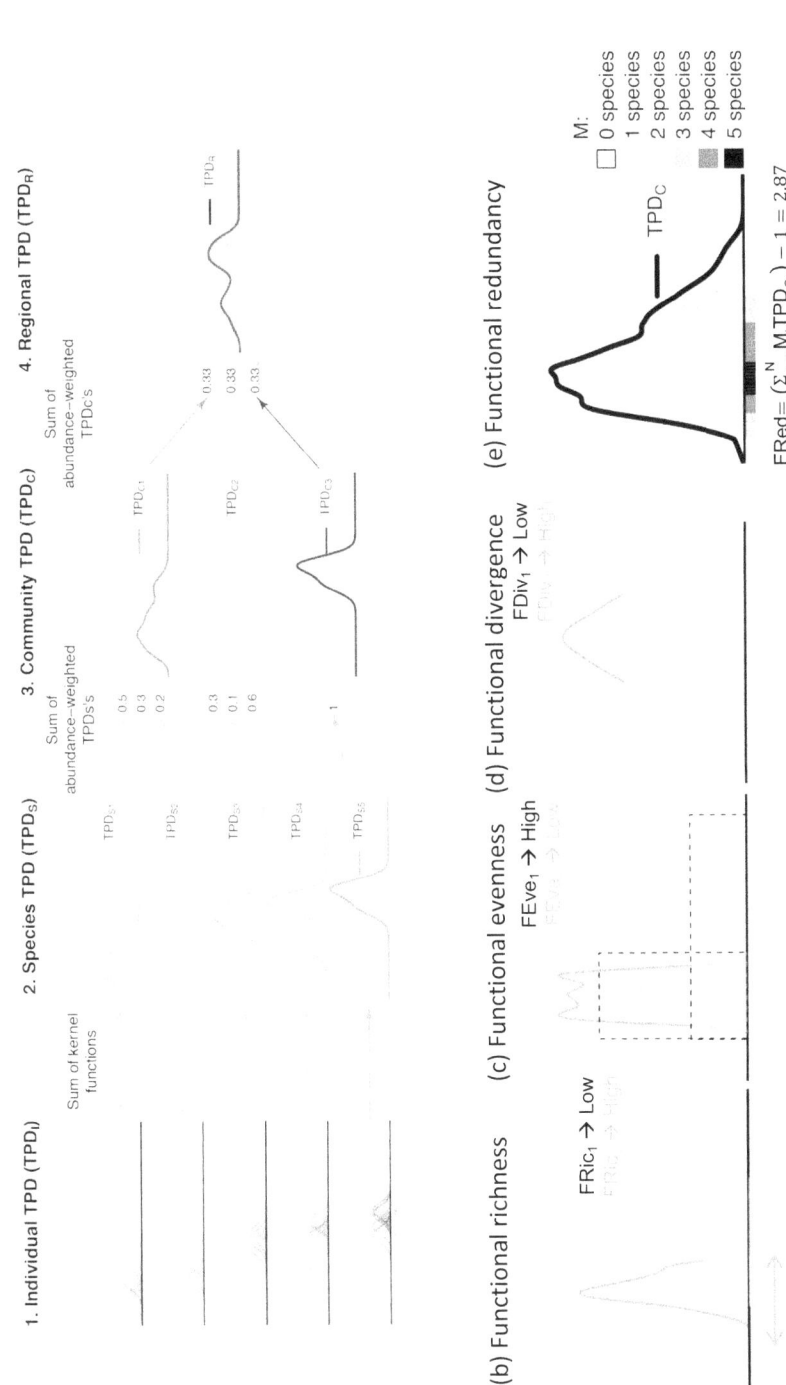

Figure 6.8 In the Trait Probability Density (TPD) framework, the traits of an ecological unit (e.g. a species, or a community, or a region etc.) are expressed as a probability distribution whose value in each point in the functional space reflects the relative abundance of the corresponding trait. For simplicity, all examples involve one trait only. By considering individuals (a1) this framework includes ITV when constructing the TPD of species (TPDS; a2). The TPDS of the species present in a community can be weighted by the species relative abundance, and combined to estimate the TPD of communities (TPDC; a3). This procedure can be repeated across hierarchical scales, leading to the estimation of TPD for any scale, for example, regions (TPDR; a4). Several functional diversity indices can be estimated within the framework, for example: functional richness (b), functional evenness (c), functional divergence (d) or functional redundancy (e). For formal definitions of these indices see Carmona et al. (2019). Adapted from Carmona et al. (2016), with permission from Elsevier. © 2016 Elsevier Ltd. All rights reserved.

Carmona et al. (2019) and use the R package 'TPD' (Carmona et al. 2019) and the accompanying R material in this book.

Take-Home Messages

- Intraspecific trait variability (ITV) is a key aspect of the evolution of species, their adaptations to environmental fluctuations and the effects of species on multiple ecosystem properties and trophic interactions.
- Important sources of ITV are genetic variation, epigenetic effects and phenotypic plasticity.
- There are different tools to quantify the strength of ITV effects on ecological patterns: for instance, to compare how strong are the differences within vs between species in a community or how strong is the effect of changes in species composition (turnover) compared to ITV along environmental gradients.
- ITV can also be used to quantify trait differences between species, for example through measures of trait overlap, which can be used to quantity functional diversity.

7 Community Assembly Rules

Why do we observe a certain variety of species in a community? Why are some communities more species-rich than others? Why are some species present in a particular community but not in others? Ecologists have long been interested in understanding what the assembly rules are that explain why certain species exist in a given community yet not others. We show in this chapter that using species traits allows a better understanding of these rules. As a matter of fact, knowledge of the principles that govern community assembly, and how they operate through species traits, is perhaps the main premise of trait-based ecology. In the current global-change context, achieving such knowledge is particularly important for building realistic models to predict biodiversity changes in the future.

In this chapter, we start by introducing the term '*assembly rules*', which we understand as any constraint limiting the number, abundance and identity of the species that we observe in an assemblage. Since we have already visited some of these ideas in Chapter 4, which deals with the effects of environmental filtering, we will now focus mostly on the effect of the interactions between species. Among these interactions, we will pay particular attention to competition between organisms, mostly because this is the process that has most frequently captured the attention of ecologists. However, although the main focus of the chapter is on competition, similar approaches can be developed to assess other effects of biotic interactions between organisms within a given trophic level, such as facilitation (Bimler et al. 2018); for interactions across trophic levels see Chapter 10. Later in the chapter, we will explain how to design null models that allow us to test different hypotheses related to assembly rules, with a special emphasis on the importance of the adequate selection of spatial scale and species pool (see 'R material Ch7' for details). At the end of the chapter we will also briefly deal with the implications of intraspecific variability on assessing assembly rules, and on how this knowledge can be applied to better predict the species composition and functional structure of communities under global change.

7.1 Community Assembly Mechanisms

The term *assembly rules* was coined by Diamond (1975), who was studying bird assemblages on the islands of Papua New Guinea and observed that certain species pairs never coexisted on the same islands. He observed 'checkerboard' patterns in community matrices, which he attributed to competitive interactions between species leading to a lower-than-expected co-occurrence of species with similar niches, and hence a similar use of resources. Therefore, assembly rules, as originally conceived, referred to *interspecific competition* and did not explicitly consider species traits. However, the existence and relevance of assembly rules based on competition was soon contested by researchers comparing the observed patterns with patterns arising from *null models* in which competition was not included. The pioneers here were Connor and Simberloff (1979), who showed that chequerboard patterns can also arise due to random colonization of communities. Later, Hubbell's neutral theory (2001) distinguished between assembly rules based on dispersal and assembly rules based on species niches. *Neutral theory* proposes that species are equivalent in their fitness, so that diversity patterns can be mostly attributed to dispersal and evolutionary processes (i.e. speciation and extinction).

In addition to neutral theory, there are a myriad of theories regarding community assembly (revised in Vellend 2016) that mostly differ in how much they focus on different processes such as dispersal, evolution, or interactions between organisms. Ultimately, all these processes can be classified into four fundamental processes, which interact to determine community assembly (Vellend 2016). These fundamental processes are *speciation* (the process by which new species arise from existing ones), *dispersal* (the spatial movement of organisms), *ecological selection* (interactions of organisms with each other and with the environment, including both environmental filtering and limiting similarity) and *drift* (random changes in species abundances). The combination of these four processes results in the patterns of biodiversity that we observe in an assemblage (as we showed in Fig. 4.4). Adopting this point of view, it seems clear that the different lines of thought presented in the previous paragraph stem from primarily focusing on one of these processes. Consequently, they can be categorized according to which processes (and scales) are given more weight. For example, consequences of speciation will become more and more important as we increase the *spatial extent* of our study, by covering different regions with different evolutionary histories. As such, Diamond's original assembly rules have a strong focus on selection (and hence on the importance of competition between species). By contrast, Hubbell's neutral theory disregards selection and tries to explain diversity by only considering speciation, dispersal and random drift (Vellend 2016). As we will see in the next sections, each of these processes calls for a different 'scale' of observation, and particularly for the definition of a specific reference species pool.

7.1.1 Defining the Reference Species Pool

As exemplified above, assembly rules can be thought of as any constraint that limits the number and identity of the species that we observe in an assemblage (Götzenberger et al. 2012). If the species that we observe in an assemblage are only a subset of a larger group of species, it follows that we first need to characterize the larger group to understand which type of species have been excluded at a finer scale. This larger group of species is regarded as the '*species pool*'. The species pool concept is a particularly slippery one because species pools are operationalized in different ways and at different spatial and ecological scales, depending on our research questions (Cornell & Harrison 2014; Zobel 2016). Selecting the adequate pool of species is crucial, since characterizing community assembly very often zooms in on examining whether or not the diversity patterns observed at a given scale are different from those expected from a reference species pool. For example, if you want to assess the assembly of plant species from a particular biogeographical realm, your species pool could include all the world's plant species. In contrast, if you are interested in knowing if *environmental filtering* limits the traits of the species that inhabit a particular habitat within a region, then your pool should include all the species that are present in that region (regional species pool).

Depending on the extent of the species pool considered we will need to use different methods to identify it, and we can focus on different *mechanisms* that cause filtering of this pool down to a smaller set of species. The regional species pool can be very roughly estimated using a list of the species present in a given region. A problem arises when we want to define which species, out of the regional species, can disperse and inhabit a given location. For example, if we want to know if *dispersal limitation* restricts which species are present in wet forest sites, then not all the species from the regional species list are relevant. Perhaps some species from other kinds of forests can live at the site of interest, but species typical of other very different habitats within the region, such as dry grasslands, are not relevant, because even if they could disperse, they most likely cannot maintain a viable population in wet forests. In such cases the so called *habitat-specific species pool* should be characterized, which should include only those species from the regional pool that can live in the habitat of interest. This requires defining the focal habitat, and then restricting the regional list of species to exclude species that cannot occur in the focal habitat. Since the distinction of sites between habitats is often not so straightforward (e.g. forest sites actually form a continuum between dry and wet forests), recent developments try to characterize *site-specific species pools* by applying much more complex methods (de Bello et al. 2016b; Karger et al. 2016).

Once we have defined the relevant reference pool for our study, we can try to assess what assembly processes are taking place. In terms of functional traits, we generally do this by examining if some aspects of the *functional structure* that we observe in the assemblages under study are different from those expected if the species in the assemblage were a (more or less) random subset of the species pool. If we find some differences, we can try to discuss them in terms of what processes are expected to drive these patterns. For example, as we showed in Chapter 4, environmental filtering can restrict the trait values that we observe at our sites with respect to the whole range of trait values that we observe in the region. Hence, the trait diversity within a habitat should be lower than the diversity in the regional species pool. There are other processes, however, that can have different signatures on trait patterns, the most commonly examined of which are biotic interactions. Before we move into providing a guide about assessing assembly rules in ecological communities, we have to discuss the potential effects of biotic interactions.

7.2 Biotic Interactions and Species Coexistence

7.2.1 Historical Perspective

Although there are different types of *biotic interactions* between species in a given trophic level (competition, facilitation, mutualism, etc.), *interspecific competition* has received the most attention. The most common expectations about the effects of competition on species composition have been highly influenced by the idea of competitive exclusion. This idea stems from the *Lotka–Volterra model*, developed almost a century ago (Lotka 1925; Volterra 1926). This mathematical model explains the dynamics of a pair of species competing for similar resources and predicts that one species will exclude the other when interspecific competition is stronger than intraspecific competition. This prediction was corroborated experimentally by Gause (1934), who observed that *competitive exclusion* happens frequently between a pair of *Paremecium* species (i.e. *P. aurelia* and *P. caudatum* – unicellular organisms that feed on other microorganisms like bacteria) competing for the same resource, under constant environmental conditions. These theoretical and experimental results suggest that *coexistence* of many species should not occur, because a small competitive advantage by one species should eventually lead to exclusion of the other species. Obviously, this is not what happens in nature, where complete dominance by a single species is relatively rare and many ecosystems like grasslands, tropical forests or coral reefs are extremely species-rich. This apparent contradiction was underscored by Hutchinson (1961) in his 'Paradox of the plankton' paper; phytoplankton species compete for very limited resources, yet many species coexist. There are several potential explanations for coexistence despite the evidence that some species have a competitive advantage over others. Most explanations point to the lack of realism in both the Lotka–Volterra model and Gause's experiments. Among other factors, both the model and experiment assume spatial and temporal homogeneity in environmental conditions and did not include dispersal. Although such conditions are hardly ever met in nature, Lotka, Volterra and Gause's findings provide a useful reference point to understand competitive interactions between species.

Despite its limitations, early ecologists studying species coexistence followed the insights from the Lotka–Volterra model and focused on quantifying the relative importance of intraspecific versus interspecific competition. One approach for this quantification is to link the difference in competitive ability to differences in resource use between species. The inspiration for this approach comes from evidence of *character displacement*. Character displacement happens when two species competing for the same resource will become more different in their use of resources when they coexist in the same area than when they live separately in different areas, which leads to the evolution of different phenotypes (Schluter & McPhail 1992). The prerequisite for character displacement is that a differentiation in certain traits results in differences in resource use. Since some characters are related to resource use and acquisition, early works started to examine these effects in terms of species traits. Character displacement is interpreted as evidence for species competition, where species differentiate in terms of traits to reduce resource overlap when they occur in sympatry (i.e. live together), which in turn leads to less intense interspecific (compared to intraspecific) competition, and enhanced coexistence. Two species of Darwin's finches in the Galapagos Islands (*Geospiza fortis* and *G. fuliginosa*) provide the classic example of character displacement (Grant & Grant 2006). In these species, beak size is associated with resource use, since which seeds the finches feed on depends on beak size. On the islands where the two species live together, they have different beak sizes. However, on the islands where only one of the species lives and the other is absent both species have intermediate-sized beaks.

These ideas contributed greatly to the ecological concept of *limiting similarity* between species (MacArthur & Levins 1967). The limiting similarity theory postulates that there is a limit in the overlap in resource use (or more generally in niches) between species that can coexist. To quantify this limit, MacArthur and Levins characterized the resource niches of species as normal distributions (see Fig. 6.1 for an illustration). They proposed that limiting similarity was the minimum separation between the peaks of the curves of two species that would allow for a third species whose peak is in the middle to coexist. They concluded that, for the three species to coexist, the distance between the mean values (i.e. the peaks of the distributions) of the two species in the extremes should be larger than their niche widths (the standard deviation of these curves). The next point, as in the case of character displacement, is the notion that species niches can be expressed in terms of traits, so that coexisting species cannot be too similar in their trait values. Of course, this is a strong assumption, because not all traits are related to the way a species uses resources or to species niche axes (in other words, not all traits are relevant for all questions, as we discussed in Chapter 2). But in practical terms ecologists use the concept of *functional trait diversity* (Chapter 5) as a way to understand possible mechanisms of species coexistence through the similarity of their traits.

The original strict formulation of limiting similarity is conceptually very attractive, since it allows us to make predictions about species coexistence based on knowledge of their traits and related resource-consumption curves. However, the theory does not hold up well when tested in natural systems, and a number of studies have shown that there is no universal maximum level of similarity for two species to be able to coexist (Abrams 1983). There are many reasons why strict limiting similarity is not apparent under natural conditions, including the multiplicity of resources, the role of temporal and spatial heterogeneity (often not accounted for in experiments), and the impracticality of quantifying species' niches as unidimensional normal distributions. Perhaps the main limitation of limiting similarity theory is that we cannot generally assume that competition for a common resource between species is the only determinant of the impact of one species on another. However, despite the lack of support for a universal fixed limit on similarity between species pairs, limiting similarity survives in the present day in the form of a less stringent concept, which is still regularly examined by scientists (Wilson 2007; Violle et al. 2011). Rather than trying to find the minimum differences possible for species to coexist, these studies assess whether the traits of coexisting species are more different than expected by chance.

7.2.2 Coexistence Theory

Keeping Chapter 4 in mind, you will understand that the idea that species need to be different to coexist clashes with the idea that species living in the same environment need to be similar in their traits. As we will see in this section, nature is more complex, and theoretical and empirical studies show that both environmental conditions and competition might act as filters on species traits, although they may act at different spatial and temporal scales (Swenson & Enquist 2009; Mayfield & Levine 2010; Kraft et al. 2015; Chalmandrier et al. 2017). This shift in our expectations about species interaction effects on species coexistence in terms of trait dissimilarity is associated with the increasing adoption of *modern coexistence theory* (Chesson 2000; Mayfield & Levine 2010), which shows that competition does not necessarily lead to coexisting species having different traits. In richer and more productive conditions, for example when a meadow is fertilized, taller plant species generally dominate, creating a convergence of species towards traits associated with greater competitive ability (e.g. greater size; Chapter 3). This means that, in contrast to limiting similarity theory predictions, competition can also lead to coexisting species being similar to each other (Grime 2006). Modern coexistence theory essentially states that coexistence among species is possible when niche differences between them are able to offset their differences in fitness (Chesson 2000; Fig. 7.1, whose details we provide later). In other words, species that have very similar competitive ability don't necessarily need to have different niches, whereas if species A has a much lower fitness than species B, A needs to have a sufficiently different niche in order to survive the battle with B. In the rest of this section, we provide a more detailed explanation of this theory.

Formally, within Chesson's (2000) framework (Fig. 7.1), two species can coexist if they are both able to *invade* (i.e. having a positive per capita growth rate) a site in which

7.2 Biotic Interactions and Species Coexistence

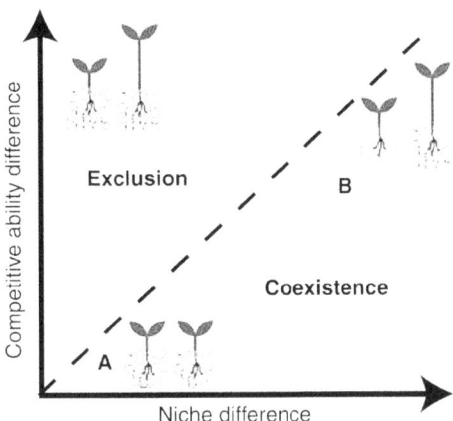

Figure 7.1 Coexistence between species based on their traits. In this example, taller species are the stronger competitors, whereas niche differences are related to their preference for different soil textures (larger particles indicate well-drained, rocky soils, whereas smaller particles indicate more water- and nutrient-rich soils). Stable coexistence (below the dashed line) occurs when niche differences are large enough to offset competitive ability differences. Coexistence is possible when very small differences in competitive ability between species are offset by very small niche differences (bottom left corner) or when large differences in niches offset large differences in competitive ability (top right corner). When niche differences are not enough to compensate for competitive ability differences (top left corner), stable coexistence is not possible. Taken from Mayfield and Levine (2010), with permission from Wiley. © 2010 Blackwell Publishing Ltd/CNRS.

other species are already established at carrying capacity. Let's imagine species A and B again, assuming that species B is a stronger competitor than A at our focal site. If the per capita growth rate of A, when competing with B, is negative even when there are very few individuals of A, then species A will eventually go extinct, without being able to invade again. Conversely, if the per capita growth rate of A is positive, then it should be able to 'invade' (experience population growth); in this case, coexistence between A and B is considered to be stable. *Stable coexistence* can only happen when the invading species has a higher growth rate when it is rare than when it is common, i.e. the growth rate of a species is dependent on its density. This *density-dependence* is a consequence of intraspecific competition being stronger than interspecific competition (Levine et al. 2017), i.e. the higher the abundance of a species is, the more its individuals have to fight against their own density.

As you can see, modern coexistence theory is most commonly explained by colonizing ability, expressed as per capita growth rate when the colonizing species is rare. Colonizing ability can be decomposed into two terms that are expressed according to the ecological differences between the species: fitness differences and niche differences. There are many different ways to formalize these ideas mathematically (Chesson & Huntly 1997; Chesson 2000; Adler et al. 2007), but, basically, they can be condensed into this general structure (considering two species: A, which is the colonizing species, and B):

Colonization ability of A on B = niche differences$_{AB}$ − fitness differences$_{AB}$

The *fitness differences* term above encompasses both the density-independent growth rates of the species (in the absence of any competition) and their sensitivity to competition. If species B has a high growth rate in the absence of competition and its growth rate does not decrease considerably with competition, it will have a high fitness. In that case, B will likely be a very strong competitor at the considered site. On the other hand, if A has a low per capita growth rate in the absence of competition and/or high sensitivity to competition, then it might be unable to invade a population of B. These differences in competitive ability between species are termed 'fitness differences' in modern coexistence theory, although some authors use different names such as 'differences in intrinsic growth rates' (Cadotte & Tucker 2017) or 'differences in competitive ability' (Mayfield & Levine 2010). Whatever the terminology, the concept remains similar: large fitness differences between species hinder coexistence.

We will borrow an example from Mayfield and Levine (2010) to show the effect of fitness differences between species in more detail. Let's imagine a community where light is limiting and there are two plant species with different heights. Since height is a trait that improves access to light, in the absence of any other factor the taller species would monopolize light and competitively exclude the smaller species (Fig. 7.1), particularly if height differences are large. However, if the two species would have very similar heights, they would have very similar fitness and it would take a long time for the species with higher fitness to competitively exclude the other. Nevertheless, if only fitness differences exist (even if they are small), the species with the highest competitive ability should eventually exclude all other species, which would not be able to invade the community afterwards. Since this does not correspond to the observed patterns in diversity in most cases in the real world, there must be some mechanism that allows for coexistence among species.

The mechanism that allows for coexistence is found within the *niche differences* part of the model in modern coexistence theory (Chesson 2000; see equation above). The underlying idea is that the more different the niches of two species, the greater is the effect of intraspecific competition compared to that of interspecific competition (see also Chapter 6). This makes sense, since conspecifics are generally expected to use the same resources in a similar way, so that the higher the density of conspecifics is, the smaller the amount of resources available per capita. By contrast, if the two competing species use resources very differently, then the growth rate of the individuals of one species should not be very much affected by the density of the other species. We can return to the Mayfield and Levine (2010) example to illustrate this (Fig. 7.1). Let's suppose the taller species prefers soils with high levels of water and nutrients, whereas the shorter species is better able to exploit well-drained, rocky soils. If both types of soils are relatively frequent in the environment, the species will segregate across soil types, and the individuals of the taller species will limit other individuals of their own species more than they limit individuals of the shorter species, which could then live in the rocky soil patches. Put in other terms, the consequence of niche differences between species is that they will experience negative frequency-dependent growth rates, which means that the average per capita growth rate of a species decreases as it becomes more frequent, due to intraspecific competition. When this happens, the average individual growth rate of a species is higher when the species is rare, buffering it from extinction or allowing it to invade when absent. On the other hand, individual growth rate decreases as the species becomes more common, making competitive exclusion more difficult (HilleRisLambers et al. 2012). The main implication is that when the niches of the competitors are sufficiently different, a species with high competitive ability might suppress itself more strongly than it suppresses a competitor species, thus allowing coexistence. It is generally acknowledged that trait differences between species should be related to their niche differences (Laughlin & Messier 2015). Niche differences can arise from species that use limiting resources differently: different tolerances to climatic conditions, different habitat preferences or different natural enemies (Adler et al. 2007; Mayfield & Levine 2010).

In a way, modern coexistence theory reconciles the ideas of environmental filtering (see Chapter 4) and limiting similarity in a unified theoretical framework. Species with traits that confer very low fitness in a given environment will be either unable to live in such conditions even under the absence of competition or be easily excluded by species with higher fitness and not very different niches. On the other hand, among the species that have traits that are better adapted, some degree of niche differentiation will be needed for stable coexistence to be possible. Thus, there are many different ways in which species can achieve stable coexistence. For example, when two species have very similar fitness, then only small niche differences will be needed for them to coexist (bottom left of Fig. 7.1). By contrast, coexistence between two species with a large fitness difference will only happen when niche differences are sufficiently high to offset fitness differences (top right of Fig. 7.1). These are the two extremes of a continuum of potential ways in which species can coexist. In general, any mechanism that reduces fitness inequalities (*equalizing mechanisms*) or increases niche differences (*stabilizing mechanisms*; Chesson & Huntly 1997) will make coexistence easier.

7.2.3 Moving Past (Pairwise) Competition

In section 7.2.1 we focused on coexistence between species pairs. However, some coexistence mechanisms emerge only when more than two competitors are interacting (Levine et al. 2017). These other mechanisms include higher-order interactions, interaction chains and intransitive interactions. In *higher-order interactions*, the effect of one species on another depends on the density of other species in the community (Mayfield & Stouffer 2017). In *interaction chains*, species are hierarchically arranged; for example, a superior competitor species (species A in Fig. 7.2a) can alleviate the competition that an intermediate competitor (species B) exerts on a competitively inferior species (species C). In certain conditions (when A and C compete for different resources, for example), the superior competitor can have a net positive effect on the inferior species by suppressing the intermediate one (Fig. 7.2b). Finally, there are *intransitive interactions* (Godoy et al. 2017), in which three (or more) species can coexist because they are not structured hierarchically, so that there is no universally

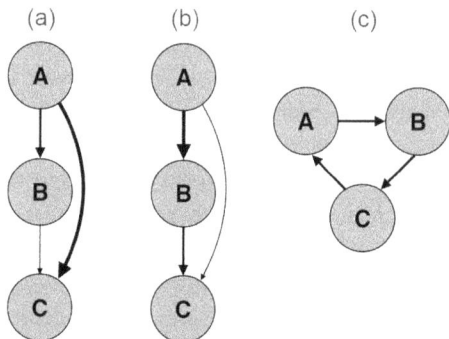

Figure 7.2 Examples of the architecture of species interactions with more than two competitors. (a) the negative effect of species B on species C can be mitigated by indirect effects of A on B. (b) If the competitive effect of A on B is particularly strong, then A can even have a net positive effect on C. (c) Under intransitive competition, there is no universally weak or universally strong competitor. Arrow width denotes the strength of the competitive dominance of the superior species over the inferior one, with the arrowhead pointing towards the inferior species. Taken from Godoy et al. (2017), with permission from Wiley. © 2017 by the Ecological Society of America.

weak or strong competitor species (Fig. 7.2c). Intransitive interactions are frequently explained using the rock–paper–scissors game as a metaphor. However, despite their potential importance for coexistence, these mechanisms and their relationships with traits are still poorly understood. Existing studies, though, show that functional trait diversity could also be essential in maintaining networks of intransitive competition (Soliveres et al. 2018).

Apart from competition, there are other mechanisms regulating coexistence that can have an effect on the trait composition of communities (revised in much more detail in Chapter 10). *Facilitation* and *mutualisms*, for example, are known to expand the fundamental niche (i.e. increase the fitness) of species (Bulleri et al. 2016; Peay 2016) and are likely to favour the coexistence of species with different traits (Liancourt et al. 2005; Maestre et al. 2009; He et al. 2013). This results in increased trait diversity (sometimes called trait divergence). Ideally most of these mechanisms could be included within the modern coexistence theory framework. For example, mycorrhizal symbiosis can increase the fitness of certain species with respect to others, allowing them to achieve dominance (McGuire 2007), but also promote niche differentiation between plants (Gerz et al. 2018), potentially favouring coexistence.

7.3 Trait-Based Community Assembly

Trait-based community assembly follows exactly the same rules as species-based community assembly, just with species swapped for traits. Historically, studies considering assembly rules have predominantly focused on different aspects of the selection process, including responses to the abiotic environment (see Chapter 4) and biotic interactions. As a consequence, assembly is commonly conceived as dominated by two apparently opposing forces acting simultaneously to determine diversity patterns: *abiotic and biotic forces*. Despite at first seeming as if these forces are mutually exclusive, the most traditional view is that they determine the trait values that can exist in a community (and therefore the identities of the species) by acting hierarchically. From this classical point of view, abiotic and biotic forces have a theoretical signature on the distribution of traits of local communities. In essence, abiotic forces are expected to restrict the range of functional trait values found in local communities (remember the concept of environmental filtering); this means that the variability in traits in local communities should be smaller than that expected if the community was composed by a random subset of the species from the regional pool. This lower-than-expected trait diversity is known as *trait convergence* (not to be confused with convergent evolution; see Chapter 8) or trait *underdispersion*. Let's consider a region with valley bottoms and mountain peaks. In this environment, plant height would be one trait (among others) that determines if a species can survive the harsh abiotic conditions (freezing temperatures, low water availability, high winds), with short plants being better adapted to such conditions. As a result, if a region encompasses different environments, only a subset of the regional pool of species and traits will be realized locally. This means that we will observe that the functional diversity of communities is lower than what we would expect if all species could live everywhere. Conversely, biotic forces should cause individuals with similar traits (similar resource niches) to compete more strongly (remember the concept of limiting similarity). Therefore, under limiting similarity, the traditional expectation is to observe higher-than-expected trait diversity (Wilson 2007), known as *trait divergence* or trait *overdispersion* (Götzenberger et al. 2012). Of course, in some cases we do not see any significant differences between the trait patterns of the regional species pool and the local community, in which case we do not have evidence for any specific assembly rule operating on traits (but we cannot disregard opposing forces cancelling each other out).

However, as we have seen before, modern coexistence theory states that competition between coexisting species does not necessarily lead to trait divergence in communities. How does this point of view affect our expectations about trait structure and the mechanistic drivers of community assembly? As we showed before, even in the simple case of two coexisting species, coexistence can be achieved in many ways. For instance, neutral theory (Hubbell 2001) assumes that fitness differences between species do not exist. In this case, community dynamics are solely ruled by stochastic variations in local species extinction, species invasion and dispersal. A less extreme scenario would include cases where small niche differences can balance small fitness differences. In this case, in order to coexist, the two species should be similar in the traits that determine competitive ability (see Fig. 7.1). At the other extreme, coexistence can arise from large niche differences counteracting large fitness differences. In this case, the two species should differ in the traits that determine niche differences. Some studies have evaluated pairwise species interactions among small sets of species (Kraft et al. 2015; Perez-Ramos et al. 2019), describing which traits or combinations of traits drive fitness and niche differences. However, real communities harbour many species, and the number of pairwise species combinations increases disproportionally with the number of species. This can make evaluating fitness and niche differences between species a herculean task. One potential way to overcome the limit of pairwise interactions is to characterize the relationship between fitness and traits of species across environments (Laughlin & Messier 2015; Laughlin 2018) so that we are able to estimate fitness differences between species based on their traits. This is a promising avenue of study (Pistón et al. 2019), but it is too early to tell how much it will improve our capacity to predict community assembly.

Moreover, modern coexistence theory also challenges the view that environmental filtering reduces the range of trait values (Mayfield & Levine 2010). Environmental conditions may act as a filter that prevents species with maladapted traits, which confer low fitness, from being present in communities. However, with simple observational data it is not possible to distinguish to what degree trait convergence is due exclusively to environmental factors or to competitive interactions (Germain et al. 2018a). In other words, it is difficult to tell if a species is locally absent because it has a higher rate of mortality than recruitment even in the absence of competitors, or if it has a positive intrinsic growth rate but competition is excluding it. Therefore, it is argued that strict tests of environmental filtering of species based on their traits, even in the absence of competition, cannot be achieved with observational data (Kraft et al. 2015).

The topic of community assembly acting on traits can be highly complex, to the point that a pure distinction of the forces that drive community assembly cannot generally be reached with certainty, particularly with observational data. But this does not mean that examining patterns of trait convergence and divergence across space and time is any less valuable. As argued by Cadotte and Tucker (2017), for most practical applications we are not primarily interested in separating pure effects of environmental filtering from those of competition. In the current context of accelerated global change, improving our understanding and capacity to predict the effects of global change drivers on ecosystems is one of the most pressing challenges for scientists worldwide. Knowing the response of trait diversity across environmental conditions is fundamental for predicting the effects of global change drivers such as climate change (Valencia et al. 2018), land-use change (Laliberté et al. 2010; Carmona et al. 2017b) or biological invasions (Loiola et al. 2018). Using the knowledge generated by studies of community assembly from a functional trait perspective can help to guide conservation and restoration activities aimed at achieving goals related to ecosystem functioning or to the resistance of communities to climate change (Mouillot et al. 2013b; Laughlin 2014a,b). HilleRisLambers et al. (2012) provide a very good overview of approaches to understand community assembly in the light of modern coexistence theory, which are beyond the scope of this book.

7.4 Assessing Assembly Rules

Following Götzenberger et al. (2012), we will focus on assembly rules associated with ecological filters: dispersal, the environment and competition-based interactions.

7.4 Assessing Assembly Rules

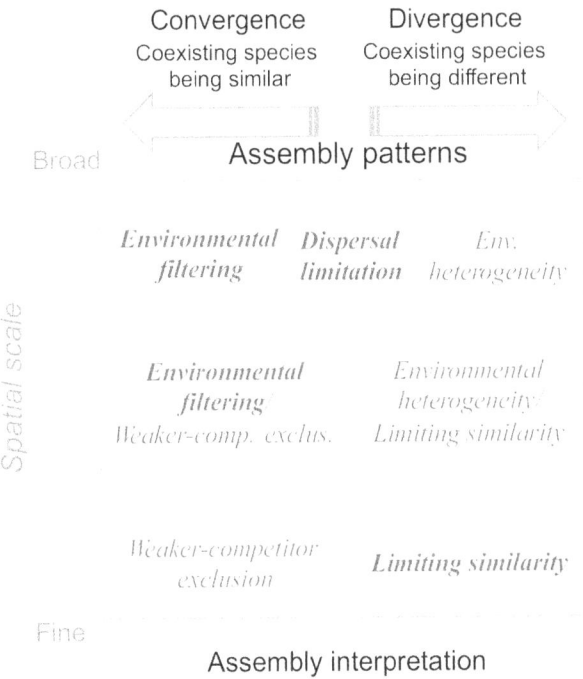

Figure 7.3 The interplay between spatial scale, community assembly patterns (trait convergence and divergence) and potential underlying assembly rules (see main text for definitions of these rules). For each given pattern, the assembly rules that are expected to be stronger at each spatial scale are shown in bold. Adapted from de Bello et al. (2013c), with permission from Wiley. © 2013 The Authors. Journal of Ecology © 2013 British Ecological Society.

If assembly rules operate, community structure should deviate from randomness (unless they cancel each other out). Testing this deviation often requires the construction of *null models*. Using this approach we create random communities from the regional pool following some selection rules and compare patterns generated by this null model with an observed pattern. Following de Bello et al. (2013c), we will consider the following assembly rules (Fig. 7.3):

- *Dispersal limitation*: species are not present at all of the sites that are ecologically suitable for them. Even if a species is able to establish and persist at a given site, species with poor dispersal abilities tend to be absent from suitable sites more often than good dispersers. Since the traits of species influence their dispersal ability (Tamme et al. 2014), filtering by dispersal limitation should result in convergence of dispersal traits (e.g. Riibak et al. 2015).
- *Environmental filtering*: even granting that environmental filtering *sensu stricto* cannot be detected with observational data, the covariation between traits and environments is widely acknowledged. In other words, species with similar traits tend to live in similar environments. As a result, environmental filtering, and its

associated patterns of trait convergence, is expected to be the dominant assembly process at broad spatial scales which include wider ranges of environmental conditions (Fig. 7.3). When environmental filtering occurs at finer scales (e.g. patches with different productivity within a site), it can result in patterns of trait divergence; this particular case has also been termed *environmental heterogeneity* (de Bello et al. 2013c).

- *Limiting similarity via competitive exclusion*: this rule is based on competitive interactions between species. As we saw before, if limiting similarity operates for some particular trait that relates to the struggle among species for a scarce resource, competition is expected to be stronger among species with similar values for those traits that should lead to trait divergence patterns. By contrast, if there is a clearly dominant phenotype, the intensity of competition will increase as trait dissimilarity increases. This should lead to the formation of competitive hierarchies in which the species that are weak competitors are excluded from communities, leading to trait convergence patterns.

Perhaps the most important point to consider from the previous definitions is that different assembly rules take place preferentially at different *scales*. Environmental filtering requires relatively broad scales to encompass sufficient environmental variation among or within sites. In turn, assembly rules associated with limiting similarity and weaker-competitor exclusion operate at finer scales. The actual spatial scales at which these processes predominantly operate are also influenced by the group of organisms studied. One main distinction between groups is how easily they can disperse over large distances (either actively or passively) and thereby escape biotic interactions by relocation. Even taking this scale issue into account, there can be further complications when assessing assembly processes with traits. First, different processes may affect the same trait in different ways. For example, plant species with heavy seeds have higher survival rates, particularly under stressful conditions (Moles & Westoby 2004; Metz et al. 2010). This should result in patterns of increasing convergence in seed mass as stress increases (e.g. as water availability decreases; Carmona et al. 2015b). At the same time, species with heavy seeds tend to have shorter dispersal distances (Cornelissen et al. 2003) and produce fewer propagules (Jakobsson & Eriksson 2000) than species with light seeds. If dispersal limitation acts as a filter, this should lead to a higher proportion of species with light seeds in the community. Filtering by environment and dispersal can therefore counteract each other, leading to random trait patterns. Detecting each of them requires carefully disentangling the effects for which they are responsible by means of appropriate analytical schemes. In this case, assessing dispersal limitation in the previous example would require first accounting for other factors that could otherwise blur the expected signals (Riibak et al. 2015). Second, different traits can be affected by different assembly processes in different ways. For example, Spasojevic and Suding (2012) show patterns of convergence for plant height and leaf area under low-productivity conditions, contrasting with patterns of divergence for traits associated with competition for below-ground resources. In general, choosing the adequate scope of any study in terms of traits, scales (spatial but also temporal) and species pools is critical to distinguish the signature of each of these rules. Only then should we be able to ascertain how different drivers (e.g. climate change, land-use intensification) affect the assembly of ecological communities. For observational data, null models are the most commonly used tool to study these questions.

7.5 Null Models

When studying if some community attribute (e.g. functional diversity) deviates from random, we need to know what possible random values functional diversity could take. This requires the generation of functional diversity values by means of null models, in which the pattern of interest is randomized while trying to fix all other relevant patterns in the dataset. For example, let's imagine that we have a set of 100 communities with different numbers of species, where we hypothesize that limiting similarity is taking place. In this case, we are interested in knowing if the species that coexist in each of these communities are more functionally dissimilar than expected by chance. We could create a null model in which we fix the number of species observed in our communities but we randomly choose (from the pool of species found in our set of communities) the identity of the species colonizing them (Gotelli & Graves 1996; Weiher et al. 1998). If we repeat this process several times for each community, for example 1000 times, then we can compare the average dissimilarity between species observed in the real communities (using, for instance, Rao) with the distribution of the values of that metric in the *null communities*. Higher values of Rao in an observed community than a null community would indicate that the species composing our community of interest are more functionally different than expected by chance (overdispersion, trait divergence), whereas lower values would indicate the opposite (clustering, trait convergence). Therefore, this strategy would allow us to tell what the chance of reaching the observed value of Rao is for that level of species richness if the species associated freely (i.e. if all species could be present in all communities in all combinations).

There are, however, some patterns in real data that have not been fixed in our null model above that could result in erroneous conclusions. Among other things, the total frequency of species (i.e. the number of communities in which each species appears) has not been maintained, so that in this null model species that are very rare in reality have the same chance of appearing in a community than very frequent species. Therefore, if the Rao value observed in one community is lower than expected by chance, we cannot rule out that this is because rare species are functionally very unique, or because dominant species have some particular traits, rather than because coexisting species have different trait values due to some competition-based assembly process. These kinds of problems can be solved by using increasingly complex null models. The downside of this is that the more constrained the null model is, the lower the number of possible randomizations is. Different authors have assessed which null models should be considered when testing different assembly rules (Hardy 2008; Mason et al. 2013; Götzenberger et al. 2016, among others). Here we provide some simple examples to illustrate how null models can be used to tackle some ecological questions.

Let's imagine we are interested in studying assembly rules along an altitudinal gradient (Fig. 11.1). We expect that only species with traits that provide resistance to freezing can survive on the mountain tops, whereas in the more benign conditions at lower altitudes species with a wider range of trait values can survive. Since our expectation is related to trait ranges, we can examine how the range of functional traits (i.e. functional richness) changes as we sample at higher altitudes as we go up our mountain. We could then try to investigate our question by simply regressing functional richness against altitude. However, an important aspect to consider is the relationship between functional richness and species richness. As we saw in Chapter 5, many functional diversity indices are related to species richness to a greater or lesser extent, and functional richness is particularly sensitive to it. This is important because it can (and frequently will!) happen that species richness reduces as we move towards less benign conditions. As a result, if we see a reduction in functional richness as altitude increases, we will not able to distinguish whether this is because mountain tops are species poor, or if it is because there is a reduction in the viable range of functional traits. In this case, we should control for the effect of species richness. For this, we can construct a null model that allows us to compare the observed functional richness with that expected given the traits of the species that we found on our gradient (which in this case would represent the regional pool of species) and the species richness of each community.

There are different alternatives for constructing such a model. Since in our study above we use functional richness, which does not depend on species abundances but only on their presence, we can use the presence–absence matrix and perform a swap algorithm on it. *Matrix-swap* null models (Manly 1995) preserve the species richness of the communities as well as the relative frequency of species in the region. By doing this, we remove the effect of any process restricting the trait values within communities (in this case, of different conditions along the altitudinal gradient), which allows us to compare the observed results with those expected under the absence of that process. Alternatively, in each community we could sample without replacement as many species from the regional pool as make up the local species richness, giving to each species a probability of being chosen that is proportional to its regional abundance (e.g. Bernard-Verdier et al. 2012).

After we define the null model, we can evaluate our hypotheses. For this, we create a number of randomizations (for example 999), and during each iteration we estimate, for each community, the index we are interested in (functional richness). By repeating this process 999 times, we obtain a distribution of 999 'simulated' (or expected) functional richness values for each community. We then want to know how our observed value of functional richness compares with the 999 null values. If, as we expect, altitude reduces the range of viable traits in communities, then the observed values of functional richness should be smaller than the simulated ones at the higher altitudes. If we do observe that, how can we test to see if there is a statistically significant effect? For this, we can compute a p-value by simply counting how many times the observed value is smaller than expected by chance. This is done by putting together our observed value and the 999 null values in a single vector and ordering that vector in ascending order. We can then conduct a one-tailed test by checking in which position (for example, in the third position) the observed value is within the ordered vector and dividing this by the total length of the vector (1000 in our example). In our case, this results in a *p*-value of 0.003 (3 divided by 1000), which indicates that the functional richness in our community is smaller than expected by chance, and therefore suggests that altitude indeed restricts the range of trait values found in the community. A more conservative approach to this would be to conduct a two-tailed test, in which the threshold of a significance of 5% would only be reached if the observed value is smaller than 975 of the expected values.

An alternative to estimating *p*-values is using the *standardized effect size* (SES; Gotelli & McCabe 2002), which is defined as:

$$SES = \frac{Observed\ value - mean\ of\ simulated\ values}{standard\ deviation\ of\ simulated\ values}$$

A positive SES value indicates that the functional richness (or whatever index we have chosen) of the community is greater than expected by chance, and negative values indicate the opposite. Because we divide by the standard deviation of the simulated values, SES values are measured in standard deviation units. This means that they can be used to assess the significance of the examined pattern in individual communities if that is what we are interested in. There are two ways to test significance using SES. A very conservative approach considers that communities with SES values greater than ~1.96 or smaller than ~−1.96 deviate significantly from random expectations (this test corresponds broadly to the one described above). Alternatively, we could simply test whether or not the set of SES values obtained for a set of communities is significantly different from zero (for example using a *t*-test across all SES values obtained). This far less conservative, but very useful, approach (Hardy 2008) clearly depends on the number of communities considered, which influences the significance of the *t*-test.

In general, and particularly if the index of functional diversity considered depends linearly on species richness, we might be interested in how the SES values change along some gradient (for example, along the altitudinal gradient of our example above). This way, if we observe that SES values decrease with altitude, we can reasonable argue that increasing altitude has the effect of increasing trait convergence in our study. Note that this approach might, for indices not strongly related to species richness, produce SES values that are tightly correlated with the observed functional diversity values (de Bello 2012). In this case, regressing directly the observed functional diversity values against altitude will result in similar results to regressing SES values against altitude.

In some cases, we will have to construct a more specific null model. For example, let's now imagine that we are interested in knowing how grazing affects functional diversity along our altitudinal gradient. Grazing can restrict the range of traits that we observe in our communities by removing species that are not able to cope with defoliation. Alternatively, it can also increase functional diversity by reducing the dominance of very competitive species with traits that confer high fitness in the absence of disturbances. We know that these alternative effects of grazing on taxonomic and functional diversity depend on productivity (Carmona et al. 2012; Rota et al. 2017), which leads us to hypothesize that the observed effect will differ between lower and higher elevations. To test this, we have sampled different sites with and without grazing at each altitude (Fig. 11.1). Since our hypothesis is expressed in terms of trait ranges within communities, we can again examine functional richness patterns. However, if we do a randomization similar to the one presented above, our null model will not allow us to test for the specific effect of grazing. This is because grazing and altitude are confounded in our dataset. In order to test for the effect of grazing, we should first 'remove' the effect of altitude as a factor that restricts the range of traits that we observe in our communities. This is done, as we explained above, by selecting different species pools for the different tests. For example, we know that plants at high altitudes tend to be shorter in height because this is a trait that allows them to cope with climatic extremes; at the same time, we know that grazing also selects for short plants, since they are better able to avoid defoliation. As a consequence, if we randomize all species in our whole dataset, the effect of grazing on plant height cannot be separated from the effect of altitude, so we might not end up finding the grazing effect. We can solve this by stratifying our null models so that the randomizations are done only among the species that are present at a given altitudinal level, which effectively fixes the effect of altitude, but not the effect of grazing. This is equivalent to saying that we are restricting the *scope* of our null model by constraining the pool of samples that is relevant for testing the effect of grazing (de Bello 2012).

As we move towards finer spatial scales, *interactions between organisms* should become an increasingly important determinant of the assembly process (Fig. 7.3). In order to detect the signature of biotic interactions, it is important to remove the effect of abiotic filters as much as possible. This will generally require substantially reducing the pool of species that are considered for tests at fine spatial scales, so that we only include those species that are potentially found at sites with overall similar abiotic conditions. This can be done in different, more or less restrictive, ways. Going back to our previous example of the grazing gradient nested in the altitudinal gradient, for each site we could simply restrict randomizations to species that are found at sites within the same altitudinal belt and the same grazing level. After doing this – and if our sampling units are homogeneous enough in terms of other fine-scale factors (such as soil characteristics) – the patterns of overdispersion or clustering that we find can reasonably be attributed to biotic interactions (de Bello et al. 2012). Indeed, this requires that we have sampled several communities (replicates) in each of the combinations of altitude and grazing. An even stricter way to test for the effect of biotic interactions, even when we do not have replicates, is making null models in which we randomize the abundance of the species that have been observed within each community (Mason et al. 2008; Bernard-Verdier et al. 2012). This randomization scheme maintains species richness and abundance distributions within communities. If competition leads to the coexistence of species with different niches (niche differentiation), then the most abundant species in a community should have different traits and we will observe overdispersed trait patterns (positive SES). Alternatively, if there is a single phenotype that confers high fitness, then the most abundant species will tend to be functionally similar and we will observe trait clustering (negative SES). This latter case, however, can be hard to detect with observational data, as the reference species pool for the null model should include the species that have been excluded by competition (Götzenberger et al. 2016). This would only be possible if the community data contain samples where the competitive exclusion is less strong, or if the definition of the habitat-specific species pool is effective enough to represent the weaker competitors.

As we can see, selecting the correct pool of samples and species for inclusion in the randomizations is one of the most crucial steps when designing a null model (de Bello et al. 2012; Götzenberger et al. 2016). Doing this in an adequate form can involve rather sophisticated methods. As we saw above with the example of mountain tops and valleys, generally, not all the species from the regional pool are potentially able to live and reproduce at a given local site. Additionally, not all species are able to withstand competitive interactions by other species. Furthermore, not all the species that are able to live in the particular ecological conditions (including both biotic and abiotic factors) of a site are actually present at the site. The set of species from the regional pool that are suited for life under the site's ecological conditions (biotic plus abiotic), but are not present there, form the *dark diversity* of that site (Pärtel et al. 2011). Dark diversity cannot be known with certainty, and can only be estimated through methods that exceed the scope of this book (Brown et al. 2019). In any case, if we are able to characterize the dark diversity of our sites, we can use species' traits to understand the reasons why these species are absent. For example, Riibak et al. (2015) compared the trait patterns of dark and observed diversities in calcareous grasslands, and found that plant species making up the dark diversity had, on average, lower dispersal ability (lower dispersal distances and lower production of seeds) and lower stress tolerance (lower height and lower S score in the C-S-R scheme we presented in Chapter 3).

7.6 Community Assembly Applications

7.6.1 Predicting Species Abundances and Trait Structure

The use of traits to predict species abundances across environmental gradients has been considered the 'Holy Grail' of ecology (Lavorel & Garnier 2002). Adequately characterizing the functional structure of communities over a wide range of environmental conditions might be one of the key pieces missing in achieving this objective. Shipley's *community assembly by trait selection* (CATS; also known as maxent) model (Shipley et al. 2006) was the first one aiming to predict species abundances along environmental gradients based on their traits. This model relies on the difference between the trait values of each species and the local community weighted mean to provide an estimation of species' abundances. This way, species whose trait values are close to the local average will be predicted to have higher abundance (Shipley et al. 2012). In doing so, the model emphasizes the role of environmental filtering over relatively large gradients, which makes trait convergence the main driver of community assembly.

A further development in this line is the *traitspace* model (Laughlin et al. 2012, 2015). The traitspace model first considers the probabilistic distribution of traits for a given set of environmental conditions and then uses Bayesian methods to estimate species abundances. By capturing the local structure of trait values (beyond the community weighted mean), the traitspace model considers both environmental filtering, with its associated effect of convergence, and limiting similarity, which would promote trait divergence. The traitspace model has great promise since it specifically incorporates intraspecific variability and can consider multiple traits (including their covariance) and environmental variables. However, it has only been tested in communities with a relatively low number of species (Laughlin et al. 2012, 2015). This perhaps reflects one of the main limitations of both methods: they cannot discriminate between functionally redundant species, which generally become more common as species richness increases (Laughlin & Laughlin 2013). One way forward is to consider more traits, particularly if they are independent, which allows better discrimination between species by reducing redundancy (Carmona et al. 2019), hence improving predictions (Laughlin et al. 2015). This raises the questions of how many and which traits are needed to optimize predictive ability, which can be investigated by examining the number of traits that should be considered until predictive ability plateaus (Laughlin 2014b).

Trait-based approaches are particularly interesting when the focus is on testing the effect of biodiversity on ecosystem functioning. In this case, we are not exclusively interested in the species composition of communities, but also in functional structure (i.e. what are the traits of the organisms that compose the community?; Díaz et al. 2007). One of the advantages of considering a trait-based approach suggested by research and observational work is that functional trait structure is much more predictable than species composition (Fukami et al. 2005; Messier et al. 2010). Convincingly showing that the functional structure of plant communities (i.e. their probabilistic trait distributions; see Chapter 6) can be predicted to a high degree from environmental information, even when this prediction is made at large spatial scales with little overlap in species identities, will be a major step forward in our ability to predict the effects of global change drivers on ecosystem functioning, and to design effective conservation (Violle et al. 2017) and restoration strategies (Laughlin 2014a).

7.6.2 Invasive Species

A particularly interesting line of research is the study of the invasive ability of species and the susceptibility of communities to invasion. Following the mechanisms outlined earlier, invading species should have a certain combination of traits allowing them to establish in the host community, i.e. to maintain a positive growth rate once they have dispersed into the community. Direct characterizations of the traits that make a species a successful invader have led to very few generalizations, because invasive ability will also depend on the environmental conditions and the traits of the local species (Moles et al. 2008). Moreover, the interactions with other trophic levels might have a large influence on the invasion success, for instance when alien invaders are free from 'enemies' that impact their growth rates in their native range (enemy release hypothesis, Keane & Crawley 2002). However, there is evidence that invasive alien species tend to rank higher in traits related to species performance than non-invasive species (Van Kleunen et al. 2010; Funk & Wolf 2016).

An alternative approach is to characterize the functional structure of the receiving communities and relate it to their susceptibility to invasion. Research in this direction suggests that communities with higher functional diversity, which should be associated with a more complete use of resources, have higher resistance to colonization (Levine et al. 2004; Lanta & Lepš 2008). However, approaches simultaneously considering the traits of the local species and those of the invaders (which can be alien or native species) can provide more detailed information about the invasion process (Carboni et al. 2016). When using observational data to answer such questions, it is important to consider that the invasion of communities by new species will also have consequences for the resulting functional diversity and composition of host communities (Lososová et al. 2015). This complication means that complete characterization of the effects of invasion needs to use comparisons of the functional structure of non-invaded communities with that of invaded communities both with and without considering the invasive species (Loiola et al. 2018; Fig. 7.4). There are not many papers studying the interplay between these two factors, so not many conclusions can be made yet, but available evidence suggests that successful invaders tend to occupy unsaturated areas of the functional space (Loiola et al. 2018; Galland et al. 2019). Therefore, being able to locate these

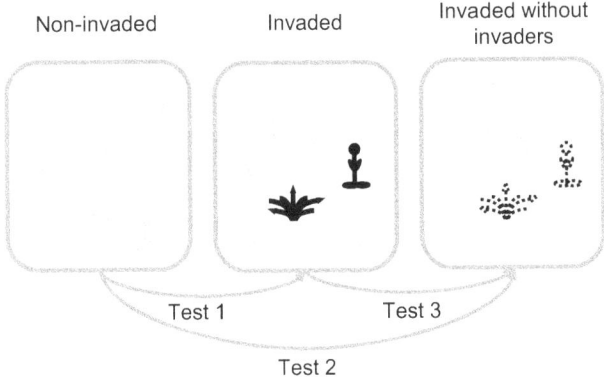

Figure 7.4 Combining different tests can help to assess the causes and consequences of invasion on the functional structure of communities. Three types of community are depicted: communities without invasive species (left), communities with invasive species (centre) and communities with invasive species in which invasive species are removed for the analyses, and only native ones are considered (right). Invasive species are indicated by a darker colour. Test 1, designed to detect the overall effect of invasion, consists of a comparison between non-invaded and invaded communities; Test 2, designed to detect if there are patterns showing a predisposition to invasion or invasive species have excluded native species occupying particular areas of the functional space, consists of a comparison between invaded and non-invaded communities considering only native species; finally, Test 3, designed to discern if invasive species occupy different parts of the functional space (which would increase the functional diversity community) or if they fill pre-existing gaps (decreasing functional diversity), consists of a comparison between invaded communities with and without considering alien species. Taken from Loiola et al. (2018), with permission from Wiley.

'holes' in the functional space occupied by communities (Blonder 2016) appears to be a good way to predict which species may be able to occupy them, which can help in guiding conservation strategies against invasive species (Laughlin 2014a; Carmona et al. 2017a) and ideally may help to predict the invasive ability of species in specific communities (Bennett & Pärtel 2017).

Take-Home Messages

- Assembly rules are any constraint restricting the number and identity of the species that we observe in an assemblage.
- Traditionally, it is assumed that environmental filters restrict the trait values that are observed in an assemblage (trait convergence), whereas limiting similarity, associated to competition, has been regarded as a force promoting trait differentiation among coexisting species (trait divergence). More recently, modern niche theory postulates that coexistence is attained when niche differences between species are large enough to offset fitness differences. This entails that competition can also lead to trait convergence.
- Whereas we can observe and quantify the degree of trait convergence and divergence, it is very hard to ascertain what are the processes (environmental filtering or biotic interactions) leading to the observed patterns without experimental approaches. However, well-designed null models that consider the effect of scale and an adequate reference species pool can help and have already provided some valuable insight.
- Understanding what assembly processes are at play in an assemblage can help us to design better restoration and conservation strategies, and to predict the impacts of climate change or invasive species.

8 Traits and Phylogenies

So far we have talked about traits as attributes of extant species, i.e. of species that we can observe in today's natural world. However, it should be clear that by treating traits of extant species, we tend to ignore, or at least neglect, that these species and their traits are the outcome of evolutionary processes. Darwin discussed how the traits that species possess evolved as adaptations to abiotic and biotic conditions. Hence, he already anticipated that phylogenetic relationships between species should affect how similar the species are to each other, stating that 'species of the same genus have usually, though by no means invariably, some similarity in habits and constitution, and always in structure' (Darwin 1859). While these ideas have never been completely forgotten, it took more than 100 years before this general idea took hold for researchers who were dealing with comparative biology, i.e. those who were interested in comparing individual species based on their traits and the environments they inhabit, also referred to as the comparative method (Felsenstein 1985; Sanford et al. 2002).

Taking into account phylogeny of species when assessing their traits has created a great deal of debate, particularly regarding the need to consider or 'correct' for phylogenetic information when performing comparative analyses across species (Harvey et al. 1995; Rees 1995; Westoby et al. 1995; de Bello et al. 2015). In this chapter we are going to see how the notion that related species should be more similar in their traits initially led to an explosion of statistical developments to test and account for this assumption (see 'R material Ch8' accompanying this chapter). Then, a second explosion of studies and methods has seen community ecologists using phylogenetic relationships as a substitute, or complement, for traits. If you are unable to follow why anyone would arrive at the conclusion that it might be reasonable to study communities from a phylogenetic perspective rather than using traits, do not worry, and bear with us. We are going to take it step by step from here, but first, as a short precursor, we will learn how we represent and describe phylogenetic information. Note that we are treating here in a single chapter something that is covered by entire books. So if you want to delve deeper into the ideas presented in this chapter, we refer you to some of the essential literature: Harvey and Pagel (1991) provided the first book-length treatment of the comparative method, and it can be regarded as a classic in the field. Garamszegi (2014) edited a book containing chapters written by many prominent scholars in the field, basically bringing the topic of Harvey and Pagel's book into the twenty-first century. Both Swenson (2014a) and Paradis (2012) cover the technical and statistical application of the matter, in conjunction with the packages *picante* and *ape*, respectively.

8.1 What Is a Phylogenetic Tree?

A *phylogenetic tree* is a way to visually represent how species are related to each other through their *ancestors* (Fig. 8.1). We can think of the ancestors as 'mother' species, where each mother has two 'offspring'. The current species are sometimes called *tips* or leaves of the tree, each of them being connected to their last ancestor ('mother' species). In tree terminology, these ancestors are called *nodes*, because they are represented in the tree as internal points in the branching structure. Furthermore, just as real trees have a root, so do phylogenetic trees, usually. The *root* is the oldest node in the tree, representing the one ancestor from which all other younger ancestors, and eventually the extant species, originated. In the ideal case that the phylogenetic tree is completely resolved, each ancestor has exactly two descendants. We also say that such a tree is completely dichotomous, as each ancestor gives rise to exactly two species. However, sometimes we lack the necessary information or certainty to specify exactly how each species is related to its ancestors, and in such cases we might have more than two species originating from the same ancestor. In terms of the topology (i.e. the shape) of the phylogenetic tree, we call such cases *polytomies*.

For a graphical representation, the tips of a phylogenetic tree can be oriented in any direction, but the most common format is to have the root of the tree on the left-hand side and the branches leading from left to right to the tips of the tree. If the tree is oriented like this, we can think of the axis that goes from left to right, parallel to the branches, as an axis of *evolutionary time*, very often given in millions of years (Myr) as a unit. Some trees lack this information, so the 'x-axis' does not contain measurable information, i.e. it does not reflect 'true' evolutionary time. In such cases the x-axis is still some representation of time, it just is not reflected in given units, or all the branches are of the same length. Other trees are time-calibrated to various degrees and with different methods, i.e. they use different methods to estimate how old some or each of

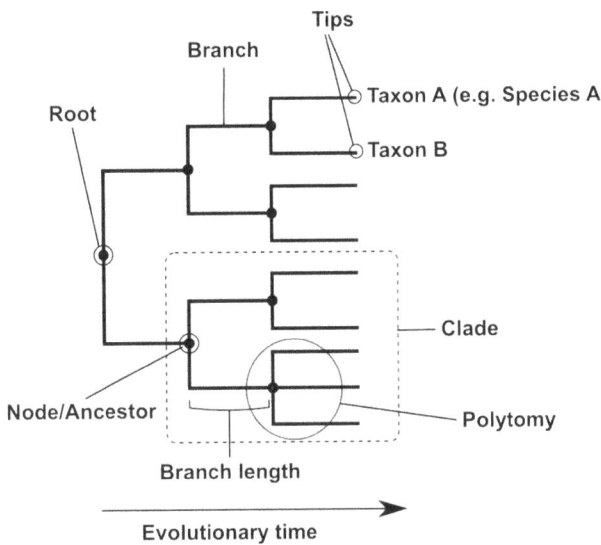

Figure 8.1 Scheme of a phylogenetic tree, i.e. a graphical representation of phylogenetic relationships between species, together with a labelling of its components.

the nodes are. You can imagine this process as taking each of the nodes in the tree and pinning them in place along the evolutionary time axis where this node has been estimated to occur. More internal nodes can also be interpreted as the origin of entire families or any other taxonomic group, so that the age of this node along the time axis also expresses an estimate of when this taxonomic group first appeared evolutionarily. More generally, a clade is defined as a group of species that originated from a single common ancestor.

8.2 Brownian Motion and Why Related Species Should Be Similar

The very basis of why we should care about phylogeny is the fact that species have evolved from each other throughout evolutionary time so that many of their traits have been inherited from their ancestors. Species do not possess certain traits only because these traits are adaptations to the current biotic and abiotic environment they live in, but also as a result of their ancestors constantly being subjected to changes in their environments, leading to extinctions for some species and to selection of newly acquired or changed traits for others. How can we depict this general idea in a phylogenetic tree? If we think of the nodes as points of speciation, then from such a point the two descendant species can be considered to depart from their ancestral species, and, in this sense, each new species can also, at least potentially, evolve a new trait value for a trait inherited from its ancestor. At the same time, the descendant species share an evolutionary history in the form of their common ancestors until this point of speciation, so they are not completely independent. If we imagine a certain trait that slightly changes its value from one generation to the next, it should be clear that the starting point for the two descendant species is the last trait value that the ancestral species possessed before the speciation event.

There is also a way to express this *evolutionary differentiation* in trait values mathematically. We can first assume, for simplicity, that the change of trait values from one generation to the next happens randomly. We will then start with a single ancestor species. Consider an arbitrary starting point for the trait under consideration, which we will set to 0 for convenience. Now every change in that trait from one ancestral species generation to the next can be expressed as a positive or negative change away from the ancestral value, i.e. away from 0 in the first new generation, and away from subsequent values in the generations that follow. To keep the model as simple as possible, and to not assume any directionality, we can draw the size of this change from a normal distribution with mean zero and a given standard deviation. At the same time, these time steps from generation to generation can be defined as 'infinitely' small, if we keep in mind how short they are in comparison to the entire evolutionary timescale under consideration.

We can depict this model (or rather one realization of it) by producing a graph that has evolutionary time on the x-axis and the value of the trait of the ancestral species at a given time on the y-axis. In 'R material Ch8' accompanying this chapter, we present code that you can use to visualize the described procedure step by step. As we draw the changes in trait values from a random normal distribution, the resulting line, i.e. the accumulated changes, can 'drift off' from the zero line. In fact, the highest probability is that all the accumulated changes arrive at a net change of 0. We can depict this by drawing a normal distribution next to the y-axis, on the right-hand side of the graph. This normal distribution is the distribution of trait values of the ancestral species after a given time if we were to repeat the evolutionary process many times, i.e. rerun the model, for instance, 1000 times. The important feature of the resulting normal distribution is that its variance becomes larger the longer we let the trait evolve (see Fig. 8.2, left panel, a case without speciation, but only genetic drift). This is because the final trait value of a single model we run is the accumulated change over time. The more time these changes have to accumulate, the bigger the change can be, even though the highest probability will always be that the net change is zero.

This simple mathematical model of evolution described above, and in Fig. 8.2, in which 'random walks' in trait values happen through time, is called *Brownian motion* (see below for the origin of these names). Now that we have seen trait evolution according to the Brownian motion evolution of a single species we can take any point in evolutionary time, and simulate a speciation event, so that a given species (in our case species X) becomes an ancestor for two new species (species Y and Z). We set that time here to zero, i.e. species X speciates into species Y and Z right at the start of our evolutionary time window. Now, let's say we wish to look at the evolution of a particular trait y. If we follow the first 'branch' originating from species X, the trait evolves for a given length of time (t_1) to a given value (y_{t1}) at which point species Y itself speciates. Hence, this trait value y_{t1} will confer the starting point for the trait

8.2 Why Related Species Should Be Similar

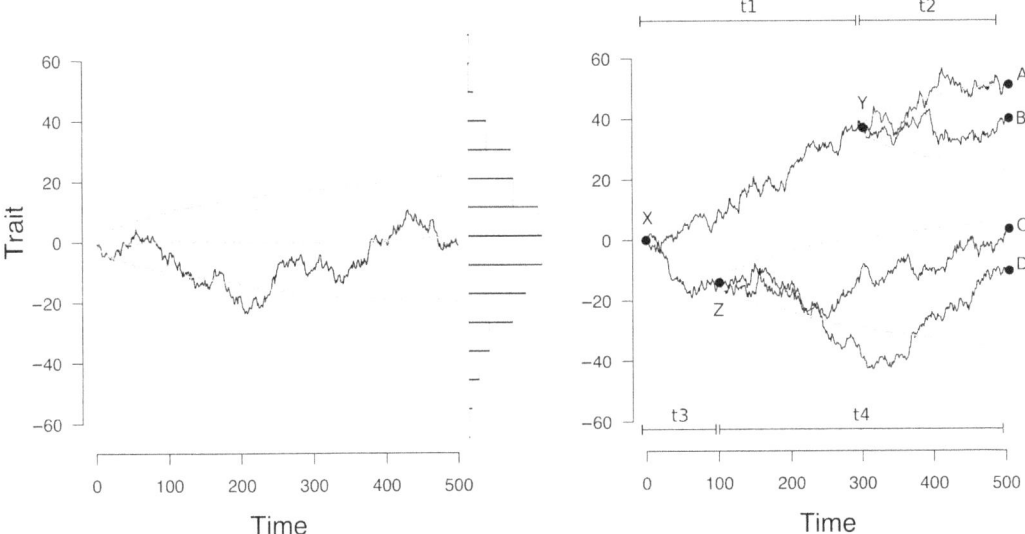

Figure 8.2 Demonstration of the Brownian motion model of trait evolution, and how it leads to similar trait values for closely related species. On the left panel, a case without speciation but only genetic drift, the black jagged lines represent a single realization of the model showing the evolution of a trait and its values through time, whereas the fuzzy grey lines represent this evolution for 500 more possible runs of the same model. The broken lines are the mean and standard deviation around that mean at any given time. From the right-hand graph, with speciation events, it becomes clear that species that have a more recent common ancestor (species A and B, with common ancestor Y) will have more similar trait values, because the time from which they could evolve independent trait values is shorter, compared to species C and D, which started their independent evolution much earlier. Note that the actual distance in trait values between values for A and B is not that different from that for C and D, but that the probability that C and D are more different is higher, depicted by the larger standard deviation. The labels t_1, t_2, t_3 and t_4 represent the time length of different speciation events and the following diversification of the species.

evolution of Y's daughter species: A and B. This trait will evolve in both species A and B by drawing, as before, incremental changes in trait values with a mean y_{t1}, but for species A and B this will be 'independent' evolution. Consequently, the direction and amount of trait change in species A is not associated with the direction and amount of trait change in species B, other than that they have the same initial trait value. After another timespan, t_2, both newly emerged species will have evolved their own trait values. Before we make our little game still a bit more complex by adding three more species, there is an important thing to note. As we know already that the possible deviation from an initial trait value is greater the longer the evolutionary time span, the differences in trait values between two daughter species can theoretically depend on how long they have had time to evolve since the speciation event from which they originated.

Now let us add species Z, C and D to the graph (Fig. 8.2, right panel). For this, we go back to the root, i.e. species X, and we can imagine a second species, Z, which evolved from X, starting at the same trait value of 0. This species follows the same pattern as its sister species, also splitting up into two daughter species. However, we assume that the time t_3 after which speciation into species C and D occurred was much earlier than the split for species A and B. What follows from this earlier split is an important distinction. Since both species C and D have had more time for their trait values to evolve, there is a higher *probability* for this pair of species to have a greater accumulated difference in their trait values. In the example the trait values between C and D may be 'similar' (as in Fig. 8.2, right panel); the important point is that there is a greater chance that, under a Brownian motion model of evolution, they will be more different than species A and B are from each other. This difference is also represented by the wider standard deviation around species C and D compared to A and B. At the same time, the fact that C and D split further back in time also means that they are less related – or more distantly related – to each other. With this realization we have recovered a simple mathematical model of why distantly related species should have dissimilar trait values, whereas closely related species should have similar values, as already anticipated by Darwin. More precisely, the model would predict that phylogenetically distant species have a higher probability of having dissimilar trait values than phylogenetically close species.

We have pretended, up to now, that this mathematical model for the evolution of a continuous trait is void of any historical background. But in fact, this particular model is called the *Brownian motion* model of evolution because of its resemblance to the model that is used to describe how gas molecules move through the air (and in fact many other similar processes). The application of this model for a particle changing its three-dimensional position randomly in infinitely small time steps is itself named after the eighteenth-century botanist Robert Brown. Brown observed a similar phenomenon (described in Brown 1828) when observing pollen grains 'jiggling' around while suspended in fluids under his microscope, like steps in random directions (hence 'random-walks'). The same concept, of traits evolving randomly in any possible direction, to which we referred above (Fig. 8.2), recalls such a movement, but on a single axis: that of the phenotype for a given quantitative trait under consideration.

8.3 Linking Evolution and Trait Filtering

The Brownian motion model of evolution described in section 8.2 results in two species that have independently evolved for a long time having a greater likelihood of being different in their trait values than two species that have evolved independently for a shorter time. At the same time, this also implies that closely related species will tend to 'conserve' the trait signal of their ancestor, hence they will show similar trait values (Kraft et al. 2007). This is called *trait conservatism* between phylogenetically close species.

While we can speak of trait conservatism, we can also talk about *phylogenetic niche conservatism*. Generally speaking, the term phylogenetic niche conservatism represents the idea that phylogenetically related species will have not only similar traits, but also similar ecological niches, i.e. similar preferred abiotic and biotic conditions, as the very traits that were inherited from a species' ancestors also define in what niche a species can live. The first use of the niche conservatism term was probably in Harvey and Pagel's seminal book (1991) but harks back to ideas already formulated by Darwin. The reasoning behind these ideas can be exemplified with speciation in a particular habitat, where given that the two emerging sister species grow, at least originally, in similar habitats means also that their traits are similar, since they inherited them from their ancestral mother species, despite slight modifications (which make them two distinct species). In this sense, based on a Brownian model of evolution, we can expect both trait and niche conservatism, i.e. closely related species are functionally similar and live in similar habitats, at least compared to distantly related species.

However, there are many examples in which distantly related species actually have evolved similar adaptations to given environmental conditions (and, by contrast, closely related species show very different traits and habitat requirements). For example, cushion plants are well adapted to arctic and alpine conditions, and this habit has evolved within several plant families. Similarly, large body size in a wide variety of different taxonomic groups of animals has been interpreted as an adaptation to colder climates, a pattern also known as Bergmann's rule (Gohli & Voje 2016). Succulent or carnivorous plants have also evolved in many distant families. In such cases we can conclude that species have evolved towards similar adaptations to the environment by following different evolutionary pathways. Sometimes the convergence in traits and trait values across different families (or more generally across distantly related species) has been referred as 'trait convergence' (Kraft et al. 2007), though *convergent evolution* is the more traditional term, and avoids confusion with the use of trait convergence in community assembly and null model contexts (see Chapter 7).

These different evolutionary modes just described are essential for linking trait evolution to trait filtering, i.e. how species adapt to different environments via a combination of traits (see Chapter 4 and following sections). To assess if, or to what extent, traits and niches are conserved, we can use different tests, all based on the definition of a phylogenetic signal, which is described in the next section.

8.4 Phylogenetic Signal

Phylogenetic signal is a rather collective term for a number of concepts and related statistical methods to address the question of whether a certain trait is evolving along a phylogenetic tree adhering to a distinctive model of evolution, in particular Brownian motion. In other words, these tests verify to what extent phylogenetically close species share similar traits, or similar niches. The two most often applied measures are Pagel's *lambda* (Pagel 1999) and Blomberg's K (Blomberg et al. 2003), and we are going to look at them in detail. The same tests can be applied to the habitat requirements of species, even though some scholars advocate against such use, arguing that derived characteristics like the niche are not heritable as such (Grandcolas et al. 2011).

We start with Blomberg's K, since its calculation is a bit more straightforward than Pagel's *lambda*. Blomberg's K has, conceptually speaking, a similar approach to an analysis of variance (ANOVA). In a simple ANOVA we can assess if the variation of a certain variable (e.g. body size) is more variable within than across different groups. You can think of Blomberg's *K* as doing something similar; it measures how much, on average, a trait varies within clades, compared to how much it varies across clades. A value of 1 refers to a baseline reflecting the case that the variance between and among clades is equal to the variance between and among clades had the trait evolved under Brownian motion. If *K* is larger than 1, the variance tends to be greater among clades than within clades, so much so that species are even more similar to each other than can be expected under a Brownian motion scenario. If *K* is smaller than 1, then variance between clades is greater than within clades. Hence, the closer *K* is to 0, the less strong the phylogenetic signal is in the trait.

Lambda, on the other hand, reflects a measure of how well the trait values of a given set of species correspond to hypothetical trait values resulting from a Brownian motion model of evolution. *Lambda* ranges from 0 (i.e. trait has no signal) to 1 (i.e. trait has high signal and has evolved as if under Brownian motion). The main distinction between *lambda* and *K* is that lambda is not defined for values larger than 1, so contrary to *K*, it cannot express cases where species are even more similar than assumed under Brownian motion. Otherwise, in most cases *K* and *lambda* would indicate similar phylogenetic signal. Note that the 'statistical' description presented above was just to aid in explaining the two indices and is not how they are calculated. We also want to mention that there has been debate over whether conservatism involves any tendency of species to be more similar than expected. A more liberal definition, and the one we use here, includes traits that evolve according to Brownian motion. In a more strict definition, only traits that are more similar among closely related species than one would expect under Brownian motion are considered to be conserved (Losos 2008; Crisp & Cook 2012).

Despite their usefulness, both these indices (and also others that serve the same purpose; Münkemüller et al. (2012) give a good overview and evaluation) have one drawback – that they try to encompass something as complex as the evolution of traits, taking place over potentially millions of years, across thousands of species, into a single number. Apart from testing trait evolution against a model of Brownian motion, other evolutionary models have become available, which also basically represent generalizations of the Brownian motion model (Pennell & Harmon 2013; see also section 8.5 below). Moreover, it is also interesting and informative to investigate how phylogenetic signal changes along the evolutionary history of species. One specific way to do this is to calculate phylogenetic signal not only for extant species, but to calculate it for different 'slices' of evolutionary time and see if and how the signal changes over time. Specifically, this can be achieved with *phylogenetic auto-correlograms*, very similar to those that are used to depict spatial autocorrelation (Paradis 2012). For this, Moran's *I*, a measure of autocorrelation, is calculated for different distance classes, i.e. in the phylogenetic case for different phylogenetic distances (see Fig. 8.6 further below). These distance classes can be defined by clades or more continuously. In the former case, this can be achieved by calculating the autocorrelation at different taxonomic levels, e.g. genera, families, orders. By using permutation tests, the significance of the autocorrelation can also be assessed. The result is a graph that represents the degree of autocorrelation along the increasing distance at the categorical or continuous scale (Keck et al. 2016). We demonstrate this method in the accompanying 'R material Ch8', alongside the more commonly used indices of phylogenetic signal.

8.5 Are Traits Brownian?

Now that we have the actual tools to measure phylogenetic signal, how well does the general idea that related species are similar hold up when we test it with data? Generally, based on our experience but also related literature, one could conclude that 'it depends'. There are dozens of studies that have examined phylogenetic signal in different contexts, for different traits and different groups of organisms. And generally, the whole spectrum from traits not showing any phylogenetic signal to traits showing very strong phylogenetic signal is represented. In the very paper that introduced Blomberg's K, the authors found a much weaker phylogenetic signal in traits that were related to behaviour, compared to traits that assessed aspects of their physiology (Blomberg et al. 2003). This study included a relatively large set of different data from different taxonomic groups and different types of traits (body size, morphology, life history, physiology, behaviour, ecology). The highest signal was found in body size. In general, though, there was a wide range in phylogenetic signal (K values) in all these trait types, so none could be categorically defined as conserved or *labile* (i.e. not conserved) across the different datasets. For plants, Pennell et al. (2015) undertook a comprehensive analysis of the three functional traits specific leaf area, leaf nitrogen and seed mass. They extended the more basic approach of measuring phylogenetic signal with a single parameter in relation to Brownian motion, and tested alternative types of trait evolution, namely *Ornstein–Uhlenbeck* (OU) and *early burst* (EB). Both of these models are basically extensions of the Brownian motion model. With OU one can imagine that instead of having entirely random increments of trait change, these changes are 'pulled' in the direction of species optima for that trait, so that the resulting trait distribution is constrained to a smaller variance around one or several means, depending on how many trait optima exist. As indicated by the name, in the EB model most of the trait changes occur during early evolutionary time after a speciation event, and the rates of change in trait values after that initial burst are much slower. Pennell and his colleagues assessed which of these three models best fitted the data of large sets of species for the aforementioned traits. They also did this at the scale of different clades, not only for the entire phylogenetic tree. This resulted in 337 'datasets' that were analysed. For the majority of these, an OU model of evolution was a better fit to the data than the Brownian or EB model. Moving from the macroscopic to the microscopic world, Goberna and Verdú (2015) provided a meta-analysis of phylogenetic signals for traits of bacteria and archaea. Very often, for these organisms, traits are in fact binary, representing information about the presence or absence of particular genes. For such traits, the authors found relatively high average phylogenetic signal, whereas for continuous traits, which expressed diverse cell or genome-related features, most data showed low phylogenetic signal, especially compared to the binary traits.

8.6 Phylogenetic Comparative Methods

8.6.1 The Comparative Method and Evolution

The *comparative method* in biology refers to the notion that in order to understand why species are different we need to compare them to each other. In the context of ecological questions this means that if we, e.g., correlate two traits across a set of species (for example to test the trade-offs discussed in Chapter 3), or if we correlate a trait with the environment that species inhabit (or more generally the niche), we are performing the comparative method. Though this sounds somewhat trivial, we will soon see how such comparisons can be complicated by the fact that species share an evolutionary history (see section 8.5), and so, statistically speaking, they are not independent.

A famous prime example of the comparative method in an evolutionary context is Darwin's finches. Darwin postulated that the different finch species living on the Galapagos Islands had adapted different beak shapes in order to use different food resources. For example, those finches with bigger beaks could crack larger and harder seeds, while tinier, pointy beaks seemed better for picking out small insects from the bark of trees and the ground. Looking at a single species, it would have been difficult to come up with this postulation, but because Darwin had taken detailed notice of the shape and size of different finches' beaks, and what various species were eating, he was able to *compare* species with each other and see the relationship between the morphological features and the occupied resource niche.

Since Darwin's observations, a lot of research has been carried out using the comparative approach in different fields of biology. What Darwin, and many other researchers after him, did not have at the time of their studies, though, was a phylogenetic tree of the species they were working with to represent their evolutionary relationships. The idea that the combination of species data with phylogenetic data is helpful (or even necessary) to draw meaningful conclusions was popularized in a seminal paper by Felsenstein (1985). He basically used the Brownian motion model of evolution to show that the correlation between two traits across a set of species can be the outcome of the evolutionary history of the species. In his rather extreme, but still valid, example, two groups of species are present, in which members of the groups evolved from one of two ancestral species that had evolved independently of each other (Fig. 8.3). If we draw a scatterplot from this data, we can see that there seems to be a relationship between the two traits. However, the correlation results from the fact that the species of the two different clades all evolved from one of two ancestors, and most of the apparent correlation comes from the fact that within these clades so much evolutionary history is already shared, i.e. what matters is the evolution of the traits before the speciation within clades. To add to this, there is no correlated evolution of species *within* clades, so that when we add this clade information to the scatterplot, we can see that the overall correlation only exists because of the independent evolution of the two ancestral clade species since their last common ancestor.

Felsenstein then introduced a test to account for phylogenetic effects: so-called 'phylogenetic independent contrasts' or sometimes simply 'independent contrasts'

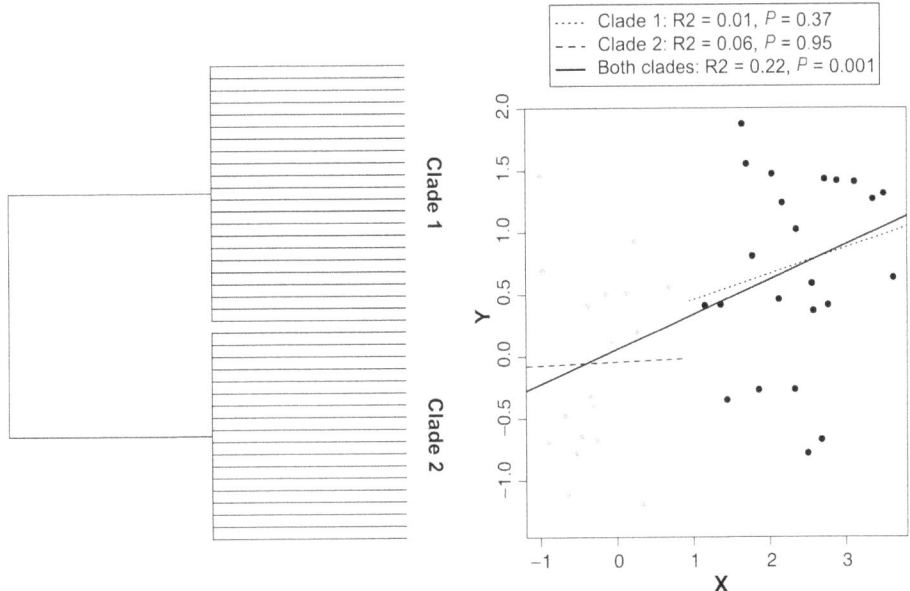

Figure 8.3 A simulation of Felsenstein's example in his seminal paper (Felsenstein 1985), demonstrating an extreme case of the potential bias when using species data for comparative analyses. X and Y can both be traits, or one of the variables can represent some niche, or an environmental condition that is part of the species' habitat. Two groups of closely related species depicted in the phylogenetic tree (clades 1 and 2) each have a common ancestor, and the trait combination of the common ancestors predicts the two-dimensional trait space of the two separate clades. In this case, the ancestor of clade 1 would have relatively high values for X and Y, and the ancestor of clade 2 would have relatively low values. This leads to two distinct clouds of points in the scatterplot of X and Y, where a relationship between the variables arises only if the data for both clades is considered. There is no relationship within each of the two clades.

(which is explained in more detail in section 8.6.2 below). This test was originally designed to remove (or 'correct') the effect of evolutionary history from correlation and regression analyses. The main original argument was that a relationship between traits that only occurs because of *phylogenetic relatedness* of species can be seen as a spurious relationship that is not interpretable in a meaningful way. From a statistical point of view this has some logic. Species traits are not independent because of their shared evolutionary history, and statistical tests should take this into account. More generally, rather than remove the effect of evolutionary history, this type of 'correction' aims to take into account the variation caused by the effect of evolutionary history. Hence whether independent contrasts are used or not will depend on the question being asked, for example if trait correlations originated across or within clades (Fig. 8.4).

Together with the statistical arguments over the non-independence of species, passionate debates over if and when to account for phylogeny in the comparative method have ensued between scholars in the past (Harvey et al. 1995; Westoby et al. 1995) and have still not faded (de Bello et al. 2015; Prinzing 2016, among others). This has certainly led to one point of view that instead of considering it a statistical necessity,

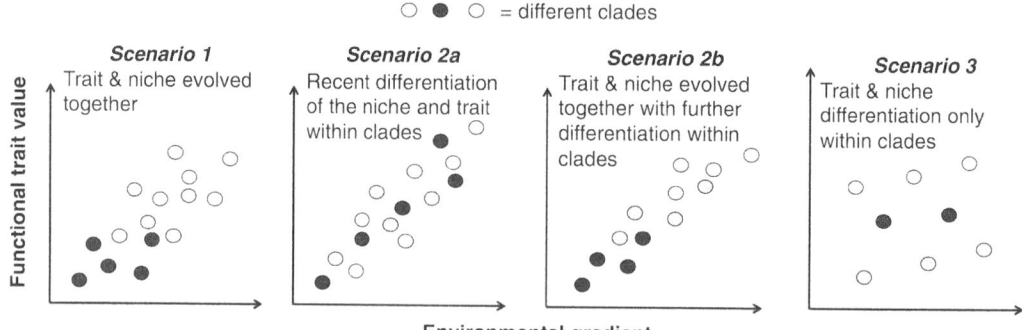

Figure 8.4 Possible results when assessing the relationship between species traits, or between species traits and environmental preferences. Each species is represented by a circle, with different colours indicating different clade membership. In Scenarios 1 and 2b, both species' niche (environmental gradient position) and the trait considered are likely conserved. In Scenario 2a neither species niche nor the trait is conserved as species from all three clades have very different values for niche and trait. In Scenario 3, only the trait is conserved within clades, but similar trait values allow inhabitation of different habitats. Taken from de Bello et al. (2015), with permission from Springer Nature.

accounting for phylogenetic relationships between studied species is a way to include *evolutionary perspectives* in a comparative analyses (where some consider this evolutionary perspective to be the only one worth studying; see e.g. Freckleton 2009). In other words, accounting for phylogenetic information, or not, basically enables different ecological questions to be answered, i.e. particularly at what *evolutionary scale* the differentiation between species can be observed. Since this field of research is large and still undergoing rapid developments, including all possible perspectives falls outside the scope of this book. We thus invite readers to dive into more specific literature such as that pointed out at the beginning of this chapter. We will, though, provide some guidance on the application and interpretation of the methods more commonly applied in the literature. We will thus give a short overview of the ways in which researchers can include phylogenetic information in their studies, if they decide to do so. Other works are available that give a much more profound introduction to the available statistical methods (Paradis 2012; Garamszegi 2014).

8.6.2 Independent Contrasts

Phylogenetic independent contrasts (PICs, or, for short, simply independent contrasts) is the method that massively popularized the idea that comparative analyses need to remove unwanted phylogenetic effects by taking into account the phylogeny of the species, while also providing a relatively easy tool to do so. At the same time there are limitations to the kind of data that can be analysed using this method. How the method works statistically is that instead of using simple correlation or regression analyses on the raw trait values of the species, differences are calculated between the values of traits for closely related sister species, resulting in the *contrasts* between the species. As mentioned above, this approach can be used to relate traits with each other or with environmental preferences (Fig. 8.4).

The substractions to generate contrasts are done not only for the tips of the tree but between any pair of sister species, be they at the tips or ancestors at internal nodes; this calculation can therefore be made between internal nodes of the tree. For this to work, trait values for internal nodes need first to be estimated, which is done assuming that traits evolved according to Brownian motion, meaning that the trait values of an ancestral species are simply the average of the two descendant species. The calculated contrasts are standardized by their standard deviations, which themselves are estimated from the branch lengths. The standardized contrasts are then used to perform a standard correlation or regression and ANOVA analyses. But because contrasts represent differences between pairs of species or nodes, the contrasts are not dependent anymore on how closely related (or not) the species (including ancestral species at nodes) are.

The use of PICs is limited for a number of reasons. First, the resulting contrasts are not assignable to species, and therefore this adds a layer of abstraction. If we plot contrasts of two traits against each other, each dot represents a difference between two species (tips or nodes), but the information on what the species trait values are is lost (Fig. 8.5). Statistically, this specific method is constrained to have a single response and a single explanatory variable, and the error term needs to be normally distributed (see below for multivariate solutions). Use of ANOVA to analyse PICs is also possible if the

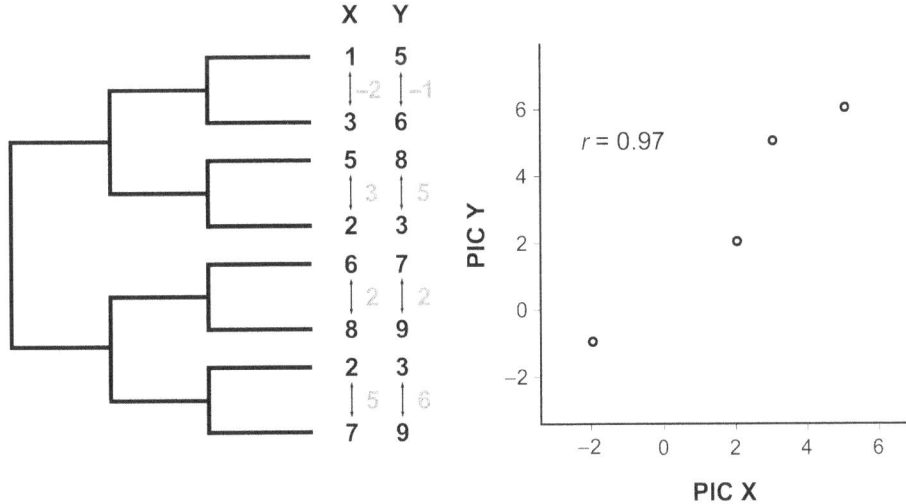

Figure 8.5 A simplified example of calculating phylogenetic independent contrasts. The values of two traits X and Y are shown at the tips of a phylogenetic tree. Contrasts are calculated by doing a subtraction of values between pairs of sister taxa, for each trait separately (shown in grey). These numbers are then used instead of the raw trait values in further analyses, in this case for creating a scatterplot of contrasts for X and Y, and calculating the correlation coefficient r. For simplicity, contrasts for nodes have not been calculated but can be and usually are, and contrasts have also not been standardized by the branch lengths, since all branches in the shown phylogeny are of the same length.

explanatory variable is categorical, but if there are more than two categories, dummy variables must be created. Methods that came after PICs all extend the toolbox of phylogenetic comparative analyses by relaxing these statistical constraints and assumptions.

8.6.3 Moving Past PICs

Having introduced the original concept of PICs, we can now introduce some more modern and flexible methods. These tools generally aim at including *phylogenetic information* in otherwise more common statistical modelling approaches that model either the relationship between traits (Chapter 3), or the relationship between species' traits and their environmental requirements (Chapter 4). These methods follow a similar approach to spatially explicit models that take into account the geographical distance between sampling units to address the potential autocorrelation. Instead of considering spatial distance between plots, the methods presented in this section take into account the phylogenetic distance (see Fig. 8.6) between species. In practice, the approaches listed here introduce phylogeny in various types of statistical models, or they use the residuals of a model in which traits are first predicted by phylogenetic differences as a way to account for the effect of phylogeny.

One method that achieves numerically identical results to PICs (under the assumption of Brownian motion) is phylogenetic generalized least squares (PGLS). Contrary to an ordinary least squares regression (OLS), PGLS allows the residuals of the model, which are assumed to be random and independent in OLS, to be correlated. Mathematically,

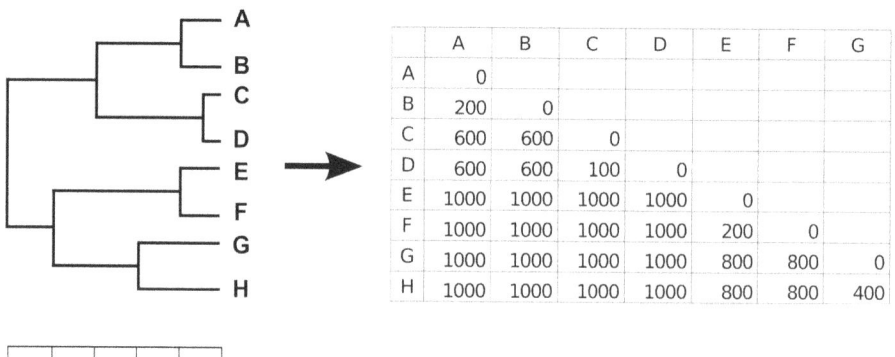

Figure 8.6 A simplified example of calculating phylogenetic distances from a phylogenetic tree. Values on the x-axis below the tree give an indication of how old nodes in the tree are. The specific way to calculate distances used here is the so-called cophenetic distance method, where the distance between a pair of species is the summation of the branches that connect them. Depending on the topology of the tree, this can lead to a comparatively large number of large distances, as is the case in the example tree (but note that real phylogenetic trees are unlikely to be as symmetrical as the one used here). Only the lower triangle of the distance matrix is shown, as the information in the upper triangle would be identical (see Chapters 3 and 5 for comparison with trait dissimilarity).

this is achieved by adding a variance–covariance matrix as a term to the model, which represents how strongly the residuals of the observations are expected to be related to each other. In the case of comparative analyses, typically the observations are species (i.e. each point in the analysis is a species). Therefore, the variance–covariance matrix represents the covariance between species, i.e. species' shared evolutionary history. This is taken into account when estimating the model's parameters, i.e. the intercept and slope of the regression line. This type of model also allows for multiple explanatory variables, continuous as well as categorical ones, and their combination. However, PGLS is only equipped to handle continuous (or under some circumstances semi-continuous) response variables. The Generalized Estimation Equation (GEE), on the other hand, can also deal with other types of response variables and, thus, is conceptually a combination of PGLS and generalized linear models (GLM). In GEE, apart from also using the variance–covariance matrix to account for phylogenetic dependence just as PGLS does, the statistical family of the distribution of the response variable needs to be specified. The most recent advances in phylogenetic comparative analysis, which enable even more complex and flexible ways to model trait evolution, have come with the use of Bayesian models (Hadfield & Nakagawa 2010; Uyeda et al. 2018).

Another useful and flexible approach is based on the concept of *phylogenetic eigenvectors* (Desdevises et al. 2003; Diniz-Filho et al. 2012). This approach allows for the inclusion of phylogeny in the analyses in a different way to those described above, as they take a little detour via principal coordinate analyses (PCoA). Note that this again resembles spatial analyses in which spatial coordinates are sometimes summarized via analyses similar to PCoA. PCoA, partially introduced in Chapter 3, transforms a pairwise distance matrix of species into axes in a multivariate space. The distance matrix is obtained from the phylogenetic tree using a cophenetic distance measure, which means nothing more than that the distances between species are defined by how distant they are in terms of evolutionary time. This is basically very similar to the distance matrix expressing trait distances as described in Chapter 3. For the phylogeny, the pairwise distances between species are based on the age of the nodes representing the last common ancestors of any species pair (see Fig. 8.6).

PCoA based on such distances produces species scores on multivariate axes, which are called phylogenetic eigenvectors. These eigenvectors provide a synthetic representation of evolutionary distances between species, with the important distinction that they are numerical values on a species level. Once these eigenvectors are produced, they can be tested against any ecological variable (e.g. the environmental preference of the species, or other traits). It is common practice to retain, for further analyses, only the phylogenetic eigenvectors that are significantly related to the variable of interest (i.e. the response variable). In the next step, the actual ecological explanatory variables of interest are tested for their relationship with the response variable. In other words, we first fit a model with phylogenetic eigenvectors alone to select meaningful phylogenetic axes, and then a model with traits and phylogenetic eigenvectors, with the latter retained from the previous model. In this way, the variation in the response variable can be explained by the main and partial effects of traits and phylogeny. For more detailed information see the accompanying 'R material Ch8'.

Eigenvectors can be used very flexibly. This is because they translate the non-tabular format of a phylogenetic tree into data that represent the phylogenetic relationships in two-dimensional table format, just as we would represent trait-by-species data. This allows their use in different univariate (see above), but also multivariate, methods, e.g. in RLQ analyses (Pavoine et al. 2011), or to decouple and combine phylogenetic and functional diversities (see section 8.8, and de Bello et al. 2017).

8.6.4 Imputing Data with Phylogeny

Though it is not strictly a case of using phylogenetic comparative methods, we also want to mention here the possibility of imputing trait data with the help of phylogeny, since this approach largely uses the same theoretical and statistical framework that we encountered in the methods above. If we assume that phylogenetically close species have similar traits, in cases where we cannot find or measure traits for a given species, we can use the trait information, for example, from congeneric species. In other words, we use phylogeny as a way to find species which, in theory, are similar to those for which we are missing trait information.

More generally, in the context of species-based data, *imputing data* refers to statistical procedures that fill 'empty holes' in species data. It can often happen that we have incomplete trait values for a set of species, especially when data have not been collected in the field but have been retrieved from biological databases. The easiest approach in such cases is of course to discard the species from the calculations, for example of CWM and FD (see Chapter 5), for the given trait considered. As discussed in several sections of the book, discarding species, especially dominant ones, can lead to problems and misinterpretations. So, when possible it might be more desirable to keep the species in the dataset, if we can provide a decent estimation of the possible trait value. To impute data can be done at very different levels of accuracy, where simpler procedures might be regarded as quite naive. For instance, one possibility is simply to replace the missing trait values with the mean of the values of all other species in the dataset. It should be intuitively clear that this is really a last-resort method that is best avoided. But given what we have learned so far about phylogenetic relatedness and how it can affect the traits of related species, it seems conceivable that we can come up with methods that should be able to do better than the overall-mean approach just mentioned. For instance, some studies impute missing data of species by checking if data are available for a different but closely related species. Such a solution implicitly makes the assumption that the related species are also similar regarding their traits; this is an assumption that we already know is not always valid. A better way to use information on species relatedness would be if we could estimate how well this assumption holds and consider this estimate when imputing the data (Swenson 2014b). This can also offer a way of saying how certain we can be about an imputed value, i.e. when the phylogenetic signal is high in the trait for which we want to impute values then we can be fairly certain about the imputed values, but the opposite holds for a trait with low phylogenetic signal (Penone et al. 2014). At the same time, other traits than the one with missing values can be used to impute data, given that there is some interdependency between traits. Various methods that integrate the correlation structure between traits, as well as the phylogenetic relatedness between species, to impute data are available. These are based on different statistical approaches (see Penone et al. (2014) for a comparison of the performance of different methods). In the accompanying 'R material Ch8' we show how one of these methods, which relies on both trait correlation and phylogeny, works in practice.

8.7 Phylogenetic Diversity and Community Assembly

In earlier chapters, we have shown that we can use functional traits of species to get an idea of how different species are regarding their *ecological strategies* (Chapters 3 and 5), and the role these differences play in the assembly and functioning of communities and ecosystems (Chapter 7). We demonstrated that one prominent methodological approach is to estimate functional diversity for a given community, to estimate how similar, or dissimilar, species are that coexist in a given defined space, such as plant individuals in a vegetation plot, or birds that live in a patch of forest. If closely related species are more similar in their traits than distantly related ones (see above), maybe we can use phylogeny *instead* of traits, i.e. as a proxy for multi-trait dissimilarity. This means that for those regions or organisms for which we still miss dedicated information about traits, tests could be done using phylogeny instead. In other words, we can theoretically use this potential resemblance between trait dissimilarity and phylogenetic distances between species to study community assembly. From a technical point of view, what we need to achieve this (in most cases) is a matrix that represents the pairwise differences between the species in our community. While we have seen how this works with single or multiple traits in Chapter 3, this is also possible given a phylogenetic tree (see sections above), as we can use the evolutionary distances between species represented in the tree to construct a species pairwise distance matrix as well (see Fig. 8.6). Under the assumption that related species have similar traits, we can potentially omit information about the traits entirely, and simply use the phylogenetic distances, forgetting about all the troubles connected with measuring traits. In this way, the information contained in the phylogenetic distances becomes some integrative measure for all potentially measurable traits that contribute to the *overall fitness* of the species.

An early adoption of the general idea to make use of phylogenetic information in a community assembly context is the application of *species/genus ratios* (Elton 1946), albeit the information here is taxonomic and not phylogenetic in the strict sense. The initial idea with species/genus ratios was that species of the same genus are more similar to each other than species of different genera, and therefore have a higher tendency to competitively exclude each other from coexisting in the same community, in line with the idea of limiting similarity (see Chapter 7). Compared to the species/genus ratio of an entire species pool, e.g. of the biogeographic area in which samples were taken, species/genus ratios at the sample level, i.e. at the scale on which species interact and coexist, should be smaller. This was indeed what was found by Elton (1946) when collating such data for plant and animal communities of Britain. His results and his interpretations thereby led to a number of subsequent publications that hotly debated the meaningfulness of species/genus ratios and ultimately led to the first use of null models in community ecology (see Simberloff (1970) for a review of the debate and a meta-study of Elton's and similar datasets). A general introduction to null models and how they are used with functional (and hence phylogenetic) diversity indices is given in Chapter 7.

Species/genus ratios make use of taxonomic information instead of phylogenetic trees, the latter usually offering better-resolved information. Before the 1990s, the low availability of phylogenies, but also the lack of communication between ecologists and taxonomists, meant that (i) the necessary data to combine phylogenetic and community information were not readily available, and (ii) ecologists simply did not see and appreciate the possibilities that could be offered by approaching community ecology from a phylogenetic point of view. This changed in the late 1990s and early 2000s, marked by the publication of the seminal paper by Webb et al. (2002), who reviewed the few then-existing phylogenetic community studies, proposed a general framework for the integration of phylogeny and community ecology, and popularized statistical tools for *community phylogenetic structure*, tests of conservatism (i.e. phylogenetic signal), and applications of null models. Together with the increasing amount of available phylogenetic information and easier access to such information for non-taxonomists, community ecologists have since then embraced the approach and adopted it in many studies of community assembly across a large span of different habitats and geographical scales.

Generally, and very similarly to interpretations of trait-based patterns (see Chapter 7), divergent phylogenetic diversity has been initially interpreted as a signal of competitive interactions leading to the exclusion of too similar (i.e. too closely related) species from coexistence, and convergent phylogenetic diversity as a signal of environmental filtering driven by abiotic factors. The important addition to that idea was that the possible resulting patterns of *convergence* and *divergence* indicate different processes depending on the conservatism of the traits responsible for the assembly processes (Fig. 8.7; Webb et al. 2002). The above-described indication of competition by divergence, and filtering by convergence, only applies when conservatism is high, i.e. if species that are closely related do indeed possess similar trait values. Hence, the connection between patterns and processes is the same as when considering functional diversity in the high-conservatism case. In the opposite case of low conservatism, to the degree that species are very similar although they are phylogenetically unrelated, filtering is actually expected to lead to a divergent pattern in phylogenetic diversity. Note that because the term convergent is also used in an evolutionary context to signify the very pattern that supposedly leads to convergent phylogenetic community structure, some authors prefer to replace divergence and convergence with the terms *overdispersion* and *underdispersion* (which have their own problems, leading to a 'jargon jungle'; see Pausas & Verdú 2010; Götzenberger et al. 2012), or phenotypic repulsion and attraction; for the same reason the term phylogenetic *clustering* is often used instead of convergence.

8.7 Phylogenetic Diversity and Community Assembly

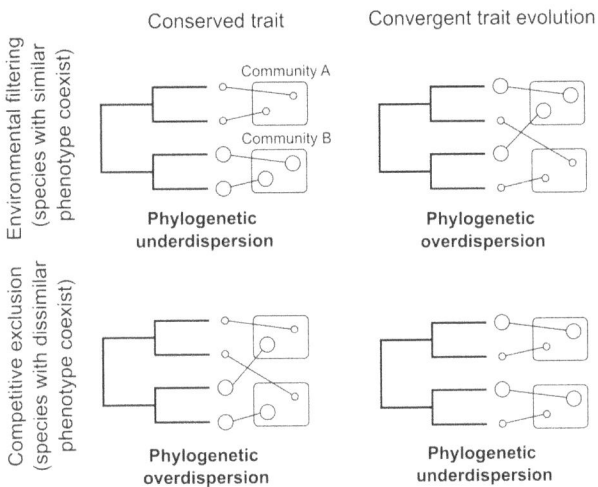

Figure 8.7 A simplified scheme of how different modes of trait evolution (conserved trait and convergent trait evolution) and different processes of coexistence (environmental filtering vs competitive exclusion) lead to different patterns of phylogenetic community structure, i.e. over- and underdispersion. The sizes of the circles next to the phylogenetic trees represent different trait values of the species at the tips of the tree.

The seemingly straightforward interpretation of potentially observed patterns in phylogenetic community structure and the inferred processes probably explain the surge in works that study community assembly through a phylogenetic lens. However, in a development not unlike the one that trait-based community assembly has undergone (see Chapter 7), the clear inference of processes from *phylogenetic community patterns* has recently been called into question (Gerhold et al. 2015). Many of the same expectations and limitations that apply to trait-based community assembly are also evident when phylogenetic-based measures of diversity are used instead. For instance, patterns of convergence and divergence are expected and have been found to vary with the spatial scale at which communities are sampled (Swenson et al. 2006, 2007; Bennett et al. 2013). Likewise, the idea that a few strong competitors can exclude weaker species from the community and thus lead to similar species in terms of their competitive traits also applies to their phylogenetic similarity when these traits are conserved (Mayfield & Levine 2010; Letten et al. 2014). Apart from the most often addressed expectations to find signatures of competition versus environmental filtering, other processes, including those that involve interactions between trophic levels, can lead to distinct patterns in phylogenetic community structure (Lortie et al. 2004; Pausas & Verdú 2010). For example, facilitation might lead to patterns of overdispersion when pioneer species facilitate distantly related species that occur at later stages in a successional system after forest fire (Verdú et al. 2009). Finally, traits rarely fall into one of the conserved/non-conserved categories in a black and white fashion, and the degree of conservatism or phylogenetic signal can also be different among different clades of the same phylogenetic tree. When we use phylogenetic distances *instead* of trait distances, we are bound to miss some information that is contained in those traits that are weakly conserved or not at all (i.e. labile), as long as they exert some influence on assembly processes. In this sense, in the example of Swenson and Enquist (2009) a clear signal of phylogeny on community assembly was not detected because different traits were acting in different directions on the assembly of species. Consequently, more and more studies are attempting to combine phylogenetic and trait information in studies of community assembly.

8.8 Combining Phylogenetic and Functional Diversities

As described in section 8.7, the phylogeny of species can potentially be used to investigate community assembly and drivers of ecosystem processes, given that a number of assumptions are met. In doing this, we assume that we can use *either* functional diversity (Chapters 5 and 7) or phylogenetic diversity (section 8.7) to assess patterns in *ecological similarity* between species. In other words, we assume that by looking through these two lenses at community ecology, i.e. functional diversity and phylogenetic diversity, similar information is provided. On the other hand, and to continue the metaphor, we already know that these two lenses are part of the same pair of glasses, and that the information they contain can be highly correlated, with many traits presenting a certain degree of trait conservatism. As a result, phylogenetic diversity is often correlated with functional diversity (Cadotte et al. 2019). This also means that if there is some potential relationship and dependency between these two kinds of information, then as well as both providing some unique information, this information may partially overlap. This could help, for example, in the case that we have measured some relevant traits, but not all; here the phylogeny could provide information about other unmeasured traits. To realize this idea, we would ideally like to combine the individual effects of the functional and phylogenetic components so that they do not overlap in a quantitative way.

Quite a number of studies have combined traits and phylogeny to the extent that they have used both the functional and phylogenetic diversity of the studied communities. Most studies so far have rather used functional and phylogenetic diversity independently, e.g. as predictors for the same processes. In some other cases, functional analyses have been complemented by reporting the phylogenetic signal of the traits used to estimate functional diversity of single traits (Kraft et al. 2007; Spasojevic & Suding 2012). As we saw in Chapter 3, one meaningful way to compute trait-based functional diversity is by using a multivariate trait approach, i.e. to estimate a compound functional diversity measure that integrates a number of traits that ideally represent different dimensions of the ecological strategy of species. The downside is that multi-trait estimates of functional diversity, and their changes along environmental gradients, can mask the response of functional diversities of single traits along the same gradients. For instance, Spasojevic and Suding (2012) found that different leaf traits responded in opposite directions (i.e. either positively or negatively) along a gradient of decreasing environmental stress, whereas no significant relationship was found between the multi-

8.8 Combining Phylogenetic and Functional Diversities

trait functional diversity and the same environmental stress gradient. In the same study, a significant positive relationship between phylogenetic diversity and the stress gradient was found; this positive relationship coincided with a positive relationship between the traits that had high phylogenetic signal and the same stress gradient. That multi-trait functional diversity did not mirror the pattern in phylogenetic signal in that study possibly meant that there were some unmeasured traits that were also responsible for the assembly along the studied gradient. The literature also includes examples where either phylogenetic diversity has stronger predictive power than multi-trait functional diversity or vice versa (e.g. de Bello et al. 2012; Craven et al. 2018).

These and similar findings have recently led to attempts to *combine* phylogeny and traits at the community level in a more complementary way. Cadotte et al. (2013) suggested an interesting approach to accomplish this more complementary treatment by combining functional and phylogenetic diversity into a single measure of biodiversity. They achieved this, simply speaking, by summing phylogenetic and functional pairwise distances of species into *functional-phylogenetic distances*. In addition, instead of simply summing the two distances with equal weight, they applied a weighting factor that allows changing the contribution of the two separate distances to the summed one. Mathematically speaking, since the weights are proportional to 1, the obtained distances are a weighted average of phylogenetic and functional distances. The idea here is that resulting diversity estimates with various weights can be tested for their relationship with an environmental gradient, and that the weighting that maximizes the explained variability represents the most likely relative contribution of trait-based versus phylogeny-based assembly along that particular gradient. This is potentially a very appealing approach. At the same time, we should not forget (from Chapter 3) that mixing dissimilarities with different properties will give stronger weight to some components. Specifically, we expect that given an equal contribution of phylogenetic and functional differences (i.e. the simple sum or average of their dissimilarity), the weight of phylogeny on the combined dissimilarity will be higher (similar to the case of combining categorical and quantitative traits described in Chapter 3). This is simply because the values for phylogenetic dissimilarity are less skewed and they will be generally greater. In some cases, they can even be similar to a bimodal distribution, because quite a number of distances can stem from species that are from distantly related clades, while within these clades species are closely related (e.g. when ferns are part of plant communities).

Another potential drawback lies in the generally observed lack of full independence of phylogenetic and functional dissimilarity (Cadotte et al. 2019). If phylogenetic and functional information *overlap* (in the case of non-negligible phylogenetic signal), their sum will overemphasize this overlap at the expense of the unique information they each provide. If our aim is to use phylogenetic information not accounted for by the traits already considered, then this drawback will be a serious limitation, especially if the traits measured are phylogenetically conserved. In the ideal case, we would like to combine traits with the information from unmeasured traits, and to combine phylogenetic information with trait information not already accounted for by the phylogeny. How can we obtain this? For this we can borrow the phylogenetic eigenvector approach to

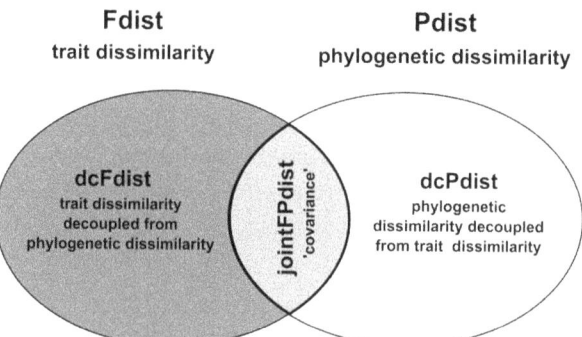

Figure 8.8 Conceptual scheme of how functional and phylogenetic distances overlap and how these components of the overall ecological dissimilarity between species can be decoupled. Taken from de Bello et al. (2017), with permission from Wiley. © 2017 The Authors. Methods in Ecology and Evolution © 2017 British Ecological Society.

disentangle (or 'decouple') the information contained in each. These 'unique' functional and phylogenetic components are separated from the overlapping component (see Fig. 8.8). At the same time, this would also enable the extraction of only the component which represents the overlap between the functional and the phylogenetic components. One way to achieve such a decoupling was suggested by de Bello et al. (2017). Basically, the approach requires two types of distance matrices for the same set of species: one phylogenetic and one functional (i.e. based on traits). Then a series of specific ordination analyses results in three additional distinct distance matrices. The two input matrices would be the same as if used independently of each other, so not different from any other application of phylogenetic and functional diversity (as also in the case of Cadotte's approach, where the weighting of either of the two matrices can be zero). The three new resulting matrices produced by this approach are as follows: decoupled functional distances, decoupled phylogenetic distances, and distances representing the overlap, called joint distances (Fig. 8.8). Decoupled functional distances are the trait-based functional distances after taking into account how closely related the species are, and the decoupled phylogenetic distances are, complementarily, the phylogenetic distances after taking into account how similar species are in terms of their (measured) functional traits.

This is all very theoretical, so how can we make sense of this new information? A first possible application is related to the question that we often have when we try to study community assembly on the basis of functional traits; did we measure all relevant traits? Ideally, we capture all traits that are important for the processes that govern the assembly of the communities we study. At the same time, we don't want *redundant traits* that capture very similar functions, while completely missing other functions or aspects of the species' ecological strategies. For instance, imagine we have measured several leaf traits that all represent the leaf economic spectrum and thus a gradient of species' leaf resource use strategy, spanning from species that have high respiration and photosynthetic rates to species that are much 'slower' with respect to these physiological processes. On the other hand, in our example dataset, we might be completely missing information on the reproductive behaviour of the species, which for plants is often captured using seed traits. If traits like seed mass are phylogenetically conserved (which they mostly are), then the premise of using decoupled phylogenetic distances (after accounting for the effect of measured traits) is that the decoupled phylogenetic component could potentially capture these unmeasured traits. Note, however, that with this approach we can of course not capture differences between species that are based on labile (i.e. not conserved) traits that we have not measured.

The second kind of application for decoupled distances is more directly related to the question of whether *differentiation* between species in terms of their traits is something that occurs within phylogenetically closely related species (e.g. within clades) or between distantly related species (e.g. across clades). We will try to approach this with a theoretical, yet concrete, example. We start with eight species that are arranged on a simple, fully resolved phylogenetic tree, and for which we have measured two traits: plant height and root type. We assume that plant height is conserved on the phylogenetic tree, i.e. closely related species will have similar height values, whereas root type is different even for sister taxa. Given this set-up of phylogenetic tree and traits, we will observe the following when we look at *pairwise differences* between species: two sister taxa (i.e. species a and b in Fig. 8.9) will have a low dissimilarity for plant height, and a high dissimilarity for root type, while they are phylogenetically very close. Taking species a and f on the other hand shows the opposite pattern, i.e. these species are phylogenetically very distant, and dissimilar in their plant height and root type (but note that their root type could also be similar, e.g. if we had taken the sister species of species f instead). In all these cases, the observed distances are non-decoupled distances. As argued in de Bello et al. (2017), the decoupled functional distances would constitute a meaningful way to capture species differences in traits, already accounting for phylogenetic information. For the purpose of examining community assembly, this approach magnifies the differences between closely related species that coexist at local scales, after they have already passed through environmental filters. In Fig. 8.7 we can see how this 'magnifying effect' is brought about; in the case of a trait with high phylogenetic signal (i.e. plant height), taking into account the close relatedness of sister species a and b leads to a higher decoupled functional distance between these two species. In other words, the two species are more differentiated regarding plant height than one would expect given their (very close) phylogenetic relatedness. If this is the case for species co-occurring at a local scale, it can be seen as an indication of coexistence through a relatively recent trait differentiation by trait assortment or character displacement within clades (Webb et al. 2002; Prinzing et al. 2008).

8.9 Evolutionary Niche Modelling

The inclusion of phylogenetic information in questions that were initially of an ecological nature means the inclusion of more and more research topics in fundamental and applied ecology. We have already read about the recently and very quickly expanding

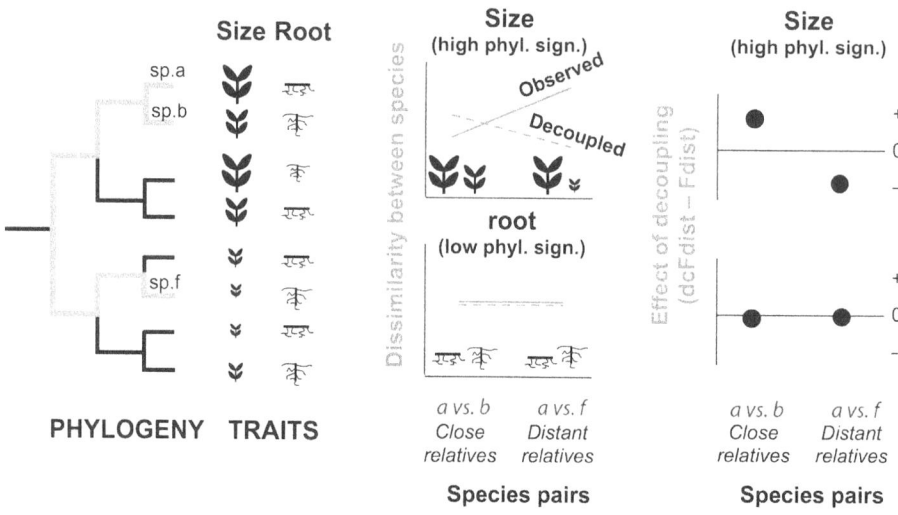

Figure 8.9 Expected trait dissimilarities between two exemplary pairs of species (a and b, a and f), comparing results with and without decoupling. A phylogenetic tree with eight species and two traits – plant size and root type – is shown. In this example, plant size represents a trait with high phylogenetic signal (i.e. 'phyl. sign.' in the figure). In the panels in the middle of the figure, the observed and decoupled dissimilarities are shown. The difference between the two (dcFdist − Fdist) is shown on the right-hand side, demonstrating the effect of the decoupling; it did not affect dissimilarities in root type between species, but led to larger dissimilarity in size between species a and b, and smaller dissimilarity in size between species a and f, compared to the non-decoupled distances. Note that for traits with high phylogenetic signal (plant size in the example) the two lines in the upper-middle panel might not always intersect and in some cases the 'decoupled' line might be horizontal. Taken from de Bello et al. (2017), with permission from Wiley. © 2017 The Authors. Methods in Ecology and Evolution © 2017 British Ecological Society.

field of *species-distribution modelling* (Chapter 4, section 4.6), which uses occurrence data of species and maps of environmental factors to predict distributional ranges of species. This methodological approach is largely based on regarding the niche of species as their integrated response to all relevant environmental (biotic as well as abiotic) factors (Chapter 4, Box 4.1). One assumption implicitly made by approximating the niche of a species in this way is that the niche is best defined by currently prevailing environmental predictors, and that these predictors, or their relative importance, do not change over time. However, as we have argued in other cases in this chapter, it is easy to see that this is an unlikely scenario. It makes much more sense to think about the niche as something that is not necessarily fixed within a species, but evolves to various degrees with changing environmental conditions across evolutionary timescales, and that the adaptations species make to these changing conditions are reflected in the change of the niche itself.

Naturally, the evolution of niches is tightly linked to the processes of *speciation* and *diversification*, and to the idea that niches of species con be conserved or labile (niche conservatism; see section 8.3). Conceptually, niche conservatism can be considered to be at one end of a niche-stability spectrum, where species 'track' their conserved niche through geographical space, i.e. their niche is rather fixed so that they follow the distribution of their niche as it changes location through time. At the other end of this spectrum, under niche evolution, species adapt to the locally changing environmental conditions, so that the niche shifts in niche space rather than geographical space as the result of adaptive evolution. At this end of the niche-stability spectrum, the processes involved would lead to a higher degree of divergence compared to the niche conservatism scenario (Pyron et al. 2015).

Apart from the aspects described above, which mostly concern the species level, there are also theoretical ideas and empirical evidence that niche evolution is affected by interactions between species. Implicit in the idea that species have a fundamental and a realized niche, biotic drivers are expected to influence where species can exist, and the modulation of the fundamental niche by coexisting species should thus shape how the niches of species evolve in a community context. As the change in niches is also reflected by changes in the phenotypic characters (i.e. traits) of the species, this line of research has been called character displacement, which is thought mainly to occur when species compete for the same resources (Germain et al. 2018b; see also Chapter 7, section 7.1).

From these theoretical considerations arise a number of possible practical approaches for studying species distributions and coexistence. We can treat the niche of species, including parameters such as niche breadth or geographic range size, just like any other trait (although we do not consider the niche as a trait in the strict sense; see Chapter 2) that we might use in phylogenetically informed comparative analyses. For instance, we could investigate if the niche is related to certain traits, and if the changes in traits are correlated with changes in the niche across the phylogenetic tree of the studied species by employing phylogenetic models on such data (e.g. Kostikova et al. 2013; Mitchell et al. 2018; see also Fig. 8.4). In addition, phylogenetic information can be used in the actual process of predicting species distributions, either in combination with traits (Ovaskainen et al. 2015), or on its own (Kaldhusdal et al. 2014; Morales-Castilla et al. 2017). These recent developments involving the inclusion of phylogenetic information in the modelling of species distributions go hand in hand with the realization that niches can be defined for taxonomic levels other than species, e.g. genera or subspecies (Smith et al. 2018).

Take-Home Messages

- Species that share a common ancestor, i.e. evolutionary closely related species, tend to share some common traits. In some cases, however, distantly related species have also evolved similar adaptations independently (convergent evolution).
- Models of evolution, such as the Brownian motion model, help to set a reference for comparison of the extent of conservatism in a trait.
- It is important to remember that the analyses at the 'species level' (Chapter 4) relating traits and species environmental preferences (niche) can, or not, take into account that species are phylogenetically independent from each other. Tests such

as Phylogenetic Independent Contrasts (PICs) have been developed to 'correct' for the phylogenetic dependence between species. We rather think that, instead of a 'correction', such tests allow questions to be answered at different phylogenetic scales.
- Phylogenetic relatedness between species can be used to compute indices of Phylogenetic Diversity (PD), similar to the ones discussed in Chapter 5 for Functional Diversity (FD).
- The PD indices can be used to assess patterns related to community assembly rules instead of, or in combination with, FD indices. New techniques allow combining phylogenetic and functional differences between species into combined indices of diversity, which can be particularly useful to account for non-measured traits.

9 Effects of Traits on Ecosystem Processes and Services

The increasing pressure on natural resources, resulting in land-use changes, over-yielding and species loss, and subsequently ecosystem deterioration, is a worldwide concern (IPBES 2019). Managing ecosystems to ensure the provision of multiple ecosystem services has thus become a key challenge for applied ecology. Traits are a focus of this effort, as they are considered the main ecological attributes by which different organisms and biological communities, through their effects on underlying ecosystem processes, influence ecosystem services (de Bello et al. 2010a; Cardinale et al. 2012; Cernansky 2017).

As we have seen in other chapters (e.g. Chapters 1 and 4), functional traits are referred to as 'response traits', given they determine which individuals and species fit to given environmental conditions. But, when a particular trait also has an effect on ecological processes, such as nutrient cycling, pollination or litter decomposition, or has an influence on the next trophic level, for instance in predator–prey interactions (see Chapter 10), then this trait can be defined as an 'effect trait' (Figs 1.5 and 9.1). Remember that *response and effect traits* are not mutually exclusive! A trait can simultaneously be a response and effect trait, thus influencing growth, survival and reproduction, while affecting ecosystem processes (see below). However, it is important to note that by no means can it be taken for granted that response traits act as effect traits, as so far it is not properly tested (for more details see section 9.4).

In plants, responses and effects are quite often expected to be associated with the same traits, or to sets of coordinated traits. For instance, plant functional traits that enable individuals to respond to environmental changes, such as those pertaining to the cycling of elements (carbon, nutrients, water), including leaf nitrogen and leaf dry

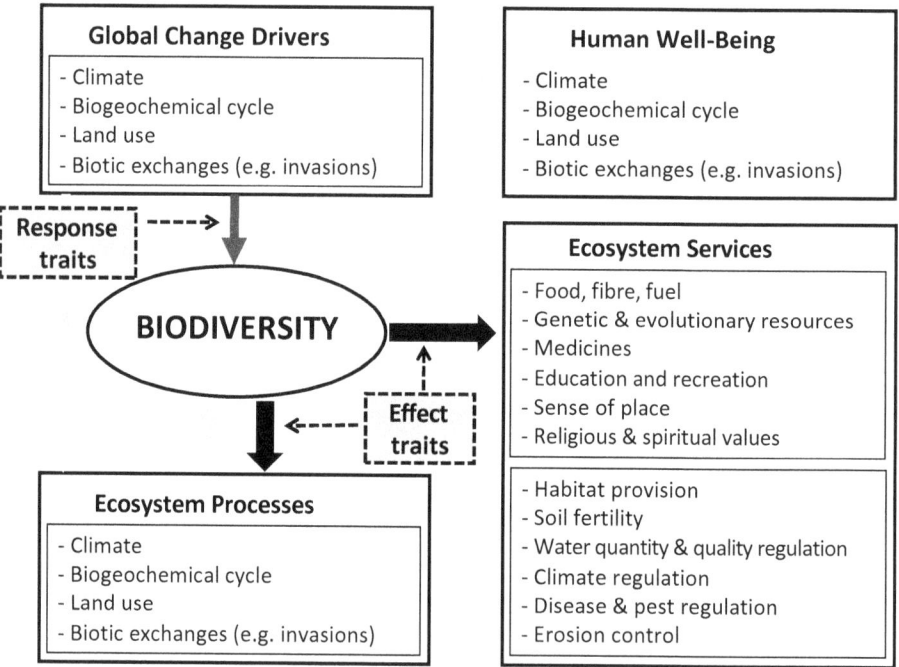

Figure 9.1 Scheme showing how biodiversity mediates the effect of global change on ecosystem processes and services, and human well-being. Adapted from Díaz et al. (2006), distributed under a Creative Commons Attribution License (CC BY). Copyright: © 2006 Díaz et al.

matter content, also impact the ecosystem (e.g. photosynthesis and net primary productivity or litter decomposition; de Bello et al. 2010a; Bu et al. 2019) and can thus be both a response and effect trait simultaneously (Hevia et al. 2017). This is usually not the case for animals, whose response traits are often disconnected from the effect traits. For instance, in an environment with fluctuating soil moisture conditions, the water loss rate of an animal affects its fitness but it does not directly influence ecosystem processes such as litter decomposition and nutrient mineralization. On the other hand, the litter consumption rate of a detritivore individual does affect leaf-litter decomposition (thus it acts as an effect trait), but it does not help to overcome drought or other stressful environmental conditions (thus it is not a response trait).

The manner in which response and effect traits are linked in organisms is central to assessing the consequences of changes in trait composition on ecosystem functions and services, both within a single trophic level (see response–effect traits framework by Lavorel & Garnier 2002) and across trophic levels (Lavorel et al. 2013; see also Chapter 10). Links between response and effect traits can be assessed by simple correlation (Gross et al. 2008; Sterk et al. 2013) or by more causal-effect procedures, such as structural equation modelling (Grigulis et al. 2013; Lavorel et al. 2013; Moretti et al. 2013) or by testing their relationship based on controlled experiments (Ibanez et al. 2013).

In this chapter we present how traits can affect ecosystem processes. We will focus on a single trait effect but also show how traits can covary both positively and negatively, creating synergies and trade-offs in the provision of ecosystem services. We then consider the mechanisms relating biodiversity and ecosystem functions (BEF), and how to integrate traits in analyses. This chapter ends by presenting the response–effect trait framework, allowing us to better understand the link between functional traits that respond to given stressors, and those that provide ecosystem processes and services. The extension of the response–effect trait approach to multiple trophic levels is presented in Chapter 10. All tools described in the current chapter are explained in detail in the corresponding 'R material Ch9' accompanying this chapter.

9.1 Links between Effect Traits and Ecosystem Processes

Within a given trophic level (e.g. primary producers, herbivores or carnivores) single traits can affect specific ecosystem processes and services, as is the case for animal ingestion rate and the amount of food consumed (Heemsbergen et al. 2004). In fact, the more food is consumed, the more organic matter is transformed into utilizable nutrients for plants and other soil organisms. The same is true for plants whose traits have a direct effect on ecosystem processes. For instance, architectural root traits, such as rooting depth, root length density and root branching, influence carbon and nutrient cycling as well as soil aggregate stability (Bardgett et al. 2014), while leaf dry matter content (LDMC) and specific leaf area (SLA) influence above-ground productivity and C-fluxes (Klumpp & Soussana 2009) (see Fig. 1.5 in Chapter 1).

Nonetheless, a review by de Bello et al. (2010a) showed that within each trophic level, specific processes and services are not only controlled by a combination of effect traits, but also some key traits are simultaneously involved in the control of *multiple processes and services*. It appears, thus, that some ecosystem services might depend on multiple traits belonging to multiple trophic levels as shown in Fig. 9.2.

Without getting into the details of multitrophic trait interactions, which will be presented in Chapter 10, it is important to note that these multiple associations between traits can results in bundles, or clusters, of associated functions and traits. *Clusters* of plant traits underlie, for example, nutrient cycling, herbivory, and fodder and fibre production. Knowing which traits link with which functions and services can help to predict changes in trade-offs among service provisions under changes of biodiversity composition. A recent review by Hanisch et al. (2020), expanding the original work by de Bello et al. (2010a), synthetized the main relationships found so far between traits and multiple ecosystem services in grasslands, focusing on trade-offs and synergies between traits and these services. These types of relationship, at present mostly limited to these meta-analysis types of data, should be enforced in studies within specific sites and regions. The data synthetized by Hanisch et al. (2020), reanalysed for this book

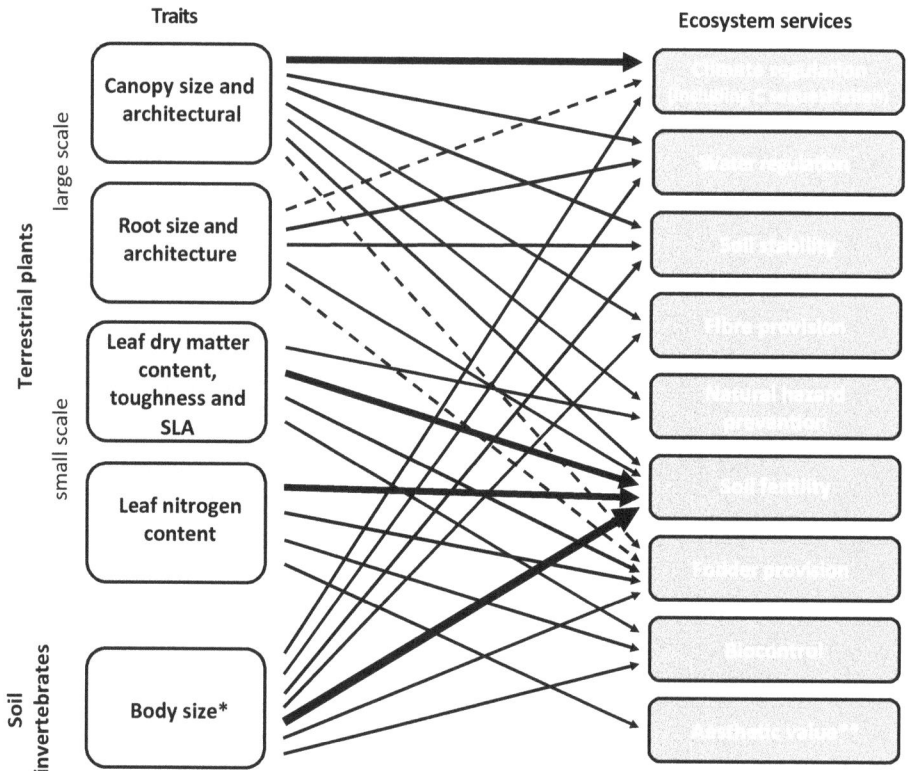

Figure 9.2 Plant and invertebrate traits and their involvement in multiple ecosystem service (ES) delivery. The width of the arrows is proportional to the number of significant trait–ES associations reported in the literature. Figure taken from de Bello et al. (2010a), with permission from Springer Nature. Copyright © 2010, Springer Nature.

(Fig. 9.3), give some hints about potential ways to assess these trade-offs among the services and the potential traits mostly associated to these services, using multivariate analyses such as PCA.

Specific trait syndromes, for example related to the leaf economic spectrum (Chapter 3), can result in *trade-offs* in several ecosystem processes and services associated to those traits (Lavorel & Grigulis 2012; Garnier et al. 2016). A shift in the composition of plant communities, and thus their associated position on the leaf economic spectrum, will likely result in a change in ecosystem processes and services as shown in Fig. 9.4. If, for instance, in vegetation a shift from a slow- to fast-growing species occurs in the community, this may cause a shift from low food quality and slow litter decomposition, to high food quality and fast litter decomposition, with an associated reduction in carbon sequestration. The dominance of traits linked with fast and slow growth will also determine changes in community temporal stability (Craven et al. 2018) with possible consequences for ecosystem functions.

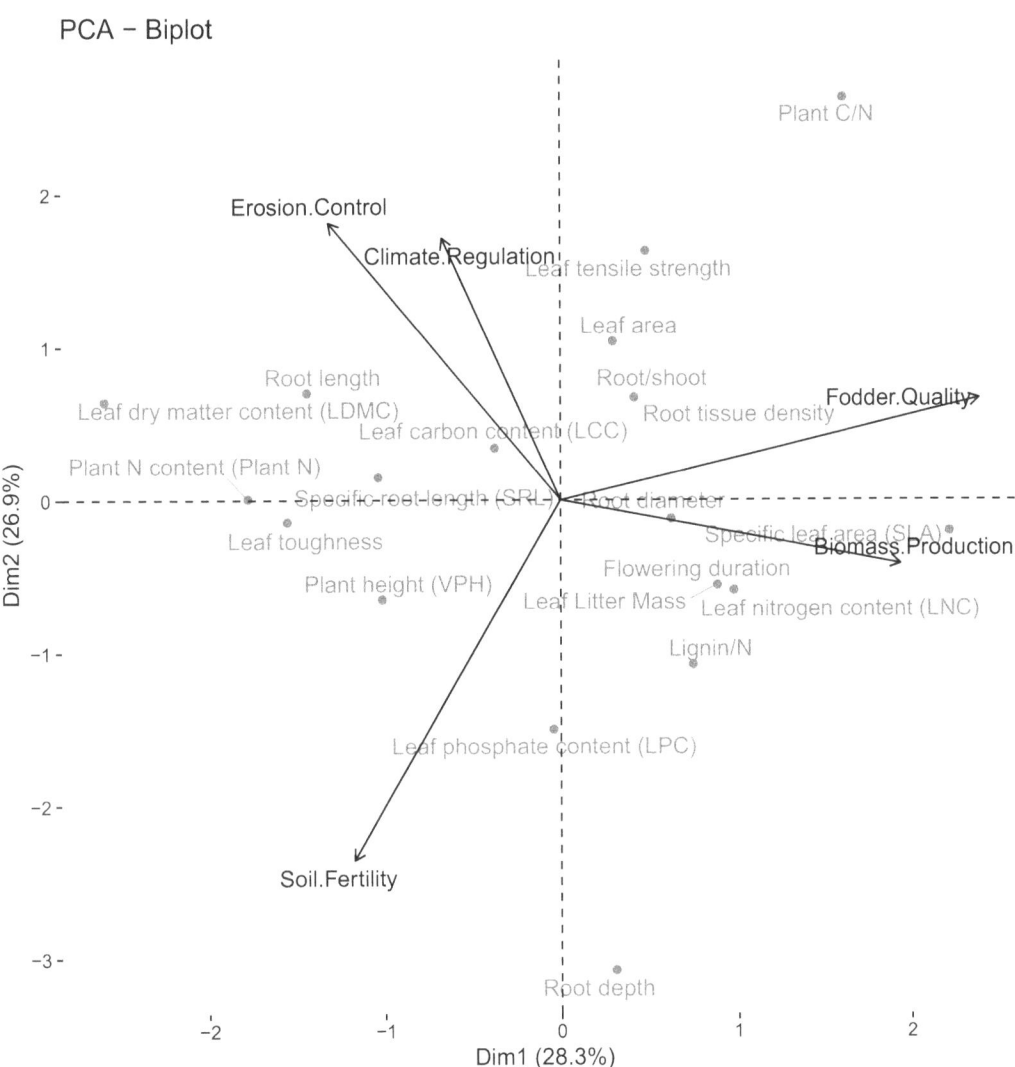

Figure 9.3 Reanalyses of the supplementary data from Hanisch et al. (2020) using a PCA analysis. The analysis, done only for visual and pedagogical purposes, implied replacing empty cells in their matrix with zeros and selecting only traits (in grey color) and services (in black color) with a greater number of entries. For a more comprehensive insight see Hanisch et al. (2020).

9.2 Assessing Biodiversity and Ecosystem Functions (BEF) Relationships

Identifying which traits are responsible for ecosystem functions could greatly contribute to our understanding of the mechanisms behind BEF relationships, and predict if and how ecosystem processes and services will change following changes in community composition (Garnier et al. 2004; Heemsbergen et al. 2004; Petchey et al. 2004; Díaz

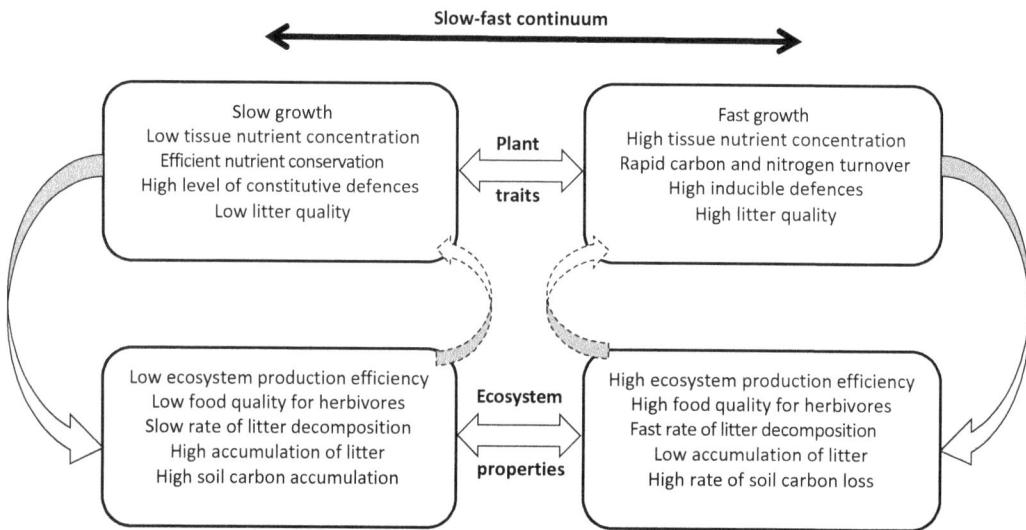

Figure 9.4 One major driver of ecosystem functioning is the difference in fundamental trade-off between slow- and fast-growing species (the 'slow–fast continuum'; see Chapter 3). Plant traits serve as determinants of the quality, quantity and fluxes of resources in the ecosystem. The effects of organism traits on ecosystem properties are represented by the large curved arrows on the outer sides of the boxes, and feedback effects from ecosystems to organisms are represented by the inner, smaller curved arrows (Bardgett & Wardle 2010; Lavorel & Grigulis 2012; Reich 2014; Ellers et al. 2018). Figure taken from Garnier et al. (2016), with permission from Oxford Publishing Limited. © OUP.

et al. 2007). Two main non-exclusive research hypotheses have emerged to explain how species traits may influence ecosystem processes (Dias et al. 2013b). The *mass ratio hypothesis* (Grime et al. 1988), also referred to as the *dominance hypothesis*, proposes that the effect of a given species on ecosystem processes is proportional to its relative abundance in the community at a given point in time. Therefore, the trait value of dominant species in a community, captured by the community-weighted mean trait value, CWM, is expected to be related to ecosystem processes rates (Garnier et al. 2004; Lepš et al. 2006; Ricotta & Moretti 2011).

Alternatively, the *complementarity hypothesis* (Tilman et al. 1996) indicates that the degree of dissimilarity in trait values between coexisting species, i.e. variation in species trait values in the community (quantified using functional diversity, FD), promotes non-additive effects on ecosystem processes. In other words, we cannot predict the effects of a single species on its own, but instead must consider the diversity of traits within the community. Non-additive effects can emerge either due to antagonistic (competition or inhibition) or synergistic (complementarity or facilitation) interactions among species, leading to more efficient utilization of resources among coexisting species (Tilman et al. 1996; Heemsbergen et al. 2004; Petchey et al. 2004; Mouillot et al. 2011). You can refer to Chapter 5 for tricks in computing CWM trait values and various FD components. It is important to remember that functional diversity can play a role in species fluctuations through time, operating on species differently, causing a decrease in synchrony and an overall increase of stability (van Klink et al. 2019). These effects can be associated with so-called compensatory effects and insurance mechanisms (McCann 2000).

To account for both the effects of CWM and FD, a mechanistic framework analysing the effect of different factors, both abiotic and biotic, on specific ecosystem properties was proposed by Díaz et al. (2007). The framework adopts a hierarchical approach, aiming at identifying the most parsimonious model combining abiotic factors, functional components of biodiversity (mostly CWM and FD for relevant traits), and key species, predicting a focal ecosystem process and service. The procedure is shown in Fig. 9.5 and consists of two steps.

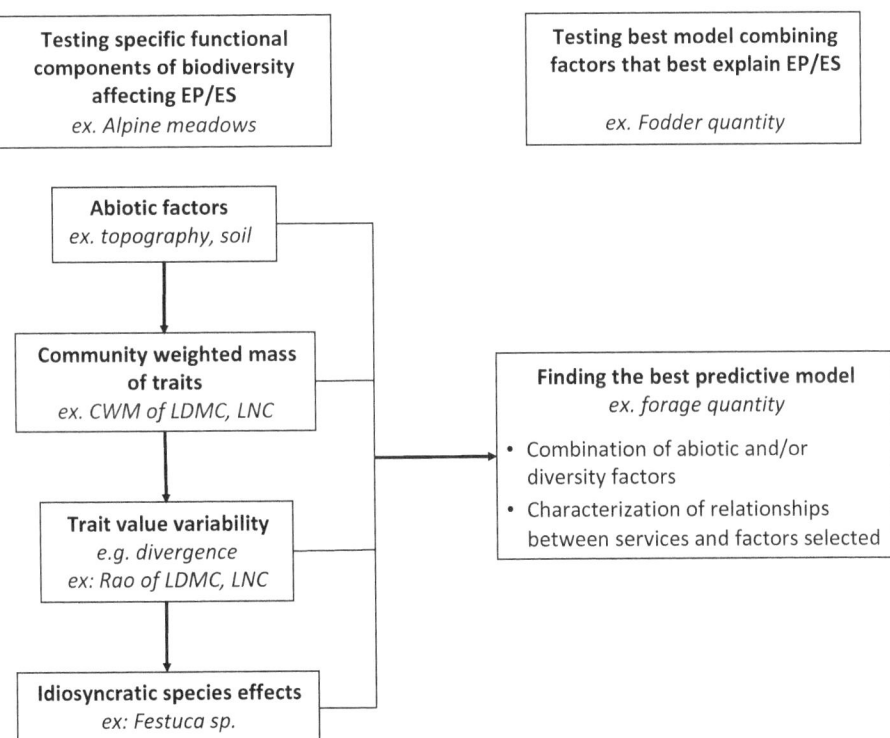

Figure 9.5 Framework proposed by Díaz et al. (2007) and simplified by Garnier et al. (2016) showing a hierarchical approach aiming at identifying the abiotic and biotic factors that best predict focal ecosystem processes (EP) and ecosystem services (ES), e.g. fodder production. In the boxes on the left, each model tests the link between EP/ES and different types of factors: abiotic factors, community-weighted mean trait values for single traits (CWM), distribution of the trait values within the each community (FD), and local abundance of each species within the community (idiosyncratic effect of single species). Significant factors are identified at each step and conserved for the final model (box on the right-hand side), which then tests the most parsimonious combination of factors predicting EP/ES. Modified from Díaz et al. (2007), with permission from the National Academy of Sciences, USA. Copyright (2007) National Academy of Sciences, USA.

First, we select predictors within each type of factor (abiotic driver, CWM value for each single trait, FD, and abundance of each species within the community) that significantly explain focal ecosystem processes (EP) and ecosystem services (ES). Secondly, the selected predictors are then combined to build a parsimonious predictive model of the focal EP/ES. By applying this framework to a study testing specific functional components affecting alpine vegetation fodder production, a key EP/ES, Díaz et al. (2007) found that prolonged delivery of green fodder was dependent upon on the combination of local abiotic factors, their effects on above-ground standing biomass, and average functional properties of the vegetation.

Ecologists have attempted to use this or similar approaches to quantify the relative importance of different biodiversity functional components on ecosystem properties and functions, mostly focusing on CWM trait values and FD-trait ranges. The studies show quite contrasting results. In a review paper, de Bello et al. (2010a) show that trait complementarity effects on ecosystem processes and services are less common than the impact of species, or functional groups and their abundance, and that they are mostly related to processes linked to primary productivity, nutrient cycling, pollination and, in particular, their maintenance through time. However, reviews could be biased by the lower number of studies considering FD effects. Also Garnier et al. (2016) and more recent studies (Lavorel et al. 2011; Ali & Yan 2017) report that ecosystem properties are primarily driven by trait dominance rather than complementarity effects. Other studies report non-significant effects of functional diversity, often in combination with both dominance and environmental effects (Díaz et al. 2007; Finegan et al. 2015). One reason for such contrasting patterns might be analytical inconsistencies, such as whether trait values are weighted or not according to their relative abundance or biomass, and the manner in which abiotic and biotic predictors are selected and processed in the models (Garnier et al. 2016).

Recently, investigations have revealed new insights into how community trait diversity drives ecosystem properties, in particular, in dry ecosystems (Gross et al. 2017; Le Bagousse-Pinguet et al. 2019). These studies suggest that species with contrasting trait values collectively exploit a greater diversity of resources and maximize multiple soil functions in drylands worldwide. In contrast, Valencia et al. (2015) found that both trait dominance (i.e. *mass ratio effect*) and diversity (i.e. *complementary effect*) were equally important drivers of multifunctionality responses to both aridity and shrub encroachment in Mediterranean dryland communities, while Chollet et al. (2014) and Peco et al. (2017) show that, beside functional attributes, local environmental conditions (such as temperature, water availability and fertilization) are important mechanisms influencing soil multifunctionality.

Summing up, research so far shows that both trait divergence and trait dominance can be important drivers of ecosystem properties, but that other factors, such as local abiotic and environmental conditions, can also affect ecosystems. The relative importance of these factors is likely to be context-dependent. In this regard, it is probable that trait dominance and complementarity effects are not mutually exclusive or independent. As such, in the next section we explain how it is possible to experimentally disentangle the effects of trait dominance from trait variation.

9.3 Disentangling Functional Traits Effects on Ecosystem Functioning

9.3.1 Designing Experiments Disentangling CWM and FD

Several attempts have been made at identifying the consequences that community changes have on the ecosystem processes which underlay key ecosystem services such as pollination, decomposition and fodder production (Beier et al. 2008; Loring et al. 2008; Carpenter et al. 2009). As we have seen in the previous section, different functional components of biodiversity, such as the mean and variation of trait values within communities, are likely to play a key role in driving ecosystem processes (Heemsbergen et al. 2004; Petchey et al. 2004; Luck et al. 2009; Mouillot et al. 2011). Up to now, however, it is still not clear what the relative role of these two functional components of communities is (Dias et al. 2013b).

Although CWM and FD express different aspects of community trait composition, they are not mutually exclusive and both explain a significant part of the variation in distinct ecosystem processes (Schumacher & Roscher 2009; Mouillot et al. 2011; Roscher et al. 2012; Butterfield & Suding 2013; Conti & Díaz 2013). The question that remains is: when, or under what circumstances, is trait mean more important than trait variation for explaining ecosystem processes? Observational and experimental studies testing the relative importance of CWM and FD for ecosystem processes have shown that it is difficult to disentangle their unique and joint contributions (Thompson et al. 2005; Díaz et al. 2007; Mokany et al. 2008; Schumacher & Roscher 2009; Lavorel et al. 2011). A major problem is that CWM and FD are often not independent (see Fig. 9.6). Ricotta and Moretti (2011) show, for example, that FD expressed with Rao diversity, one of the most widely used FD indices (see Chapter 5), and CWM are mathematically related, despite the fact they describe complementary aspects of community trait composition (Laughlin 2011; Mouillot et al. 2011). This will necessarily leave uncertainties about whether community effects on ecosystem functions are due to CWM or FD.

Let's try to understand why FD and CWM cannot be considered independent. Imagine a simple case in which the trait is nitrogen-fixing abilities in plants. In this case when CWM is highest, i.e. all plants can fix nitrogen, FD is zero. In the similar case when CWM is lowest, i.e. no plants can fix nitrogen, FD is also zero. FD is maximized when 50% of species are nitrogen fixers and 50% are not, i.e. with intermediate CWM values. As such, in general terms at highest and lowest values of CWM it is impossible to obtain a high FD, and FD is maximized at intermediate CWM values. This generally produces unimodal relationships between CWM and FD (Fig. 9.6), but in some empirical data the relation can be also almost linear (Dias et al. 2013b). It is thus unclear how a classic statistical approach can prove sufficient to assess the relative effect of two non-independent variables.

Traditional BEF experiments do not solve the problem of covarying CWM–FD values, because these experiments are designed to create a gradient of species richness, using species randomly selected from a species pool, irrespective of trait values for the selected species (Petchey et al. 2004; Meier & Bowman 2008; Mouillot et al. 2011).

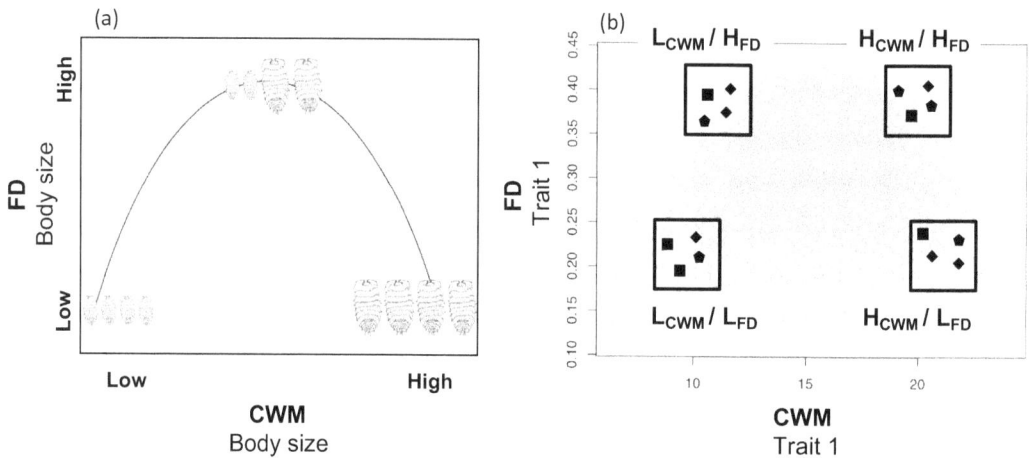

Figure 9.6 Relationship between CWM and FD for a given trait. (a) Hypothetical example of the relationship between CWM and FD for carabid beetle body size. When CWM approaches the lower (low) and upper (high) bounds of CWM body size range (the ends to the right and left of the curve), FD (here expressed as Rao diversity, but valid for other indices in Chapter 5 as well) will necessarily be low, while at an intermediate level of CWM trait value, FD body size will reach its potential maximum value at the top of the hump (Dias et al. 2013b). (b) A hypothetical example of the relationship between CWM and FD for a given (simulated) quantitative trait (trait 1). The dots represent the result of 5000 simulated combinations of CWM and FD for a given quantitative trait, resulting from the possible assemblages of 8 species out of a pool of 30 species. In order to disentangle the combined effect of CWM and FD shown in (a), four quasi-orthogonal CWM–FD areas (grey quadrats) of low–low (LL), low–high (LH), high–high (HH) and high–low (HL) can be selected. In each quadrat, four unique assemblages of distinct species combinations (symbols) are randomly selected.

However, new types of experiments are now being performed that try to disentangle the effect of CWM from FD (Tobner et al. 2016; Galland et al. 2019). We argue that to properly achieve this, it is necessary to design experiments with species assemblages reflecting an almost orthogonal design of CWM and FD effect trait values, as suggested in the framework by Dias et al. (2013b). Basically, the framework proposes plotting the possible combinations of CWM and FD (out of the potential species pool) for the focal effect trait (of a model taxa), while potentially controlling for other community parameters, such as the total biomass, total density and species richness. You will notice that the plotted points will follow a hump-shaped line as shown in Fig. 9.6a.

In order to control for the correlation between CWM and FD, and to disentangle their effects, it is necessary to identify areas insuring the most orthogonal combination of 'low' and 'high' CWM and FD values. From each area, a given number of species assemblages should be selected for use in the experiment. Fig. 9.6b shows the results when this framework is applied to an experiment with four isopod species, in which the potential effect of CWM and FD leaf litter consumption rates (a key effect trait for litter decomposition) are studied (Bílá et al. 2014). A simplified version of the framework was used by Finerty et al. (2016) in a litterbag experiment showing that mass effects mainly drive litter decomposition, irrespective of the presence or absence of an exotic species in leaf litter assemblages, as long as exotic species and native species have similar trait values.

9.3.2 Analysing Biodiversity Experiments

A quantitative approach for disentangling the effects of dominant species and complementarity on a given ecosystem function (EF) was proposed by Loreau and Hector (2001). Their method allows us to additively partition the contribution of the *selection effect* given by the dominance of species in a species mixture, and that of the *complementary effect* emerging from positive species interactions. This approach, whilst often applied, has also been widely questioned. It assumes that the EF (e.g. decomposition rate or pollination success) in a given mixture of leaves or flowers is equal to the average of the same EF in the monocultures of the component species. This is the expected amount of EF under the null hypothesis. The difference between the observed and expected mixture EF is the so called *net diversity effect*. In other words, a positive net diversity effect occurs when mixtures perform better than the averages of the corresponding monocultures. A positive *complementarity effect* occurs if species yields in a mixture are on average higher than expected based on the weighted mean monoculture yield of the component species. This can be due to a *selection effect*, i.e. the presence of a given species improves the outcome of the mixture (for example, dominant species producing most of the biomass), *niche complementarity* and facilitation (i.e. in the presence of species A, one or more of the other species in the mixture perform better than in their respective monocultures). The approach by Loreau and Hector (2001) can be summarized as follows (the corresponding algorithm is provided in the accompanying material):

Net diversity effect = Selection effect + Niche complementarity

An advantage of this additive partition approach is that it provides absolute measures of biodiversity effects, thereby allowing quantitative comparisons of the contributions made by net diversity components. Some criticisms, however, have been raised by Pillai and Gouhier (2019), who argue that the expectation of EF based on neutrality does not account for the non-linear abundance–ecosystem-functioning relationships widely observed in nature, and that much of BEF research is based on a trivial and circular expectation of over-yielding arising naturally from species coexistence. Therefore, the additive partitioning approach is likely to overestimate the positive effects of biodiversity on ecosystem functioning. In order to avoid such inflated biodiversity effects, the authors propose to at least account for that portion of the increased ecosystem functioning observed in mixtures that is merely the consequence of species coexistence. Nevertheless, this recent hypothesis still needs to be empirically tested, as stated by Loreau and Hector (2019), to continue building on the major findings from hundreds of experiments carried out during the past quarter of a century.

Going back to the *net diversity effect* approach, you should now choose an approach that allows you to determine the relative importance of selection effects and of niche complementarity by relating the two components to different functional metrics

(Cadotte 2017). The general hypothesis is that CWM will be related to selection and dominant species effects, while FD is related to niche complementarity (Díaz et al. 2007). Mixed results appear in the literature, however (Tobner et al. 2016; Cadotte 2017). Complementarity effects on the relative litter consumption rates, for single or different leaf species in combination, were found in a controlled litter decomposition experiment using two (De Oliveira et al. 2010) or more species of detritivores (Heemsbergen et al. 2004). Also Deraison et al. (2015) found complementary effects on plant biomass production in a controlled herbivory experiment using combinations of grasshopper species with different mandible strengths (Ibanez et al. 2013). In contrast, using experimental data from plant assemblages both in field plots and greenhouses, Cadotte (2017) reported positive net diversity effects on productivity driven by both selection and complementarity effects: the former was maximized in species mixtures with low functional diversity and with tall plants, while the latter was strongest in communities with high functional diversity across multiple trait axes. Also Finerty et al. (2016), using a litter bag experiment with functional homogenous versus different species assemblages found contrasting patterns, with mass effect mainly driving litter decomposition rate, while complementarity effect modulated the mass effect by increasing decomposition rate in mixtures with low mean decomposability, and decreasing mass loss in mixtures with high mean decomposability. Conversely, other authors found that selection effects primarily drive ecosystem functions (e.g. Bílá et al. 2014; Tobner et al. 2016). Bílá and colleagues tested the net diversity hypothesis by disentangling the effect of the relative importance of dominant species and their traits from the combinations of functionally different species using combinations of four isopod species in a mesocosm experiment and found that mass (selection) effect explained litter decomposition of a single leaf species (Fig. 9.7). Also Tobner et al.

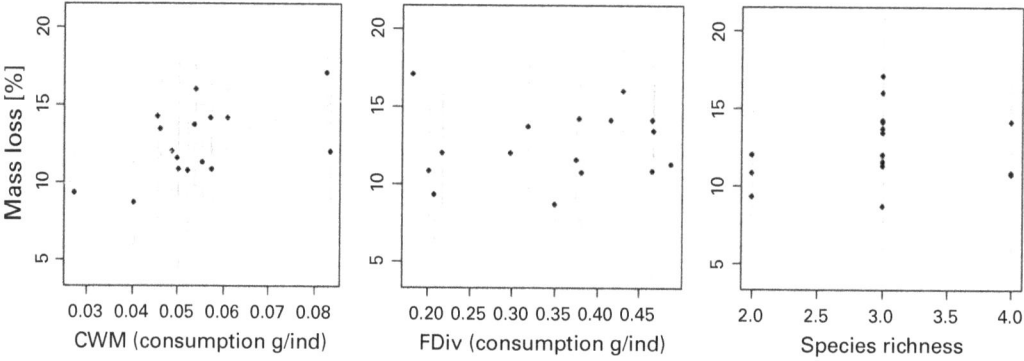

Figure 9.7 Relationships between all explanatory variables tested individually, that is, isopod community-weighted mean (CWM) consumption rate (consumption g ind-1), functional diversity calculated as functional divergence (FDiv) of consumption rates by isopods (consumption g ind-1), and species richness and leaf litter mass loss (%). Only CWM showed a significant positive effect on leaf litter mass loss (linear regression, $P = 0.032$), whereas FDiv and species richness were not significant (linear regression, $P > 0.1$). Taken from Bílá et al. (2014), distributed under a Creative Commons Attribution License (CC BY). © 2014 The Authors. Ecology and Evolution published by John Wiley & Sons Ltd.

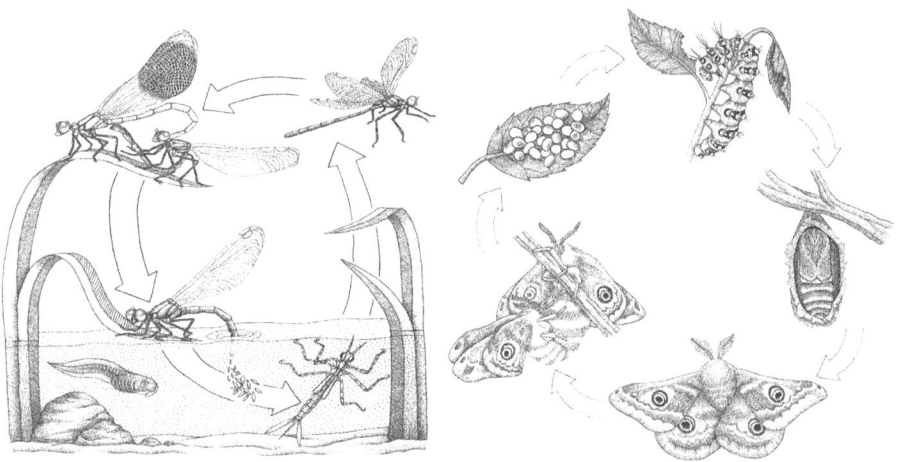

Figure 9.8 Life cycles of the dragonfly and the butterfly, and their ontogenetic habitat and diet shift from the larval stage to adult. In the case of the dragonfly, both larvae and adults are predators, but while the larvae hunt underwater, the adults hunt in flight. In the case of the butterfly, the larvae are herbivores and the adults feed on nectar. Figure created by created by Luís Gustavo Barretto.

(2016) found that tree growth in an early-stage mixture plantation experiment was mainly driven by selection effects due to the dominance of some deciduous species and the competitive suppression of most evergreen species in mixtures during the first four years.

An interesting question that remains is whether selection and complementary effects can also emerge from individual differences within the same species (i.e. intraspecific trait variation; Chapter 6). It is, for instance, known that the diet of consumers can vary significantly during their ontogeny, changing with body size, metabolism and physiology. Take, for instance, longhorn beetles. While the larvae of this family of coleopterans live in dead wood and feed on sap and microorganisms in decaying wood, the adults feed on pollen. The same happens to butterflies (the larvae are herbivores while the adults feed on nectar) or dragonflies (both carnivores, but with larvae living under water and the adult flying; Fig. 9.8).

This is called ontogenetic niche shift, an aspect that is largely overlooked in trait-based approaches (see Chapter 6) but can influence key ecological processes. For example, Fontana et al. (2019) found that leaf litter mixtures enhance decomposition beyond the additive effect of each single plant species, especially in treatments with mixed-body size classes of the detritivore *Oniscus asellus* (Isopoda). This result is explained by a shift in the diet of individuals with different body sizes, where small individuals fed on softer litter material. These results suggest that interspecific litter diversity and intraspecific consumer diversity impact net diversity effects.

9.4 The Response–Effect Trait Framework

In 2002, two plant ecologists, Sandra Lavorel and Eric Garnier, presented the response–effect trait framework (Fig. 9.9). This allows us to link shifts in community composition resulting from species response to environmental drivers, with the effect of community compositional changes on ecosystem processes.

A key idea of the response–effect trait framework is that if response and effect traits within a community are somehow linked, the impact of stress on ecosystem processes can be directly inferred (Chapter 1). This is because a shift in species composition due to changes in environmental conditions (which can be ideally predicted from specific response trait/s; Chapter 4) can have consequences for an ecosystem process if the effect trait/s are also modified (Lavorel & Garnier 2002). If we can identify the response traits an organism uses to maximize their fitness, and the effect traits they possess through which they influence ecosystem processes we can ideally evaluate the direct impact of environmental change on ecosystem processes via biodiversity.

The link between environmental changes and changes in ecosystem functions occurs when the two response and effect traits are the same or are correlated (Suding et al. 2008). For instance, high-precipitation events, hence high soil water content, might increase nutrient mineralization and additional N uptake by plants, resulting in higher N content of leaves (Wright & Westoby 2002). As some plants compete for this additional N more efficiently than others, plants will become more or less dissimilar

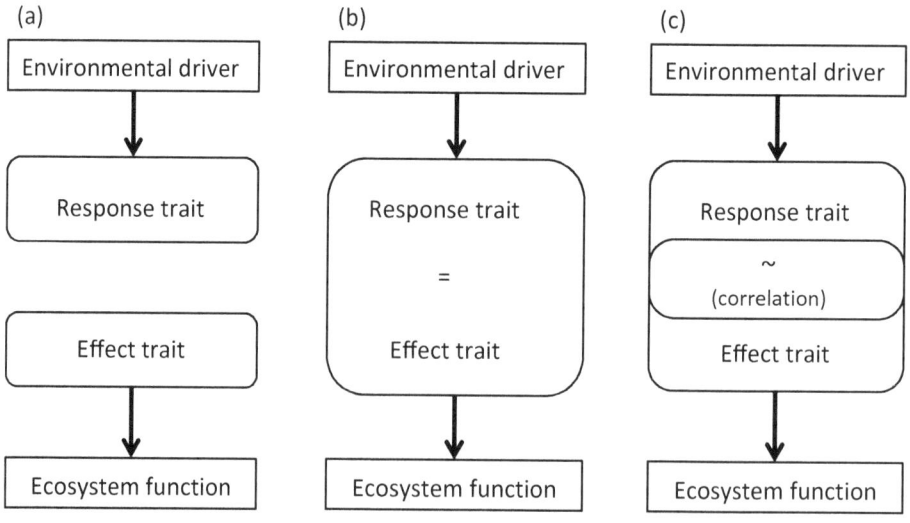

Figure 9.9 Conceptual response–effect trait framework, where a community response to environmental conditions is the result of species response traits, with their effect on ecosystem functions mediated by effect traits. There are three options: (a) no links between response and effect traits; (b) response traits directly affect ecosystem functions (response trait = effect trait); (c) correlation between response and effect traits (response trait ~ effect trait through evolutionary trade-offs). Based on Lavorel and Garnier (2002) and Lavorel et al. (2013). © 2013 International Association for Vegetation Science.

in N content (Elser et al. 2010). This might influence their sensitivity towards leaf herbivores, making species with high leaf N-content more prone to herbivory given herbivore performance is N-limited (Lu et al. 2007; Aqueel & Leather 2011). The level of herbivory, in turn, affects primary productivity, a fundamental ecosystem process (Brathen et al. 2007). Hence, leaf N content is both a response trait as it increases the growth of a plant, as well as an effect trait as it determines palatability to herbivores, which in turn affects primary productivity (Kurokawa et al. 2010).

Response and effect traits might also be correlated, which means that the response of the community to stress overlaps with the effect on the ecosystem. It results in a predictable positive or negative stress effect on the process (Lavorel & Garnier 2002). If we go back to the isopod-litter decomposition example, data show that species that are sensitive to drought (or insensitive to inundation due to a trade-off) have a rather minor effect on litter decomposition compared to less drought-sensitive species (Dias et al. 2013a). This is because drought-sensitive species are small and, as a consequence, have a low absolute consumption rate and a small effect on litter decomposition. Conversely, large-bodied species are less prone to drought stress but more sensitive to inundation. Therefore, if soils become flooded due to prolonged precipitation events, this will have a negative effect on litter decomposition because we lose the large, inundation-sensitive species which have a high consumption rate.

There are also examples where traits responding to an abiotic factor or environmental change do not link with any traits affecting the target ecosystem process. For instance, most soil fauna groups are sensitive to drought, and the reaction of soil fauna communities towards dry spells can be inferred from interspecific differences in water loss rates determining drought tolerance (Dias et al. 2013a). Earthworms are an important soil ecosystem engineer as their tunnel-digging habit affects soil pore space and structure; however, their drought tolerance is not strongly related to body size, but instead to their soil vertical stratification (Felten & Emmerling 2009; Taylor et al. 2019). However, large-bodied species have a stronger effect on soil bioturbation than small-bodies species, suggesting that drought tolerance and bioturbation do not correlate in a predictable way. Nonetheless, as mentioned earlier (Chapters 2 and 3), traits never act alone; they are often correlated with other traits. We therefore recommend checking for such correlations and testing for links with the target effect traits.

Several authors have applied the response–effect trait framework in various contexts and using different taxonomic groups. Most work has been done on plant responses to stress and how this affects primary productivity (Garnier et al. 2007; Suding et al. 2008; Minden & Kleyer 2011; Pakeman 2011; Sterk et al. 2013; Solé-Senan et al. 2017). While there are only a few such studies on vertebrates, e.g. birds in apple orchards along a landscape homogenization gradient in northern Victoria, Australia (Luck et al. 2012), there are plenty of examples for invertebrates, such as dung beetles under increasing grazing pressure across Scandinavia (Piccini et al. 2018), grasshoppers under different management regimes on Alpine pasture in France (Moretti et al. 2013), earthworms in Brazilian Amazon deforestation (Marichal et al. 2017), and collembolans in mesocosms mimicking the loss of life history groups (Eisenhauer et al. 2011). There have also been recent efforts to use this approach for the first time on microbes such as ectomycorrhizal fungi in forest ecosystems (Koide et al. 2014; Yang et al. 2019) and bacterial biofilms (Lennon & Lehmkuhl 2016) with promising results, but not yet extensively tested in BEF studies (but see Piton et al. 2020). See also Chapter 10 for applying this framework across trophic levels.

It is crucial to define the stress and ecosystem process of interest as precisely as possible when applying the response–effect trait framework, and when assessing its applicability as a predictive tool. Only then can the appropriate response and effect traits be selected by the three-step approach presented in Fig. 2.2 at Chapter 2 (adapted from Brousseau et al. 2018). While trait values can be retrieved from a database, precautions must be taken when using them to predict response–effects on ecosystems (see Chapters 1 and 2). It is possible that, at present, existing trait databases do not cover extensively potential effect traits.

We would like to enter an additional word of caution. Often the impact of functional traits on ecosystem processes is assumed but not properly tested. In plants, at least for certain processes, functional traits are seemingly often involved in stress reactions, as well as having ecosystem impacts, whereas this is not the case for animals. For instance, a plant's response to increased herbivory, i.e. via an increase in secondary metabolites (Rosenthal & Berenbaum 2012), has an afterlife effect on litter detritivores as these compounds also affect litter palatability (Hättenschwiler et al. 2005). It would be good scientific practice to validate this assumption beforehand! The link between response and effect traits can be evaluated using structural equation modelling. Or, at least, the link could be assumed when two focal traits are correlated and their correlation is biologically meaningful, or when a physiological, biomechanical or behavioural link has been reported in the literature, while the predictability of the framework can be tested by laboratory or field experiments. As we have seen in Chapter 2, there are basically two options to select traits: hypothesis- or pattern-based. In the approach outlined above, the traits are selected based on hypotheses, using our knowledge of stress reactions and their impacts on ecosystem processes. However, sometimes corresponding traits are not known. In these cases one can take a selection of traits with suspected information about response and/or effect and test these traits across stress gradients. Often, the traits that respond to stress show a consistent pattern in their underlying mechanisms.

There is an increasing number of studies testing the applicability of the response–effect trait framework, especially for plants, and by extension to animals (e.g. Moretti et al. 2013; Schmera et al. 2017) and multiple trophic levels (Lavorel et al. 2013; see Chapter 10), however the value of the framework still remains open for debate. We cannot yet adequately determine which type of stress, model organism and/or ecosystem process the framework can be applied to, so that it can function as a predictive tool.

One open question is to know whether, and to what extent, the effect traits influence response traits, and therefore have an indirect effect on the growth, reproduction and survival of the individual. While the positive feedback of effect traits on functional traits is easy to imagine for a few well-known keystone and environmental engineering organisms, such as top predators and large herbivores, for the majority of the organisms such a link is completely unknown or has been investigated very little (e.g. Ellers et al. 2018).

9.4 The Response–Effect Trait Framework

Take-Home Messages

- Response and effect traits allow us to improve our mechanistic understanding of the response of communities to environmental changes and how these can influence ecosystem processes and related services. A trait can simultaneously be a response and effect trait, but this double role cannot be taken for granted and has to be tested; in some cases response and effect traits are dissociated.
- Two main non-exclusive hypotheses have been proposed to explain how traits influence ecosystem processes: the *mass ratio hypothesis* (the dominant trait in the community, expressed by CWM) and the *complementarity hypothesis* (the variation in trait values in the community, expressed by FD). CMW and FD might affect ecosystems simultaneously, but their relative importance is likely to be context-dependent.
- To empirically disentangle the roles of CWM and FD in affecting ecosystem functions, one needs specific designed experiments with species assemblages reflecting an almost orthogonal design of CWM and FD trait values. Thereby, it is possible to disentangle these two components and to quantitatively assess the non-additive contributions of species interactions to service provision by calculating the net diversity effect based on the contribution of the single species to species mixtures.
- The response–effect trait framework allows us to link changes in community composition in response to environmental drivers, with the effect on ecosystem processes.

10 Response and Effect Traits across Trophic Levels

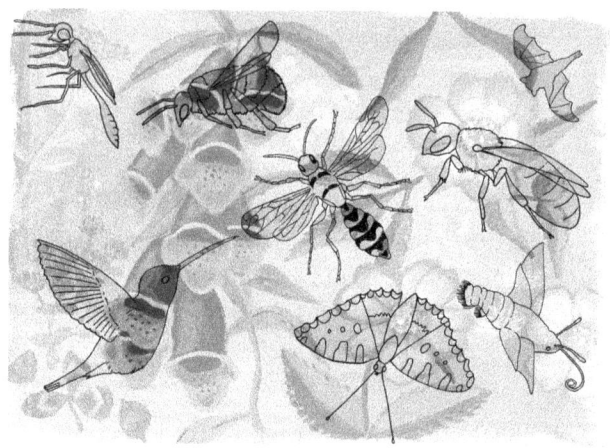

Trait-based approaches have been increasingly used to understand and predict both species response to environmental changes and their effects on ecosystem processes. As we saw in Chapter 9, the response–effect trait framework (Lavorel & Garnier 2002; Suding et al. 2008) has been proposed as a tool to unify both response and effect perspectives. By identifying the linkages between response and effect traits, it is possible to predict whether shifts in community composition due to environmental changes are likely to promote changes in effect traits with consequences for ecosystem functioning. So far, this approach has been applied mainly within communities of organisms belonging to the same taxonomic or functional group, like plants, pollinators or detritivores (Larsen et al. 2005; Gross et al. 2008; Chapter 9). However, these groups of organisms do not stand alone in nature. They interact with organisms at different trophic levels and, through these trophic interactions, they structure community assemblages, which, in turn, provide key ecological processes and services (Schmitz 2010).

Working simultaneously with organisms belonging to different trophic levels presents evident challenges for ecological research. The first challenge is to have the basic biological knowledge about taxonomical groups as distinct as plants, vertebrates, invertebrates, fungi etc., which are often involved in intricate interactions that are essential for maintaining biodiversity and providing ecosystem functions. The second challenge is to identify and quantify the realized species-specific (or even individual–individual) interactions, which need to be distinguished among hundreds, or even thousands, of potential interactions that can take place in a single community. These challenges are old companions of community ecologists, but novel techniques (e.g. barcoding and other molecular diet markers) are providing interesting new perspectives in this field. The third challenge, which we analyse with particular attention in this chapter, is specifically defining trait-matching between organisms from different trophic levels, i.e., selecting traits linked to trophic interactions and including them in different types of analyses. Multitrophic interactions have been the core of different analytical frameworks, but only recently have traits been directly considered. In this chapter, we will explore trait-based approaches applied to trophic interactions. We discuss how these approaches can help us to predict species interaction and resulting ecological processes and services by recognizing general rules driving species' feeding associations and their consequences for maintaining biodiversity and ecosystem functioning. See the 'R material Ch10' for further details.

10.1 Multitrophic Controls on Ecosystem Functioning

As we have shown in the previous chapter, there is strong evidence that species' traits determine, to a great extent, the rates of ecosystem processes and the services provided by them. While much emphasis has been placed on plants due to their prominent role as biomass primary producers in most ecosystems (Garnier et al. 2016), the majority of ecosystem processes result from the combined actions of organisms from different trophic levels (Kremen et al. 2007; de Bello et al. 2010a; Lavorel et al. 2013). For instance, organic matter decomposition and nutrient mineralization depends on the traits of plants, microorganisms and soil invertebrates. In an extensive reassessment of the contribution of traits to the provision of ecosystem services, de Bello and colleagues (2010a) introduced the concept of *trait–service clusters*, which result from the associations between multiple traits and multiple ecosystem services, and they showed that these clusters operate across different trophic levels. They specifically showed that the provision of ecosystem services depends on the traits of species belonging to different trophic levels, as in the case of pollination, biological control, water purification and biogeochemical processes. At the same time, a given trait often influences more than one service. Similarly, Harrison et al. (2014) used network diagrams to show that the delivery of several ecosystem services depends on an intricate set of associations with the traits of different types of organisms.

Even in processes in which plants predictably exert a strong control, the functional composition of other organisms can further substantially drive and modulate ecosystem functioning. For instance, plants exert a strong control over soil biogeochemical processes due to, among other traits, the chemical and physical traits of plant litter (Cornwell et al. 2008; Freschet et al. 2012; see also the extensive field experiments across five terrestrial and aquatic ecosystems by Handa et al. 2014). On the other hand, Grigulis et al. (2013) showed that both plant and microbial traits jointly affect ecosystem properties. In that study, considering microbial functional traits in addition to plant traits substantially increased the explained variation of soil organic matter content, potential N mineralization and potential leached N. Additionally, species at different trophic levels often show strong contrasts in their biology (e.g. sessile vs mobile organisms, or endothermic vs ectothermic organisms) and, therefore, are expected to respond differently to environmental changes (Concepción et al. 2017). This increases the risk of disrupting interactions and mismatching between trophic groups, with detrimental effects on the provision of ecosystem processes (Berg et al. 2010). In this way, shifts in ecosystem processes and services that rely on multitrophic interactions might not be properly predicted from the response of a single trophic level to environmental changes. For instance, temperature-induced changes in phenology were shown to disrupt the oak–winter-moth–great-tit association (Visser & Holleman 2001). Earlier winter-moth egg hatching as a response to climate changes leads to a mismatch with the bud burst of oak trees. This results in the starvation of winter moth larvae, which are an important food source for great tit hatchlings. A reduced recruitment in great tit population due to the mistiming between oaks and moths can potentially compromise the service of biological control provided by this bird species (Mols & Visser 2002). Also, in a long-term warming experiment covering nine growing seasons in Swedish Lapland, plant and soil fauna showed marked differences in sensitivity to environmental changes. While vegetation composition and structure show no response to warming (Richardson et al. 2002), the abundance of soil micro-arthropods decreases sharply along with a shift in the community-weighted mean values of traits related to drought tolerance (Makkonen et al. 2011). Such changes in the soil arthropod community can strongly affect soil processes (Berg & Bengtsson 2007; David 2014), even if the plant community remains unchanged. Berg et al. (2010) show that plants, herbivores, macro-detritivores, predators and microbivores exhibit marked differences in sensitivity to temperature changes and potential dispersal capability. Such differences in responses to environmental changes are expected to lead to disruption of community interactions, with potentially strong compromises to biodiversity support, ecosystem functioning and service provisioning. Therefore, it is important to consider different trophic levels and their interactions to improve predictions about the provision of ecosystem services under environmental changes.

Predicting shifts in interactions due to a changing environment and subsequently mapping the consequences for ecosystem functioning remains a challenge. There is increasing hope that combining a multitrophic perspective with a trait-based approach can improve our ability to predict species distribution in space and time and, thereby, identify mechanisms that drive biotic control over ecosystem service delivery (Lavorel et al. 2013; Moretti et al. 2013; Schmitz et al. 2015; Brousseau et al. 2018). Classic studies on trophic interactions have focused on quantifying 'who eats whom' and 'how much'. In this chapter, we present recent approaches attempting to use traits to elucidate mechanisms behind trophic interactions, both at the species and community levels, and their consequences for ecosystem functioning (Schmitz 2010; Ibanez et al. 2013; Lavorel et al. 2013; Schleuning et al. 2015; Brousseau et al. 2018).

10.2 The Multitrophic Response–Effect Trait Framework

The response–effect trait framework (see Chapter 9) assesses the effect of environmental change on ecosystem functioning through community dynamics (Lavorel & Garnier 2002; Suding et al. 2008). The framework was originally proposed for a single trophic level, but an extension of the response–effect trait framework including multiple trophic levels was developed by Lavorel et al. (2013). Such an approach, as described here, can help in moving from qualitative to quantitative predictions of ecosystem services relying on trophic interaction and their responses to environmental changes. Unlike studies on food webs and interaction networks, this approach does not require quantification of complex species-to-species trophic interactions. Different trophic levels are represented as compartments of the ecosystems for which it is possible to calculate aggregated functional properties, such as community average trait values and functional diversity (see Chapter 5), that reflect both their responses to the environment and their effects on the ecosystem. While this can be considered a methodological advantage, as specific interactions are hard to quantify, such a simplification can also be considered a limitation of this approach because grouping species in trophic levels neglects important properties of food webs that can influence ecosystem stability (see section 10.3).

The main novelty of the multitrophic response–effect trait framework is to identify traits responsible for controlling multitrophic interactions, which are called response and effect trophic traits. In the case of response trophic traits, we focus on those traits through which organisms depend on another trophic level, such as proboscis length in pollinators in response to the type and depth of flowers in vegetation. In the case of effect trophic traits, we focus on the traits of one trophic level that affect other levels, such as flower corolla length affecting the type of pollinators that can feed on the nectar of the flower (Ibanez 2012) or prey cuticula toughness affecting which carnivores can predate on them based on the proper mandible strength (Brousseau et al. 2018). It is important to notice, though, that the classification of a given trait as a response or effect trophic trait depends on which trophic level is directly affected by the external driver (as for a single trophic level, as explained in Chapter 9). Therefore, the main external drivers need to be identified in order to determine which trophic group is expected to change in functional composition and, by doing so, affect the other trophic level(s) (Fig. 10.1). For instance, if predator functional composition changes due to species invasion, prey cuticula toughness could be considered as a response trophic trait enabling some species to escape from new predators. Nevertheless, from an evolutionary perspective, defining which trait is affecting and which is responding to the other trophic level is difficult, as some traits involved in trophic interactions are shaped by coevolutionary processes (see section 10.3 for the concept of matching traits, which is embedded in an evolutionary perspective). The response and effect trophic traits can be used to identify and quantify functional links, which can connect primary producers and consumers (Lavorel et al. 2013). In this way, it is possible to identify how shifts in one trophic level, due to environmental changes, can affect other trophic levels and their

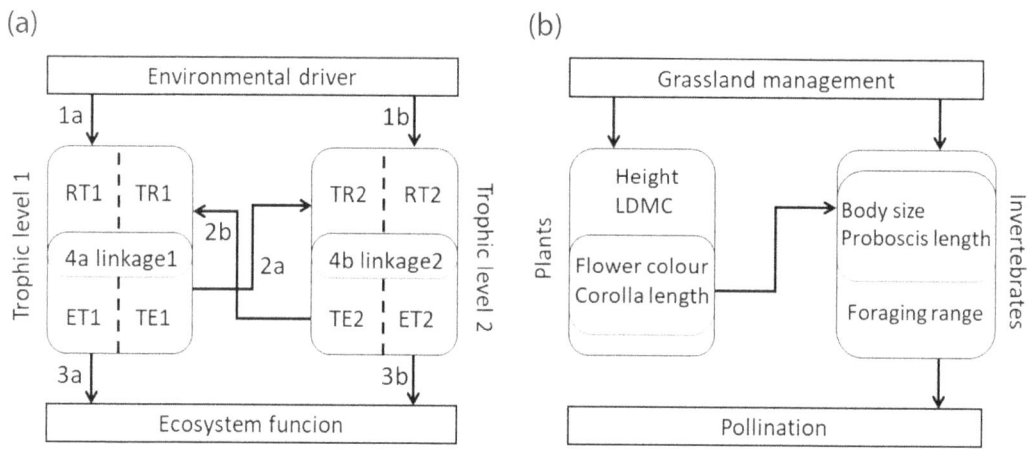

Figure 10.1 Multitrophic response–effect framework of Lavorel et al. (2013) to investigate trait linkages across trophic levels. (a) General framework in which two trophic levels respond to an environmental driver and affect ecosystem function. ERT: environmental response traits; TET: trophic effect traits; TRT: trophic response traits; and EET: environmental effect traits. Numbers refer to analytical steps described in the text. (b) Example of pollination as a function of trophic interactions responding to grassland management. Adapted from Lavorel et al. (2013) and Moretti et al. (2013), with permission from Wiley. © 2013 International Association for Vegetation Science.

consequences for the provision of ecosystem services. See the 'R Material Ch10' for two approaches linking traits to species interactions, one at the species level and another at the community level.

Fig. 10.1a illustrates a conceptual example of the response–effect trait framework with two trophic levels. Organisms in each trophic level respond to an environmental driver through their environmental response traits (ERT; 1a and 1b). The first and lowest trophic level affects the second and higher trophic level via its trophic effect traits (TET; these effects are perceived by the second trophic level via its trophic response traits, TRT; 2a). Conversely, the second trophic level can feed responses back to the first level, thus also affecting the first trophic level via trophic effect traits (2b). Both trophic levels can affect the ecosystem function via their environmental effect traits (EEF; 3a and 3b). Possible linkages between the response and effect traits within trophic levels will indicate the likelihood of the environmental driver impacting ecosystem functioning (4a and 4b) if the two trophic groups are strongly interacting. This framework has a modular nature, and new trophic levels can be added or removed depending on the environmental driver and ecosystem function of interest.

In their work proposing the multitrophic response–effect trait framework, Lavorel et al. (2013) use an exemplary case study to show how the framework can be used to identify traits linking grassland management intensity to pollination efficiency (Fig. 10.1b). In the example, the intensification of management leads to communities dominated by species with lower plant height and leaf dry matter content, as well as a reduced contribution of legumes to the vegetation. Such shifts, including in the phylogenetic composition of the vegetation, promote changes in the colour and length of flower corollas present in the vegetation, which are known to influence the community of pollinators. This linkage between plant environmental response traits and trophic effect traits due to phylogenetic constraints can have important consequences for the provision of pollination services that are not predicted by the direct effects of management intensity alone. Similarly, Perović et al. (2018) use the multitrophic response–effect trait framework in a qualitative way, framing a literature review to identify traits and trophic interactions related to the provision of biological control of pests in agroecosystems. They show that management intensity is intrinsically related to an enhanced disturbance regime, favouring communities dominated by small, generalist arthropod predators. At the same time, management intensification leads to a simplification of flower traits related to attraction, accessibility to floral resources, blooming period and nutritional suitability of the provided resources. This homogenization of floral traits has negative effects on most groups of natural enemies of herbivores, including hoverflies, hymenopterans, dipteran parasitoids and spiders, which partially rely on floral resources (Nyffeler et al. 2016). In this way, the environmental response traits of both plants and natural enemies to management intensification are strongly linked to the provision of biological control.

From the above-mentioned examples, it is clear that identifying linkages between response and effect trophic traits is essential to understand and predict how shifts in biodiversity due to environmental changes will affect ecosystem functioning across trophic levels. A recent review shows that it is possible to identify a few 'key functional traits' that act as both response and effect traits for plants, invertebrates and vertebrates (Hevia et al. 2017). If response and effect traits covary (i.e. trait syndromes) they can transmit the community response signal to the ecosystem. This opens the possibility to focus on a limited number of traits when monitoring biodiversity to identify potential shifts in ecosystem functioning. However, the studies compiled by this review focus on each of these taxa separately, limiting the range of possible inferences that can be drawn on how trait shifts in one trophic group can influence other trophic levels and the consequent effects on biodiversity and ecosystem functioning. The next important step is to identify traits involved in trophic interactions that also act as response and/or effect trophic traits.

10.2.1 Adding Numbers to the Framework

Adding numbers to the multitrophic response–effect trait framework (Fig. 10.1) could be achieved in various ways. Moretti and colleagues (2013) provide the first quantitative operationalization of the framework using a dataset on plant and grasshopper traits for modelling the effects of contrasting grassland management regimes on productivity. They upscaled plant and grasshopper traits to the community level by calculating two community trait metrics for each trait at both trophic levels: community-weighted mean trait value (CWM) and functional diversity (FD), using Rao diversity for this purpose (see Chapter 5 for more metrics details). To add numbers to the framework, they performed partial analyses using different subsets of variables in a step-by-step manner to untangle relationships between land use, plant and grasshopper traits and their contributions to primary productivity. This is a good example of how to use a multistep analysis comprising different uni- and multivariate regressions for identifying direct and indirect effects between the different components of the trait framework across trophic levels.

This analytical approach comprises four steps, as indicated in Fig. 10.1a. First, Moretti and colleagues (2013) identified the environmental response traits of both plants and grasshoppers that significantly react to changes in management regimes. Second, they identified trophic trait linkages by testing for grasshopper traits that respond to plant trophic effect traits. This was done in a multivariate manner using a redundancy analysis for selecting which plant trait metrics (TET) determined the assemblage of grasshopper traits. In turn, the selected set of plant traits was tested on each grasshopper trait metric (TRT) to identify which grasshopper trait specifically responded to vegetation compositional changes. The third step consisted of testing which plant and grasshopper trait metric (EET) directly affected plant productivity. The fourth step, the identification of linkages between environmental response and effect traits, consisted of establishing trait metrics that both respond to management regimes or other trophic levels and affect productivity.

The multistep approach for selecting the responsive trait metrics for the different trophic levels can then be further tested for cause-effects using structural equation models (SEM) (Lavorel et al. 2013), which are a powerful tool for testing complex hypotheses accounting for multiple causal relationships (Shipley 2000). Fig. 10.2 shows the path graph of the relationships between the selected variables in the above example with plants and grasshopper traits affecting primary productivity. The hypothesized direct and indirect interactions, linking environmental changes to ecosystem processes

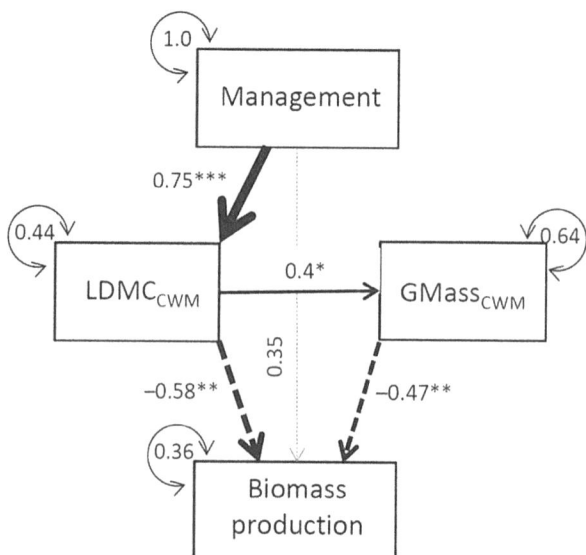

Figure 10.2 Path diagram showing the hypothesized causal relationships between management, plants and grasshopper traits and primary productivity. Community weighted mean (CWM) leaf dry matter content (LDMC$_{CWM}$) and grasshopper body mass (GMass$_{CWM}$). Taken from Moretti et al. (2013), with permission from Wiley. © 2013 International Association for Vegetation Science.

through community dynamics, can be tested using SEM. Taking the example of Fig. 10.2, grassland management only affects primary productivity via changes in community-weighted mean leaf dry matter content (LDMC). LDMC shows a direct negative effect on primary productivity (standardized path coefficient $\beta_{(LDMC,\ productivity)} = -0.58$). Additionally, higher LDMC leads to an increase in the community-weighted mean body mass of grasshoppers ($\beta_{(LDMC,\ GMass)} = 0.40$), which, in turn, negatively affects primary productivity ($\beta_{(GMass,\ productivity)} = -0.47$). Such an indirect effect of LDMC on productivity via grasshopper body mass can be calculated as $0.40 \times -0.47 = -0.19$. Therefore, the total negative effect of LDMC on productivity ($-0.58 + -0.19 = -0.77$) is stronger than what could be estimated by evaluating the shifts in plant community composition alone. This illustrates how including multitrophic interactions can also increase our accuracy when modelling the effects of environmental changes on ecosystem processes. See the 'R Material Ch10' for more details on how to use structural equation models to test the multitrophic response-effect trait framework.

We do not claim that this is the only approach for adding numbers to the framework in Fig. 10.1; surely, different ones can also be delineated using species-level analyses, as opposed to community-level analyses (see Chapter 4). Complementary laboratory and field experiments can be very useful for testing which traits affect species interactions. For instance, Brousseau et al. (2018) showed that predator biting force and prey cuticular toughness are better traits to describe pairwise predator-prey interaction in soil arthropod communities as compared to the often used predator-prey size ratio.

10.2.2 Intraspecific Trait Variability in Trophic Cascades

Most of the studies using the response-effect trait framework focus on how trait differences between species can affect ecosystem functioning after species filtering (Díaz et al. 2013; Lavorel et al. 2013; Moretti et al. 2013; Perović et al. 2018). However, organisms can respond to a given environmental driver by shifting their trait values, i.e. by intraspecific changes within and between generations (see Chapter 6). This source of trait variation can be both plastic, when individuals or genotypes change their trait values under new environmental conditions, or genetic, when new environmental conditions filter out individuals with given trait values from the population, leading to a shift in the average value of the selected hereditable trait. There is some evidence that intraspecific trait variability can substantially affect ecosystem functioning via alterations in the result of community interactions (Schweitzer et al. 2004; Miner et al. 2005; Schmitz 2008; Palkovacs et al. 2009), which highlights the importance of considering this source of trait variation in multitrophic response-effect trait frameworks. We thus recommend considering the tools described in Chapter 6 to compute indices needed for the framework described in the previous section.

Most traits show some degree of intraspecific trait variability. Animal behaviour, for instance, can be very plastic, allowing individuals to respond almost immediately to changes in the environment (Wong & Candolin 2015). Schmitz (2008) shows that grasshoppers shift their diet from the preferred plant species to other species that offer better protection against predation. This shift in grasshopper food preference has a substantial impact on plant diversity, vegetation composition and, consequently, on primary productivity and nitrogen mineralization. Behavioural traits are involved in many aspects of individuals' interactions with the environment and other species, such as maximizing forage gains, minimizing predation risk and adjusting to or seeking suitable abiotic conditions. Therefore, the plasticity of animal behaviour is likely to play an important role in cascading effects through trophic interactions (Schmitz et al. 2015). The phenological traits of both plants and animals are also very plastic, commonly showing rapid response to environmental changes. However, the often distinct phenological responses of different trophic levels (Berg et al. 2010) are likely to disrupt trophic interactions, with detrimental effects to biodiversity maintenance and ecosystem processes (Visser & Holleman 2001).

Rapid evolutionary changes can also cascade through trophic levels, affecting ecosystem functioning. We now know that contemporary evolution is much more common than previously thought (Post & Palkovacs 2009; Schoener 2011). Although there are few complete descriptions of a causal chain linking trophic interaction, phenotypic selection and their consequences for ecosystems, some study cases show a potentially strong impact of evolutionary changes on ecosystem functioning (Rudman et al. 2017). Palkovacs et al. (2009) experimentally tested if predation pressure can shift life history traits in a prey population, which in turn determine changes in ecosystem processes, using Trinidad guppies as a model system. They show that high predation pressure selects guppy phenotypes that reach maturity faster and at smaller size. This phenotype contributes nearly double the amount of N and P to the nutrient pool via excretion, leading to a significant increase in primary productivity as compared to low-predation populations. The growing selection pressure of anthropogenic drivers on organism phenotypes urgently calls for appreciation of contemporary evolution's effects on ecosystems and services they provide (Rudman et al. 2017).

10.3 Response and Effect Trophic Traits in Interaction Networks

The study of plant–animal interactions has traditionally focused on contrasting scales of investigation. On the one hand, detailed studies including one or a few focal plant species provided knowledge about the detailed mechanisms behind different plant–animal interactions, including pollination, seed dispersal and herbivory, as well as on coevolutionary processes shaping such interactions (e.g. Johnson & Steiner 1997). On the other hand, studies with an ecosystems perspective often aggregate species into relatively coarse trophic levels, allowing the study of fluxes of energy and matter through these different compartments of the ecosystem (de Ruiter et al. 1998). These different scales of investigation can be merged by using an interaction network approach, which allows upscaling of the study of pairwise interactions to account for whole communities. Building community-wide interaction networks (e.g. food webs, pollination and seed dispersal networks) represent an important line of research to investigate factors determining the structure of such networks and their consequences for the persistence of biodiversity and for ecosystem functioning (Pascual & Dunne 2006; Bascompte & Jordano 2007).

In interaction networks, species are seen as nodes which are linked to each other by one type of interaction. Plant–animal interactions are often depicted as bipartite networks which comprise two sets of nodes that interact between, but not within, sets (Fig. 10.3). The bipartite representation explicitly illustrates the reciprocity of the interactions, helping to describe and understand the structure of plant–animal interaction networks (Bascompte & Jordano 2007). Much attention has been paid to the importance of network structure to network robustness, i.e. the probability of secondary species' extinctions after species loss (Memmott et al. 2004). For instance, in some networks, species with few links have a subset of links of other species (nested networks), while other networks are divided into relatively independent sub-networks (modular or compartmentalized networks). In highly nested networks, specialist species are functionally redundant, as they interact with a subset of the species interacting with generalists. On the other hand, modular structure leads to complementarity of species belonging to different modules, while species are redundant within modules (Lewinsohn et al. 2006; see also Fig. 10.3). Cascading extinctions are less likely in modular networks, as effects of species losses are expected to be limited to the original compartment. Nevertheless, the stability of the network strongly depends on the sequence of species' extinction. As we will discuss below, traits can play an important role in determining both the structure of the network and the sequence of species'

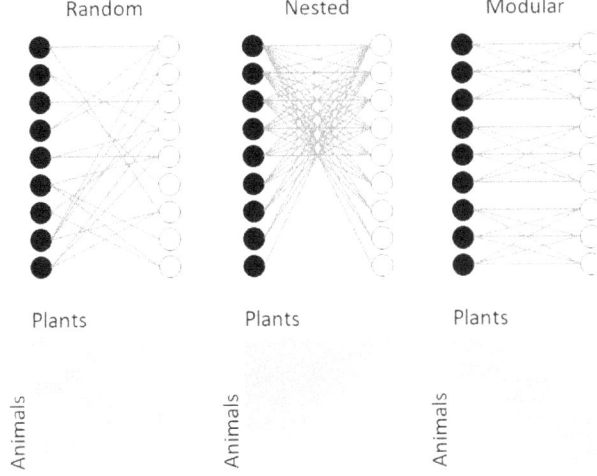

Figure 10.3 Different network structures represented as bipartite networks (top panels), where plants (black symbols) interact with one or more animal species (white symbols), and corresponding interaction matrices (bottom panels). In interaction matrices, plants are represented in columns and animals in rows. Interactions between a given plant and animal species are represented by a shaded cell in the intersection of the respective line and column. Modified from Lewinsohn et al. (2006), with permission from Wiley. © OIKOS. Published by John Wiley & Sons Ltd.

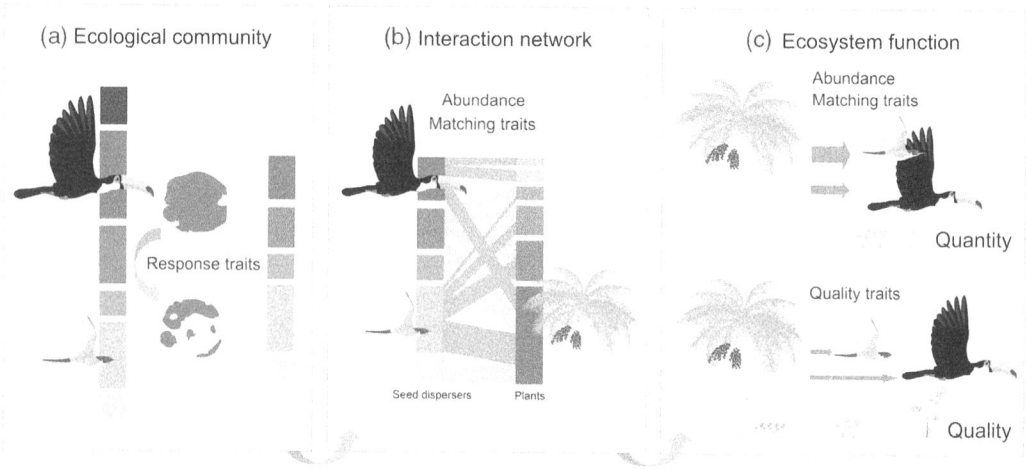

Figure 10.4 Scheme of three sequential processes that are influenced by different trait types: assembly of an ecological community, structuring of an interaction network and provision of an ecosystem function. Interactions between plants and seed dispersers are used as an example. (a) Community assembly is the result of environmental filtering and, therefore, is determined by species' environmental response traits. (b) Species density and matching traits between species pairs determine the interaction frequencies and network topology. (c) While species abundance and matching traits determine the quantitative component of the interaction (i.e. seed removal), quality traits determine the qualitative component of the interaction (e.g. dispersal distance). Both quantitative and qualitative components ultimately shape the services provided by seed-dispersers to plant species. Taken from Schleuning et al. (2015), with permission from Wiley. © 2014 The Authors.

extinction and, therefore, are determinant for the robustness of interaction networks and the stability of ecosystem functions they provide.

Despite the widely accepted idea of plant–animal interactions being strongly constrained by both plant and animal traits, the fields of functional ecology and interaction networks remain mostly disconnected. Recently, Schleuning et al. (2015) proposed an extension of trait-based concepts to plant–animal interaction networks. They define two types of traits that are important for the structuring and functioning of interaction networks: *matching traits* and *quality traits* (Fig. 10.4). Matching traits are those that modify the interaction probability for plant–animal pairs, strongly determining the structure of the network. The matching between interaction partners can be modulated by morphological traits (e.g. fruit size vs mouth-gape size, corolla size and shape vs body size and shape; leaf traits vs mandible traits; Ibanez 2012, 2013; Bartomeus et al. 2016); chemical traits (e.g. fruit chemical and nutritional composition and interactions with frugivores; Sebastián-González 2017) and phenological traits (e.g. flower and fruit set period vs activity period of pollinators and herbivores; Visser & Holleman 2001). Quality traits are related to the quality of the service provided by an animal for a plant species. These traits determine the per-interaction effects of animals on a given ecosystem function. For instance, the body size of bees is positively related to pollination efficiency (pollen deposition per flower visit; Larsen et al. 2005). It is important to notice that there is a correspondence between the concepts of matching and quality traits and the multitrophic response–effect trait framework. We can consider matching traits to include both trophic response and effect traits.

The framework proposed by Schleuning et al. (2015) extends the classical (Elmqvist et al. 2003; Naeem & Wright 2003; Suding et al. 2008) and multitrophic (Lavorel et al. 2013) response–effect trait frameworks by explicitly incorporating the structure of the interaction network. For this, it is necessary to recognize and integrate three sequential processes, all of which are influenced by different trait types (Fig. 10.4). First, the presence and the abundance of potentially interacting species within a community can be restricted by environmental filtering (see Chapter 4). This process is modulated by response traits to restrictive environmental conditions like tolerance to drought or extreme temperatures. Second, the formation of an interaction network is modulated by morphological, chemical and phenological traits, increasing the interaction probability of pairs of species with matching traits and restricting the interactions of pairs of species with non-matching trait combinations (forbidden links, e.g. anti-predatory features, such as warning signals or defensive secretions; Olesen et al. 2011). This strongly determines the structure of the interaction network. Third, the ecosystem functions provided by plant–animal interactions are determined by both quantitative (how frequent interactions are) and qualitative (how good interactions are) components. In this way, the product of quantitative (matching traits) and qualitative (quality traits) effects determines the functional importance of a species in the ecosystem. As in other response–effect trait frameworks (Naeem & Wright 2003; Lavorel et al. 2013), the linkages between response and effect traits have fundamental consequences for the stability of the system. However, depending on the structure of the interaction network (and the linkages between traits), the effects of environmental changes can be either reinforced or buffered when cascading through trophic levels.

The sequence of species loss due to environmental changes strongly depends on species' environmental response traits, while the consequences of species loss for ecosystem functioning depend on response-matching-quality relationships (Schleuning et al. 2015). When traits related to the ability to withstand harsh conditions are negatively correlated with, or are the same traits determining, the effect on ecosystem processes, we expect a strong effect of species loss on ecosystem functioning. For example, large seed-disperser birds are more susceptible to extinction due to land-use changes or hunting (Galetti et al. 2013), while body size is also positively related to fruit removal (Muñoz et al. 2017) and long-distance dispersal (Díaz et al. 2013). Therefore, land conversion and increasing hunting pressure can cause the loss of large seed dispersers and strongly compromised long-distance dispersal. This impact can be enhanced in nested networks where large species are generalists and interact with many plant species (Fig. 10.5; see Correa et al. 2016 for an example of seed dispersal by fish). However, when large species are specialists (Farwig et al. 2017) or the seed-dispersal network has a modular structure (Donatti et al. 2011; Fig. 10.5) the initial effects of species loss can be buffered. To the contrary, when (effect) quality traits are positively related to response traits, we can expect higher stability of ecosystem processes facing species loss. For example, Bommarco et al. (2010) showed that large bees are more

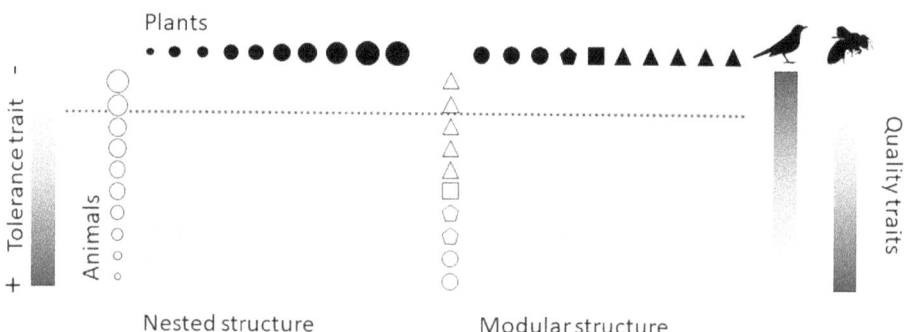

Figure 10.5 Possible consequences of a correlation between environmental response traits (e.g. tolerance traits to abiotic conditions), matching traits and quality traits. Matching traits of animals (open symbols) and plants (closed symbols) determine the structure of the interaction network. The two interaction matrices with animals as rows and plants as columns display realized interactions in grey. The dashed line shows a hypothetical scenario of environmental change in which only very sensitive animal species (above the dashed line) are extinct. This leads to contrasting consequences for the two networks. The loss of sensitive animal species has a strong impact on the plant community in the nested network, while the modular network is more resistant to the extinction of sensitive species due to redundancy within modules and complementarity across modules. Figure inspired by Schleuning et al. (2015).

resistant to fragmentation as compared to small bees, while large bees also display higher pollination efficiency (Larsen et al. 2005) and foraging ranges (Greenleaf et al. 2007). In this way, fragmentation could lead to extinction of small bees with little consequence for pollination rates at the community level.

Developing tools to quantify the direct contribution of traits for shaping the structure of interaction networks (Rafferty & Ives 2013; Bastazini et al. 2017) and pairwise interactions (Spitz et al. 2014; Krasnov et al. 2016) is fundamental for identifying matching traits and to further integrate the study of interaction networks with trait-based approaches. These analytical tools can be used for testing trait structures, as hypothesized in Fig. 10.5, for instance, with higher species trait similarity within network modules and lower similarity between different modules. Despite the notorious direct role of matching traits in structuring interaction networks, response traits can have an important indirect effect on network structure by determining the occurrence, timing and abundance of species that can potentially interact. Phenological overlap is a necessary condition for interaction to take place, and species abundance influences encounter probability and, consequently, frequency of interaction (Blüthgen 2010; Bartomeus et al. 2016), shaping network topology and interaction strengths. Such abundance-driven effects have long been recognized as important factors structuring interaction networks (Blüthgen 2010) and are often referred to as neutral mechanisms (Krishna et al. 2008). While abundance-driven effects can be considered neutral when strictly focusing on species matching within a community, the occurrence and abundance of species are, to a great extent, determined by species' traits in response to their environmental context, relying, therefore, on niche differentiation (Schleuning et al.

2015). This broader perspective is central for further integrating the interaction network and trait-based approaches.

Methods for comparing networks along environmental and resource gradients are advancing quickly (Pellissier et al. 2018) and have a strong potential to contribute to fully addressing the role of traits in shaping species interactions. The dissimilarity among networks can be divided into two components, one attributable to the turnover in species composition and the other to the turnover in interactions (Poisot et al. 2012). In this way, explicitly addressing the role of traits in these two components would allow researchers to infer interactions in the context of environmental change. For that purpose, it would be necessary to take the following steps: (1) identify species response traits that determine community assembly and species abundance patterns, (2) partition the relative contribution of species abundance and matching traits to determining interaction frequencies and network topology, (3) identify matching and quality (effect) traits that determine, respectively, the quantitative and qualitative components of interaction outcomes, ultimately describing the magnitude of services provided through the trophic interaction, and, finally, (4) identify linkages between those different types of traits. As exemplified above, by identifying the linkages between response, matching and effect traits, one should be able to infer the stability of the community in providing ecosystem services when facing environmental changes.

10.4 Perspectives

As presented in the previous sections, trait-based frameworks can help us to understand species' trophic interactions and their consequences for processes within ecosystems. However, for many animal species, individuals often perform different activities (e.g. feeding and resting) or spend part of their life cycle in distinct ecosystems (i.e. ecosystems are not closed but open). This leads to more complex situations in which trophic interactions in one ecosystem can affect interactions and processes in another ecosystem. In the 1960s, Margalef (1963) already hypothesized that mature ecosystems, such as forests, can 'exploit' ecosystems at earlier successional stages because consumers from mature ecosystems move to nearby early-successional ecosystems for foraging, creating a preferential flow of nutrients and energy from early- to late-successional ecosystems. More recently, the meta-ecosystem perspective (Loreau et al. 2003) has stressed the importance of accounting for both gains and losses of nutrients and energy across ecosystem boundaries. In the interface between forests and streams, such reciprocal subsidies (from aquatic to terrestrial ecosystems and vice versa) can make up to 25.6% of the annual energy budget for birds and 44% for fishes (Nakano & Murakami 2001). In addition, predation in one ecosystem can limit the abundance of species in nearby ecosystems, thereby promoting cross-ecosystem trophic cascades (Sabo & Power 2002; Knight et al. 2005; see example in Fig. 10.6). To expand the study of trophic interactions beyond ecosystem boundaries, we still have important questions to answer: When/where are cross-ecosystem interactions more likely to occur? Which species traits can promote ecosystem subsidies?

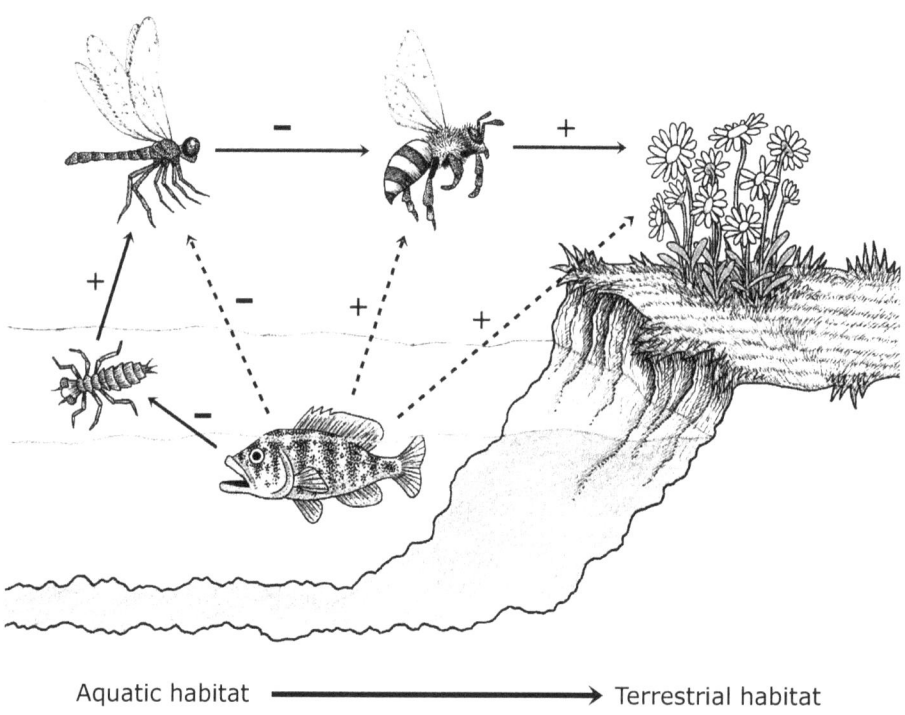

Figure 10.6 Diagram showing species interactions across aquatic and terrestrial ecosystems quantified in experimental ponds by Knight et al. (2005). Because dragonflies have an ontogenetic habitat shift, aquatic trophic interactions cascade to the terrestrial trophic chain, affecting plant–pollinator interactions. Fish predation on dragonfly larvae results in a lower abundance of adult dragonflies. Consequently, a higher abundance of pollinators and higher number of flower visits are observed around ponds where fish are present. Solid arrows indicate direct interactions, and dashed arrows indicate indirect interactions. Drawing by Luís Gustavo Barretto after figure 1 in Knight et al. (2005), originally created by S. White and C. Stierwalt. Copyright © 2005, Springer Nature; adapted and redrawn version, permission via RightsLink permission and additional email communication.

Species with complex life histories showing complete metamorphosis often present a strong ontogenetic niche and habitat shifts (aquatic to terrestrial, below ground to above ground, or herbivore to predator lifestyle). Such species can be important agents of cascading interactions across ecosystems. The classification of ontogeny and the feeding guild of each life stage is an important first step for detecting key species in this process (Moretti et al. 2017). Behavioural traits related to daily movement patterns or migration can also promote cross-ecosystem interactions. Even random movement patterns of species using different ecosystems can result in substantial transport of nutrients. via excretes and carcasses. from richer to poorer ecosystems (Doughty et al. 2013; Wolf et al. 2013). Considering behavioural traits in differential uses of distinct ecosystems can further highlight species promoting cascading effects across ecosystems

(Stevenson & Guzmán-Caro 2010). Identifying and validating traits that interconnect ecosystems via species interactions and movements is an important next step for further advancing landscape ecosystem ecology.

Another challenge we face is to understand how traits determine non-trophic interactions. Most studies on species interaction focus on competition or trophic interactions. Still, other types of interactions like facilitation, symbiosis and ecosystem engineering can substantially affect biodiversity maintenance and ecosystem functioning (Wilby 2002; Brooker et al. 2008; Powell & Rillig 2018). The extension of the response–effect framework to non-trophic interactions (other than competition; see Chapter 7) can help increase our capacity to predict the effects of environmental changes on ecosystem functioning.

Take-Home Messages

- The response–effect trait framework originally proposed by Lavorel and Garnier (2002) can be expanded across different trophic levels, thus assessing how functional traits control species interactions and the consequences of these interactions for ecosystem functioning.
- The concepts of 'trophic effect and response traits' are introduced to assess how traits within a trophic level affect other trophic levels.
- Incorporating trait-based concepts into plant–animal interaction networks can help identify both niche and neutral mechanisms driving interactions networks and the resulting ecosystem services.
- Intraspecific trait variability can modulate species interactions and the ecosystem processes resulting from them. This can be especially important when traits involved in interactions (e.g. phenology, body size, food preference) show substantial shifts (i.e. via plasticity or selection) due to environmental changes.

11 Trait Sampling Strategies

One of the most important aspects of designing any study on functional traits is the acquisition of the trait data. In Chapter 2 we presented some suggestions about how to select and acquire traits, including using literature and trait databases. We cannot always rely on databases, however. This applies when a significant proportion of our target species are actually not included in any database, but also when the question being asked requires that we account for intraspecific trait variability, i.e. the differentiation of trait values within species (ITV; see Chapter 6). In other words, even if databases contain valuable trait information for our species, sometimes this is insufficient given the context of our study. Trait information from databases sometimes reflects the traits measured in a given location well, but this is not always the case (Cordlandwehr et al. 2013). This is because, as explained in Chapter 6, trait values are not fixed within species, and individuals of a given species often differ in their trait values because of genetic variation and phenotypic plasticity. Trait values from databases can be problematic when the geographical, environmental or biotic context in which the traits were measured are not similar to that of our particular study.

When trait values are absent or we want to explicitly consider ITV within our study area, it is imperative to measure traits *in situ*. Indeed, after deciding which traits one is interested in, and becoming familiar with their measurement and standardization (see Chapter 2), one frequently faces the question of how many and which individuals of each species need to be measured to get trait values that address a given question. This chapter will address these questions as we explore *optimal sampling strategies*. Most of the examples provided here are based on plants.

However, the same principles apply for other taxonomic groups as well, although sometimes with additional complications like phenotypic differences within the individuals' life cycle and strong male–female differences. We will start this chapter with a sort of 'game' that we usually 'play' with our students to show the important aspects of trait sampling.

11.1 The 'Ambitious Supervisor' Exercise

We often ask our students to play the ambitious supervisor exercise to familiarize themselves with the design of trait sampling strategies. Imagine that your supervisor has a grant to examine changes in plant traits along an altitudinal gradient, assessing possible consequences of climate change on vegetation composition. The idea of the project is to examine differences in species and community traits between low and high altitudes but also changes within species (see Chapters 4–7 and related questions to support such a scientific objective). Specifically, as we saw in Chapter 6, we expect to observe changes in trait values caused both by species turnover along the altitudinal gradient, i.e. species adapted to the environmental conditions experienced at a specific altitude, and by ITV, i.e. individuals adapting to the altitude (Lepš et al. 2011).

Your supervisor has already sampled community composition in a previous project, so that species compositional changes along the gradient are known. So let's look at the characteristics of this previous hypothetical sampling, keeping in mind that it is based on a sampling scheme similar to that already discussed for null models in Chapter 7. The sampling design, in this case, would include a total of 40 plots of 5 × 5 m each, with five plots (replicates) at each vegetation belt (Fig. 11.1). Let's assume there are eight vegetation belts at different altitudes (say around 1300, 1500, 1700, 1900, 2100, 2300, 2500, 2700 m above mean sea level). Species cover data are available for each of the 40 plots, for a total of 150 species. To keep it simple, we presume there are around 20 species per plot, and that generally the six most abundant species in each plot account for more than 80% of the total species cover. Let's also keep the turnover within each vegetation belt (within each altitude) small, so that for each belt there are some 30 species in total. Along the altitudinal gradient some species will be present at one belt only, but in this hypothetical example 50 species are present at more than two altitudes. You can see this as a realistic example, but please consider that very often the sampling design includes much more variability, with a higher number of species both across the environmental gradient and across replicates under the same environmental conditions.

Your supervisor, who, for simplicity, we can assume did not study this book, asks you to present a sampling design to collect trait information for the project. He or she has some ideas on which trait(s) to measure (Chapter 2) though. Let's make a conservative estimation that the measurements of these traits will take 20 minutes in total for each individual measured (including field and lab time). For example, measuring

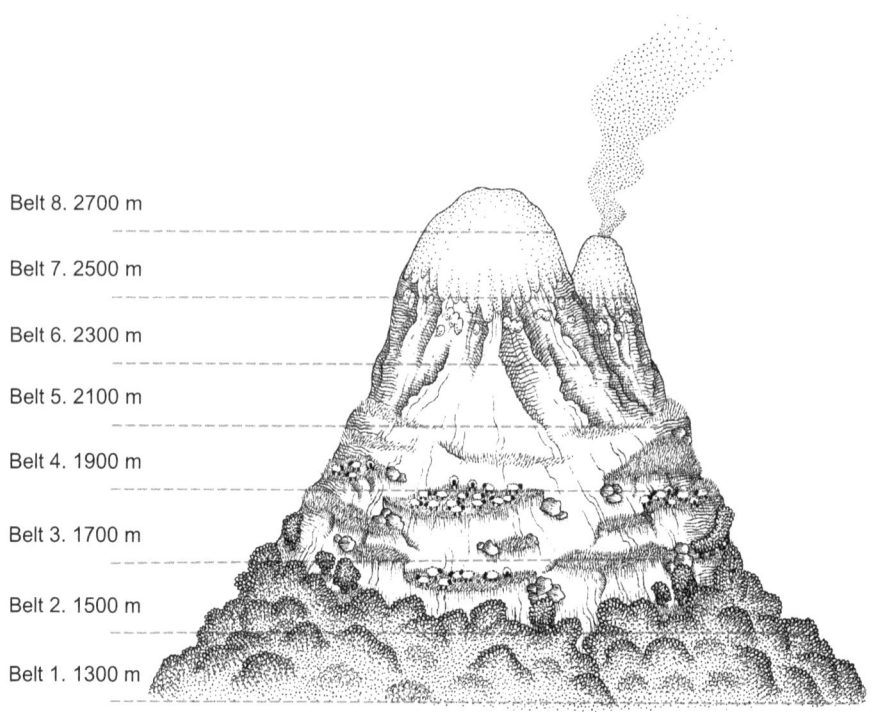

Figure 11.1 The trait 'game': sampling traits across an altitudinal gradient divided into eight altitudinal belts. In each belt, we have sampled species composition in five different plots. Figure created by Luís Gustavo Barretto.

specific leaf area – very commonly measured in plant studies – will take some 10 minutes per individual plant (collecting leaves, storing them for transport, scanning them, estimating leaf area in the image, preparing the leaf for drying, weighing the dry leaves, entering the data into the computer). For many traits the measurements will take more time.

Now that we have all the elements more or less straight, we are ready to play the game. What are you going to propose as a sampling design to your supervisor? How would you conduct this sampling campaign to meet the objectives of the project, publishing some nice papers, while still having a personal life with some free time? Think about it for a while. Close this book or put it aside; take a sheet of paper, write down your options and try to figure out how many species and individuals you would like to measure. Take 5–10 minutes, come up with some possible scenarios and then, only then, read the following text. You can also start reading the first example if you are unsure of the game we are prescribing, and then (when facing the problem) close the book and try playing again.

OK, let's see what the different options are. To please your supervisor, one option would be to sample 'everything everywhere'. How much sampling would this imply?

We know that there are 20 species per plot on average. So the sampling, for a trait-like specific leave area, could be:

'everything everywhere' strategy = 20 species × 5 plots × 8 altitudes × 10 individuals = 8000 individuals (8000 × 20 minutes ≈ 2700 person-hours)

Notice that we just suggest, on this occasion, sampling 10 individuals per species, which some standardized sampling schemes suggest for several traits (Chapter 2). However, the number of individuals measured per species could be much higher.

Obviously, this sampling strategy will impress your supervisor but it might not excite the people you work with very much. It is not very realistic to sample 8000 individuals in one season, unless you have a lot of resources in terms of people, transport and technicians (and hence sufficient funding). In our experience, a team of four people can easily sample around 100–150 individuals per day, or more. However, sampling is only the first step, followed by a number of other procedures (storage, cleaning, rehydration etc.) before the actual measurement of specific leaf area. Of course it depends on the type of organisms, number and type of traits measured, accessibility of the study sites etc., but in the end it all comes down to time constraints.

A solution would be to reduce the number of altitude levels, focusing on the climatic extremes. But your supervisor has already sampled the vegetation at each altitude, so we will assume that we have to respect the altitude scheme. A more reasonable solution is to forget about the five replicates per altitude and to sample plant species '*by environmental levels*', i.e. per altitude. This will make your project less ambitious (hence, feasible), but still produce high-quality trait data. As we mentioned above, the main objective of the project is to examine changes in traits associated with climate (hence, altitude), and not so much intraspecific variability within similar environmental conditions. We thus could hypothesize that the main driver of trait variability would be altitude with no large trait changes within altitudes (or at least we can say that this is not part of the questions currently being addressed). Accordingly, we do not need to sample each species in each replicate plot at each altitude, but instead sample species at least once at each altitude. Then, we can just sample each species in at least one of the five plots by altitude (e.g., randomly chosen). Since we have 30 species per altitude on average (see above), this would come down to:

'by environmental levels' strategy = 30 species × 8 altitudes × 10 individuals = 2400 individuals (2400 × 20 minutes ≈ 800 person-hours)

This is much more reasonable! But still, 2400 individuals in one season means quite a lot of work for a small/medium research group. It is feasible, but still requires a dedicated group of researchers to spend several weeks measuring specific leaf area (perhaps even imposing some unacceptable phenological changes in particular species).

What other possible solutions do we have to reduce the workload? Keep in mind that, at each step, simplifying the sampling scheme means that you need to give something up. For example, you can try to convince your supervisor that there is no need to sample all species, as you might 'only' sample the most dominant plant species. Here we assume that (i) trait filtering operates mostly on dominant species (Chapters 4 and 5), that (ii) species are dominant because their traits fit a given condition better than rare species (Shipley et al. 2016), that (iii) the assembly processes are the result of interactions between dominant species (Chapter 7) and even that (iv) ecosystem functions and services largely depend on the most dominant species (Chapter 9). The sampling strategy based on dominant species is supported by studies such as Pakeman and Quested (2007), Pakeman (2014) and Majeková et al. (2016a), which suggest that having traits for plant species accounting for *80% of the total abundance* in a plot can give a relatively good approximation of the functional structure of the plot. Indeed, you can combine this strategy with the sampling '*by environmental levels*' mentioned above. So, let's say that, out of the 30 species per altitude, there are around 10 species which are dominant at each altitude level (remember we mentioned above that six species accounted for most of the species cover in each plot). Now the sampling strategy could be:

'**dominant by environmental level**' strategy = 10 species × 8 altitudes × 10 individuals = 800 individuals (800 × 20 minutes ≈ 270 person-hours)

This is much less work to measure specific leaf area! The potential downside now is that in this '*dominant by environmental level*', it is quite possible that you actually sample 'only' some 40 species out of the total 150 plant species in the region, because the dominants in two contiguous altitudes might be the same. The sampling strategy just described is interesting if you do not care about what is happening to the non-dominant species. But sometimes one is interested in these, especially if you want to assess how environmental filtering occurs across species.

In that case you might try to sample more species. To do so, while keeping a similar sampling size, you need to sacrifice something, which is likely the trait variability within species (which is accounted, indirectly, in the '*dominant by environmental level*' scenario, because a single species could be dominant in more than one altitude). So you can try to convince your supervisor that the trait changes within species across altitude are of less interest compared to changes in species composition. If you want to focus on the effects of species' compositional changes (turnover; Chapter 6), which is important if you have a lot of turnover in species between altitudes, then you need to consider as many species as possible. Following this, you could for instance sample each species in just one altitudinal level, for example where they are the most abundant. This sampling strategy is actually very often described in the literature, and assumes that the trait values within a species are 'fixed' (Lepš et al. 2011), and the differences between conspecifics are less important than the differences between species (Garnier et al. 2001). In other words, interspecific differences in traits overrule ITV. If you apply this sampling scheme, then it means that we sample species only in one vegetation belt, for example where it is more abundant. In this case we can then sample up to all 150 species. This would be summarized as:

'**fixed species**' *strategy* = 150 species × 10 individuals = 1500 individuals (1500 × 20 minutes ≈ 500 person-hours; almost half if we further select only 80 of the 150 species)

This 'fixed species' strategy is still a bit too ambitious and eventually you can combine it with the approach of sampling only dominant species, as described above, until you reach a good compromise of sampling as many species as possible while keeping the sampling effort within limits. You could then also use the 'fixed species' approach to sample 800 individuals, as in the 'dominant by environmental level' strategy, but at least covering more species, in this case 80 species.

But let's now consider that your supervisor is not happy at all with the 'fixed species' approach or with the idea of restricting sampling to the dominant species. He/she thus insists you need to sample more species, while considering ITV as well. So, in principle we need to increase the number of species, but, to keep the numbers of trait measurements within reason, you need to reduce something else. One option is to reduce the number of individuals that you sample for each species in each belt, sampling, for instance, 5 instead of 10 individuals. As we will discuss below, this strategy will reduce the precision of your estimations, but it will retain information for species within belts, thus improving accuracy:

'reduced by environmental levels' strategy = 30 species × 8 altitudes × 5 individuals
= 1200 individuals (1200 × 20 minutes ≈ 400 person-hours)

Of course, you can even choose a mixture of the previous strategies. Since the dominant species are potentially the main determinants of the functional structure of the communities, you could choose to sample them a bit more intensively within belts (let's say 5 individuals per dominant species per belt). At the same time, you could still collect a reduced number of individuals of the non-dominant species within each belt (3 individuals, for example). The sampling would be as follows:

'more for less' strategy = (10 dominant species × 8 altitudes × 5 individuals) +
(20 non-dominant species × 8 altitudes × 3 individuals) = 880 individuals
(880 × 20 minutes ≈ 300 person-hours)

As you can see, you sample many species with this method, even taking ITV into account, although possibly not very accurately (especially for non-dominant species).

The examples shown within this exercise, overall, reflect many discussions we have had with students during our courses and with various colleagues. The examples are by no means exhaustive but reflect different alternatives, including some extreme strategies that can be employed when planning an apparently simple sampling scheme. There are indeed different strategies to sample traits, satisfy your supervisor, reviewers and, most important, satisfy your own biological interest in understanding species differentiation. No strategy is in principle better than others, but some will better reflect the 'true' changes in traits than others, obviously depending on your research focus and type of data. In the exercise above we did not consider, although we could have, cases where sampling could be organized phylogenetically (for example to compare congeneric species along gradients; see tests in Chapter 8; Defossez et al. 2018), nor a sampling organized within (and across) functional groups, functional types, which has already produced so many interesting studies on trait trade-offs (Wright et al. 2004).

11.2 Accuracy and Precision

After having played a bit with the notion of different sampling strategies, let's discuss some theory behind these ideas. In functional trait studies we are usually interested in estimating different aspects of the trait structure of biological communities (Chapter 5) or populations (Chapter 6). To do this, we need reliable estimations for trait values of the individuals composing these communities or populations. If we were able to measure a trait on all the individuals present in our study system, we would be able to attain a 'perfect characterization' of this trait structure. However, with very few exceptions (e.g. Baraloto et al. 2010; Paine et al. 2015), such a level of sampling intensity is unfeasible for most research groups and most projects, due to the limitations in time and resources. Hence, we are again facing the estimation problems we explored above. For a given sampling effort (i.e. the number of individuals that we can reasonably expect to be able to measure) our aim is, implicitly, to maximize the similarity between our estimations of the trait structure and the 'real' trait structure.

The conceptual framework of trait sampling dealt with in this chapter is based on two important concepts for statistical estimations: accuracy and precision. *Accuracy* is the proximity of our variable estimate, i.e. our trait of interest, to the real value of that variable in our model system. *Precision* is the degree to which different measurements of that variable differ between them (Fig. 11.2). As a general rule we can say that a measurement system is better the higher its precision and accuracy are. Let us illustrate this point with a very simple example: imagine you want to estimate the average height of a local plant population, but you cannot measure all individuals. The aim of our trait measurements is to approximate average height with the highest level of precision and accuracy. It is quite easy to know where to select the individuals in this case: just go to the population in question and measure plant height. If you end up measuring individuals from another population (or using data measured in another population) with, say, much higher levels of soil fertility – which generally results in taller individuals – your height estimation for the target population is going to be inaccurate (Fig. 11.2). So, the first conclusion is that measuring traits of a local population can avoid the bias that would occur when using trait data obtained from populations growing under different environmental conditions.

Then, of course, even if we measure traits only in one site, there will also be a certain degree of intraspecific height variability in our focal population. Hence, estimating the average height of the population by measuring a single individual is probably going to result in an unreliable estimation (we will discuss, though, some cases in which this might even be preferred over the absence of local measurements). Accordingly, measuring heights for an increasing number of individuals will increase the precision of our average height estimation. The larger our sample size is, the more precise our estimation will be. It is important to remark that precision and accuracy are independent. For instance, extending the previous example, you could measure a very high number of individuals in a population other than the one you are actually interested in, e.g. with taller individuals in fertilized conditions. The resulting estimation will be of high precision but low accuracy.

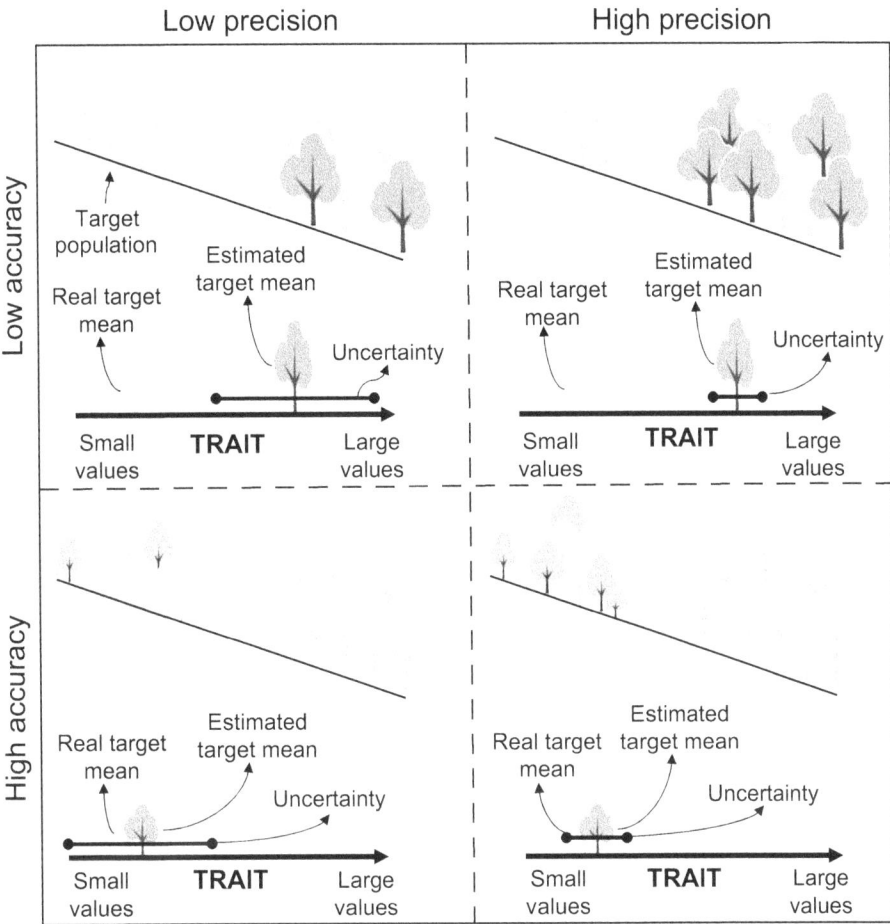

Figure 11.2 Illustration of the precision and accuracy concepts. Our goal is to estimate the mean plant height for a target population, represented by the trees on the top of the slope. Plant height takes different values in response to environmental factors, so that different populations will differ in their average height (trees in the bottom of the slope represent a population of the same species with taller individuals than the target population). Our estimate of the mean plant height of the target population is based on averaging across a series of selected individuals in which we measure plant height (darker trees). The level of accuracy indicates how far our estimation of plant height is from the real average value in the target population. Accuracy of our plant height estimation can be improved by making sure that we select individuals for trait measurements that grow in the same environmental conditions as the target population (see improved accuracy in lower panels compared to upper ones). Precision of our plant height estimation is an indicator of our degree of certainty about the average trait value we are trying to estimate; precision increases as we measure more individuals (see improved precision in right-side panels compared to left-side ones). Note that accuracy and precision are independent, so that if we select individuals that are very different from the ones in our target population, we will not improve accuracy simply by measuring more individuals (e.g. low-accuracy, high-precision case).

11.3 Trait Variation at Different Scales

The example in Fig. 11.2, illustrating the concept of accuracy and precision, is deliberately simple, as it only deals with the issue of trait estimation for populations of a single species. One might think that this is not much of a problem, since it should not involve an unreasonable number of individuals. But this perception can be misleading, as trait variability occurs within species across *multiple spatial scales* and environmental conditions. For instance, Albert et al. (2010) partitioned the variability of trait values across three hierarchical ecological scales: subplot (3 individuals within 1–10 m^2), plot (3 subplots within 2500 m^2) and gradient (7–18 populations for 16 species across strong altitudinal gradients). They show that plant trait values within a species are largely variable among populations, but also within populations and even within individuals sharing the same local conditions. Leaf trait values can differ a lot within a single individual, especially among canopy strata or when comparing shadow and sun leaves (Messier et al. 2010). In Chapter 6 we discussed how to partition variance across these scales using different methods. Seeking high accuracy could drive us into an escalation of trait measurements that can make a study impossible to perform. There is a necessary trade-off between the level of detail (is the smallest level considered the branch, the individual, or the population?) and the overall sampling effort (number of replicates for each level), especially when considering multiple species at a time.

But not all hope is lost. There are *protocol handbooks* for measuring traits in a standardized way (Cornelissen et al. 2003; Pérez-Harguindeguy et al. 2013; Moretti et al. 2017; see Chapter 2), which try to solve this issue by providing recommendations aimed at homogenizing the conditions in which traits are measured. An important aspect of this standardization is related to the selection of *which individuals* to collect and *how many individuals* need to be measured to answer a great many ecological questions. A typical recommendation is to select adult, healthy individuals, to remove the part of the variability within a species due to ontogeny or natural enemies. In the case of plants, leaf traits (such as specific leaf area) can be measured in leaves coming from different areas of the canopy, which can differ in sun exposure, in turn affecting leaf trait values. In this case, handbooks recommend selecting undamaged mature leaves exposed to direct sunlight (Cornelissen et al. 2003; Pérez-Harguindeguy et al. 2013). These recommendations reduce the need for intraspecific and intra-individual replication, allowing for a greater number of species. We recommend following these practices whenever possible, although obviously they do not apply when the aim is to quantify the full extent of intraspecific trait variability, for example due to different leaf positions in the canopy (as in Messier et al. 2010) or the variability due to ontogeny (Moretti et al. 2017), among other possible ecological questions.

Let's assume we are not interested in trait variability within populations. Despite standardization efforts, there is still substantial intraspecific variability among populations, which implies that several individuals within a location must be measured to have a reliable (i.e. accurate and precise) estimation of the local average trait value. Then how should we proceed? Measuring traits for several individuals (as suggested in some handbooks; see Appendix 1 in Pérez-Harguindeguy et al. 2013 for plants) in each population, across locations that differ in conditions, may lead to sampling and measuring a huge number of individuals. However, depending on our sampling design, it is not always viable to measure traits in so many individuals. How do we proceed then?

11.4 Sampling Strategies

We will now further expand some ideas of the previous exercise, while also developing some new concepts. Assuming that you cannot draw on available trait databases, a trait characterization of our communities might imply thousands of individual trait measurements, even if only one trait is involved; this increases quickly (exponentially in some cases: Carmona et al. 2016; Blonder et al. 2018) with the number of traits, species and environmental conditions under consideration. Hence, measuring functional traits in order to characterize the functional structure of communities can be much more challenging than sampling individuals of one or a few species. The remainder of this chapter will mostly focus on how to design this kind of sampling.

The main idea to keep in mind is that selecting our sampling strategy depends, first of all, on the *study objective*. Unfortunately, there are so many different aspects and potential cases that it is virtually impossible to give a general, fail-safe recommendation to follow for all cases. However, one simple and intuitive concept is that if you want to consider ITV across environmental conditions, you may have to measure individuals of each species in each environmental condition (i.e. the '*by environmental levels*' strategy). This means that you will try to improve accuracy (by not assigning trait values measured in different environmental conditions), but this comes at a cost in terms of precision (decreasing the number of individuals measured per species and altitude, because you have to divide the total number of trait measurements that you can perform between a greater number of environmental conditions). If you want to consider ITV within each of the environmental conditions in which a species grows, the number of individuals per species and environment will be higher. If, on top of this, your communities are very species-rich, you may have to further reduce the number of measurements for each individual species, resulting in a yet lesser precision of your trait estimates.

The best advice we can give you is to rely on common sense; to acquire as much information as possible about your study system before you start measuring traits (for example, in the exercise above, knowing the structure of the communities in advance); and to discuss your sampling designs and alternatives with colleagues. Below we further suggest a step-wise approach to come to a valid sample design.

11.4.1 Starting Point: Setting Limits

Every trait-based study starts with a properly stated research question which should help to determine which trait(s) you need to measure, where and how often (Chapter 2). At this point it should be clear that the key to designing a feasible sampling strategy is to decide which and how many individuals of each species to measure to attain an estimation of the actual trait values as precisely and accurately as possible. You have to consider several factors when designing a strategy. First, which traits to measure? Although trait selection is not the objective of this chapter (see Chapter 2), it is evidently a very important step in developing your sampling strategy. Some traits are relatively fast and inexpensive to measure, while others are not, which can profoundly impact the maximum number of individuals that we are able to measure within a study. Some traits are known to be more variable than others (Cornelissen et al. 2003; Siefert et al. 2015) and this might mean you need greater within-species replication. Measurement time and trait variability should be considered beforehand to avoid unpleasant surprises. Dividing the total available time (or resources) by the average time (or resources) spent measuring each individual gives you an idea about the number of individuals you are able to measure. It is probably wise to decrease this number by a certain proportion to account for unexpected events; better safe than sorry!

Once you know the maximum number of individuals that you can afford to measure for a given trait, you are ready to move to the following steps. Before continuing, you have to answer an important question: is the number of measured individuals sufficient to answer your research question? If the answer is no, then there are some alternatives. The first one is to abandon the idea of performing this particular trait-based study, or reformulate the question into one that can be answered with a smaller number of measurements. Another alternative is to reconsider the factors that determine this maximum number of trait measurements. Assuming you have carefully selected traits for the problem at hand, this can only be done by increasing the total available time or resources (for instance, by getting new people to collect and measure traits). This process should be done iteratively until you are happy with the number of individuals that you are going to sample. Now, it is time to decide your sampling strategy, which basically consists of deciding how to distribute these individuals both between and within species, and sampling units, similar to the earlier trait exercise.

11.4.2 Literature on Sampling Strategies

Although the optimum trait sampling is relatively recent and understudied in trait-based ecology, there is still interesting literature on this topic. In these studies, simulation methods are used to find out the strengths and weaknesses of different sampling strategies. Ideally, studies dealing with the issue of comparing the best sampling strategy in a given system should have a complete, or quasi-complete, estimation of the real trait structure of communities in order to gauge the optimal sampling strategy. This is done by sampling a large number of individuals of each species in each studied plot and then considering this as the optimal sampling (if resources were unlimited). Afterwards, different realistic trait sampling strategies (each considering only a subset of the total number of individuals) are simulated. Some of these strategies are based on constructing *regional databases* in which some individuals of each species are selected, either along the whole study area or in the plot in which the abundance of the species is maximized (*'fixed species'* strategy). Then, the average trait value of these individuals can be assigned to all the conspecifics in the dataset. Other strategies are based on different random selections of a varying number of individuals in which the traits are estimated (as in the '*more for less*' strategy). There are even *taxon-free* strategies that do not require species identities. The goal of doing these simulated strategies is to have an idea of what the results would be if we selected specific individuals to measure traits on. Then, for each of these simulated strategies, we can compare the values of different aspects of community functional metrics (Chapter 5) with the 'real' ones (estimated considering all the individuals). This process is analogous to calibration in metrology; we are trying to estimate the accuracy and precision of the measurements provided by each sampling strategy. Let's go back to Fig. 11.2 to better illustrate the main idea highlighted here. The complete sampling allows us to know the real plant height in our population, which is not known in most real-life applications. Then, the figure proposes four strategies based on two main factors: the number and the location of the selected individuals. A higher number of individuals gives a higher precision, while measuring individuals in the population of interest yields higher accuracy. Indeed, the number of potential alternative sampling strategies in real life is much larger than the simplified version from Fig. 11.2, which is what makes this kind of study interesting.

If you are interested in knowing more about the range of potential sampling strategies along with deeper explanations about each one, we suggest reading Lavorel et al. (2008), Baraloto et al. (2010), Gross et al. (2013), Carmona et al. (2015a) and Paine et al. (2015). In any case, these exercises allow you to rank sampling strategies according to their resemblance to the real functional trait structure, as well as the amount of sampling effort each requires. Unfortunately, these papers only cover a few particular cases, and they cannot include all the existing heterogeneity in research questions and target habitats, or the different interacting aspects that determine the best sampling strategy in each case. This hinders reaching general conclusions and guidelines. Keep in mind that these are case studies that will surely have some degree of disconnection to your particular questions; further studies expanding these analyses to different conditions (and organisms) are very much needed. We recommend that you also look for examples in the literature that more closely resemble your study system to get an idea of the most appropriate sampling design. Regardless, we can extract some valuable lessons from these papers; the next section is an attempt to summarize them, focusing on different criteria.

11.4.3 'Length' of the Environmental Gradient

As we mentioned in Chapter 6, changes in community functional structure along environmental gradients are caused by two different processes: *species turnover* and *intraspecific trait variability* (Lepš et al. 2011). We can expect that, in general, intraspecific trait adjustments of species to local habitat conditions will be more important when species turnover is small, assuming that turnover is small when there is little environmental variability (Auger & Shipley 2013). In other words, the relative importance of ITV compared to interspecific trait variability due to species turnover may decrease as spatial scale increases, as environmental variability will also increase (Albert et al. 2011). This has significant consequences for your sampling strategy. On

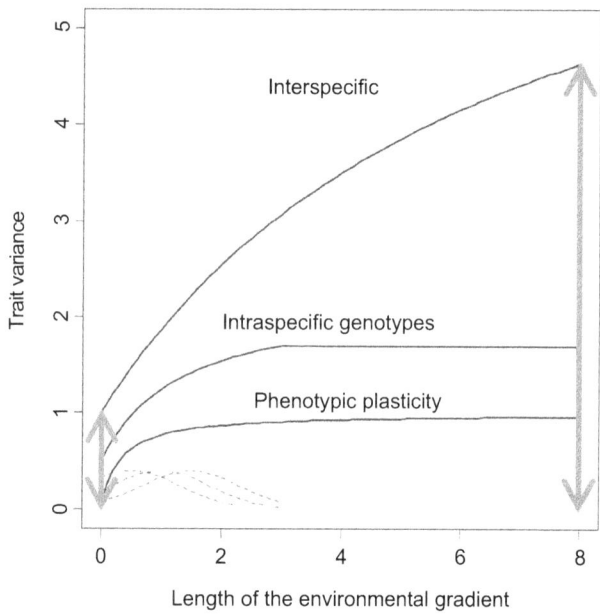

Figure 11.3 The importance of intraspecific trait variability (composed by the combination of phenotypic plasticity and differences between intraspecific genotypes), compared to the effect of changes in species composition (interspecific variability) along environmental gradients. Trait values change along environmental gradients, so that the x-axis represents both the length of the environmental gradient and the traits of species (with each species' trait values being represented by a normal distribution). In general, the longer the environmental gradients considered (x-axis), the smaller is the relative contribution of ITV to total trait variability. The x-axis can represent, for example, the length of a first axis in analyses like Detrended Correspondence Analysis (DCA). In a very short gradient, we can expect that a large part of the total trait variability will be due to ITV (left arrow). By contrast, when the considered environmental gradient is long (right arrow), the relative contribution of ITV should decrease. Figure adapted from Auger and Shipley (2013), with permission from Wiley. © 2013 International Association for Vegetation Science.

the one hand, if your gradient is long enough (in terms of spatial extent and/or environmental heterogeneity), it will encompass sufficiently different conditions as to have dissimilar communities at the extremes of the gradient. The species that occur at the ends of the gradient will probably differ significantly in their average trait values (Fig. 11.3), which means that a large portion of the differences in trait structure between communities will be due to species turnover. Hence, interspecific trait differences will account for most of the functional differences between communities. Consequently, the importance of estimating the extent of species turnover along the gradient *before sampling trait values* cannot be stressed enough. If species turnover is strong then it is often assumed that intraspecific trait variability can be ignored as its effect is small, as in the '*fixed species*' strategy outlined above. On the other hand, if your gradient has a small spatial extent and/or little environmental heterogeneity, the species compositions of your communities are not likely to differ much. But even in these cases we can still observe community functional differences along the gradient, which to a large extent will be due to shifts in trait values between conspecifics. In other words, ITV can cause changes in community trait structure between different environmental conditions even with restricted species turnover (Fig. 11.3). This means that if your study is done on a local scale (for example, assessing change in a community after the application of fertilization), then intraspecific differences will play a crucial role, and should be accounted for.

It is not easy to know beforehand how intraspecific trait variability due to adjustment of trait values to local conditions scales to interspecific trait variability due to species turnover. One potential option to assess the relative importance of the two sources of community trait variability is to perform a *prospective sampling*, for instance by first sampling community composition and collecting traits only in the most extreme conditions along your gradient. Then you can check if trait variability within species is substantial, as well as the amount of species turnover. If species turnover is large, or if the trait values of species that are present at both extremes do not change much, then it should be safe to use a single trait value per species for the whole gradient. If this prospective sampling is not feasible, we recommend checking available studies in similar systems to get an idea of an optimal sampling strategy.

Let's examine potential cases in a bit more detail. The whole world is the largest spatial scale one can imagine from an ecological point of view. Some studies have tried to characterize functional patterns at this scale. For instance, Díaz et al. (2016) analysed the global variability in several plant traits, such as plant height, leaf area or seed mass, identifying groups of species clumped together within a multidimensional trait space. For this analysis, they gathered trait information for more than 46000 species retrieved from the TRY database (Kattge et al. 2011). Selecting such a huge number of species, from all kinds of habitats, implies that most of the potential trait variability at a global spatial scale is under consideration. Since their research question focused on trait differences between species and given that on this spatial scale interspecific differences are expected to be, by far, the main source of trait variability (Albert et al. 2011), including intraspecific trait information would have made very little difference. Accordingly, the authors used a fixed trait value for each species. However, even studies considering global trait patterns can benefit from including more than one value per species and trait, especially when trait values can be linked to the particular environmental conditions in which they were measured. Fortunately, trait databases frequently include information on the geographical location where individuals have been sampled, which can be used to retrieve climatic information, and sometimes more detailed measurements on other parameters. Wright et al. (2017) took advantage of this information, and analysed how climate drives plant leaf size at a global scale. They did not simply use a single trait value per species, but focused on site-based trait values. This means that if a species' leaf size was available for more than one site, they were able to pair these local measurements with the local climatic conditions. If we shift our focus from species to communities, there are studies available that analyse changes in functional diversity patterns across large spatial scales, such as whole continents (Lamanna et al. 2014). Again, this encompasses a large variety of environmental conditions, which translates into large variability in trait values between species in the dataset. There is little to be gained by considering ITV in these cases, as communities along such long latitudinal gradients will share few species. At smaller but still relatively broad spatial extents, strategies that ignore intraspecific changes in species trait values along the gradient are still able to capture the main changes in functional structure. However, this can depend on the trait. For instance, Gross et al. (2013) show that in an extensive aridity gradient, ignoring intraspecific trait variability was a suitable strategy for leaf area and leaf thickness, but not so much for specific leaf area or plant height, whose patterns of variation along the gradient were not fully captured by considering fixed trait values.

When we work on finer scales, such as localities, ITV often becomes important. This is not a trivial aspect of your study, since it determines whether traits collected from databases can be used to characterize the functional structure at your site(s). If, for instance, your study site is located in the boreal zone, but the trait values obtained from a database are based on individuals collected in the temperate or Mediterranean zone, you can imagine that there might be a discrepancy between real trait values and those used. There is some evidence that the ranking of species according to trait values does not change substantially regardless of whether local trait values or database-extracted trait values are used (Kazakou et al. 2014). However, this depends on the trait and most likely also on the communities that are sampled. This assumption might only hold if trait values extracted from the database are from species that were collected in the same latitudinal zone. Yet, other studies are more sceptical (Cordlandwehr et al. 2013). As we explained above, spatial scale plays a key role in our decision to include or exclude ITV: the finer the scale, the less reliable your estimations of functional structure using fixed trait values from databases will be. Moreover, even averaging trait values at the species level, i.e. using a single average trait value for our species in a dataset, is not necessarily the best option in some cases. For example, trait values, such as plant height or leaf size, can change with mowing and fertilization treatments within a meadow, creating large intraspecific trait variation (Lepš et al. 2011). At local scales, ignoring ITV associated with shifts in environmental conditions can result in significant misestimations of community functional structure (Carmona et al. 2015a). Studies analysing the functional differences between coexisting species are likely to particularly benefit from using locally collected trait data (Mason et al. 2011; Albert et al. 2012; Le Bagousse-Pinguet et al. 2014).

Sampling individuals at the plot scale and measuring traits for each species is the most demanding scenario (remember the *'everything everywhere'* case in the game). As we saw above, if you need to sample many individuals of each species in each plot to have a fair idea of the plot average trait values, this is going to have a huge impact on the total number of individuals you must measure. In this case, you can evaluate the importance of intraspecific trait variability at the plot scale compared to total trait variability. Such evaluations can be done using the approach proposed by Violle et al. (2012; Fig 11.4). In this approach, total variability is partitioned in interspecific and intraspecific components, the latter being defined as the ratio between within-species trait variance and total variance, as discussed in Chapter 6. In general, if intraspecific trait variability contributes significantly to the total trait variance within a sampling unit

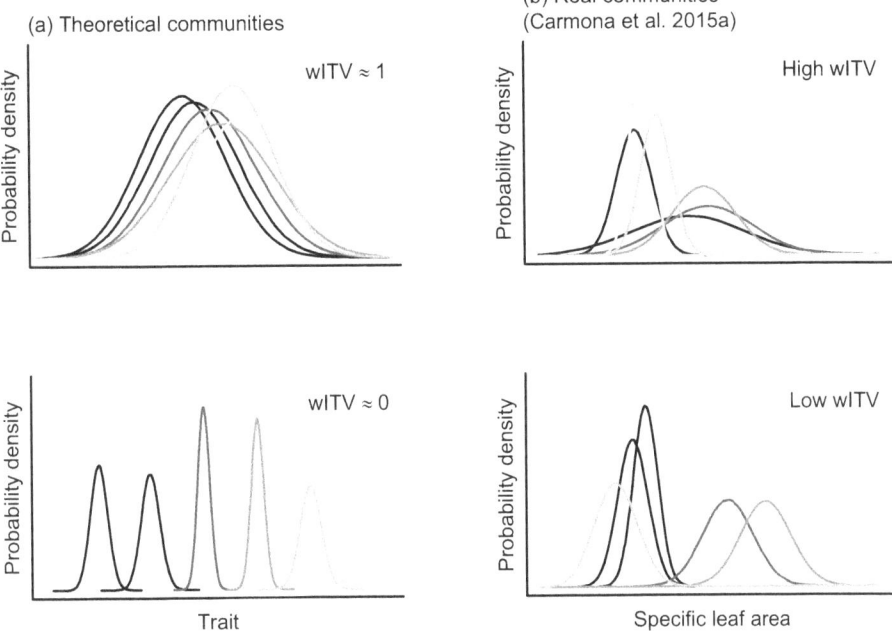

Figure 11.4 Schematic representation of the influence of the average variance within single species relative to the total variance of the sampling unit in trait sampling (wITV; see Chapter 6). (a) Accurate characterizations of the functional structure of sampling units in which the variability within species is high (wITV close to 1 due to high intraspecific trait variability; top panel) will require greater sampling intensities than those of sampling units in which the variability within species is low (wITV close to 0; bottom panel). (b) Examples of two communities from Carmona et al. (2015a), with contrasting values of wITV for the trait-specific leaf area. Note the much smaller overlap between species in the bottom panel, which is reflected in the wITV values.

compared to interspecific differences (top panel of Fig. 11.4a), you should select a greater number of individuals per species to adequately estimate local average trait values (Albert et al. 2015). What does this imply for your sampling design? As an example, let's examine data from Carmona et al. (2015a), in which grassland species were sampled along a local environmental gradient, with traits of the dominant species being measured at several points along the gradient. Within each of these plots, we can calculate the proportion of total variance in trait values accounted for by within-species trait variances (following Le Bagousse-Pinguet et al. 2014). Intraspecific trait variability in the dataset accounts for, on average, 33%, 29% and 37% of total variance of plant height, leaf area and specific leaf area, respectively, but there is substantial variability between quadrats. This means that the study from Carmona et al. (2015a) presents a case with a relatively high proportion of intra- to interspecific trait variance, which implies that you need to sample a relatively high number of individuals of each species in each plot. Lastly, studies using probability density functions of trait values (e.g. Mason et al. 2011; de Bello et al. 2013a; Micó et al. 2020; see Chapter 6) generally require sampling a higher number of individuals per species and plot.

11.5 Species' Abundances and Missing Values

In this section we deal with the issue of species' abundance distribution within communities and how this can affect trait sampling. Moreover, we will address the problem of missing trait values for species in your analyses.

Firstly, an important reminder that some trait-based studies do not focus on species identity but treat all individuals as coming from the same species. In a taxon-free sampling strategy, individuals are randomly collected, irrespective of species identity, and used to characterize the functional trait structure of the community (Lavorel et al. 2008). Taxon-free sampling strategies have a clear advantage when compared with species-based sampling: they do not require the ability to recognize species. This is a non-trivial advantage if you do not know your species. This approach can therefore be quicker, more economical and used by people with little taxonomical training. Taxon-free sampling strategies permit the characterization of many aspects of the functional trait structure, including community average trait values (Ricotta & Moretti 2011), trait value range, the fraction of the functional trait space occupied by your communities (Carmona et al. 2016; Blonder et al. 2018), partitioning of functional diversity across spatio-temporal scales (de Bello et al. 2009; Lamanna et al. 2014), or decomposing beta trait diversity into nestedness and turnover components (Villéger et al. 2013; Martello et al. 2018). However, they come with a trade-off. Without composition data, commonly used species diversity indices cannot be applied. You implicitly renounce all the questions related to species identities, or species richness, such as what drives species coexistence and community assembly, the functional redundancy of species, the dissimilarity between species, the functional uniqueness of species, the relationship between functional and taxonomic diversity, or attempts to predict the species identities and/or abundances under some environmental conditions. In addition, taxon-free trait information cannot be included in *public trait databases*, which reduces its value for the overall scientific community. Therefore, although the objective of the study might allow the use of taxon-free approaches, this decision must be carefully considered in advance, given the potential limitations that the data will have should you wish to address additional questions other than those you originally posed.

As we discussed above with the sampling game, the distribution of abundances, or *abundance ranking*, of species within communities is an important aspect of identifying the best sampling strategy. In this vein, Baraloto et al. (2010) and Paine et al. (2015) performed a complete species characterization of trees in tropical rain forests plots, measuring ten traits in all trees >10 cm diameter at breast height, in nine 1-ha plots. They then compared the simulated trait mean and variance obtained with different sampling strategies to the measured trait mean and variance for each trait and plot. Interestingly, they observed that taxon-free sampling strategies describe community trait structure more efficiently than some strategies based on species identity in which traits were obtained from databases or where only the most abundant species were sampled (Paine et al. 2015). Additionally, the strategy consisting of sampling a single individual of each species in each plot performed well generally. In contrast, Carmona et al. (2015a) used data from a steep topographical gradient in a Mediterranean grassland with relatively little change in species identity across communities. The authors of this study examined two somewhat related questions: first, which sampling strategy yields the most accurate and unbiased estimates of community average trait values and functional diversity (using Rao's quadratic entropy)? Second, which sampling strategy best detects patterns of variation in the functional structure of the communities along the gradient? The results show that using the trait values from individuals collected in the corresponding plot was clearly the best strategy. For a comparable sampling effort, measured in terms of numbers of sampled individuals, this strategy outperformed the alternatives, yielding unbiased estimates of trait structure.

The dissimilarity of optimal sampling strategies between these above studies reflects, to a great extent, the different abundance distributions in the communities sampled in each study, among other possible factors. In the Paine et al. (2015) study, taxon-free sampling strategies outperformed both strategies based on the relative abundance of species (such as the 'dominant by environmental level' strategy) and strategies that completely ignore abundances (such as the 'one individual per species and plot' strategy). This may be because the sampled communities were particularly *species-rich*. Working with high local species richness has two consequences that most likely determine the relative efficiency of sampling strategies. First, in species-rich systems, even the dominant species will have a low relative abundance in the community, which implies that their trait value will not be such a strong determinant of the aggregated community trait value at the plot level. Second, sampling one individual of each species per plot might, paradoxically, not differ substantially in terms of total effort from sampling all the individuals, since many species will be present as one or very few individuals. However, in most ecosystems, only a few species dominate in the community. For example, even in species-rich grasslands, there are a few *dominant species* that account for the majority of the total abundance (e.g. Lavorel et al. 2008; Carmona et al. 2015a). In ecosystems like these, sampling a reduced number of individuals of the most abundant species will likely yield reliable estimations of the plot trait structure (Majeková et al. 2016a).

Species abundances also interplay with missing trait data for species and functional trait metrics in a complex way. Take community weighted mean (CWM) trait values as an example: if a single dominant species in your community accounts for 80% of the total abundance, and if you are able to characterize the trait values of that species correctly, then your estimate of CWM will likely be quite reliable. This immediately raises a question: what proportion of the species present in a community do we need to characterize to attain a reliable estimation of functional structure of that community? The answer seems to depend on the type of trait, qualitative or quantitative. Pakeman and Quested (2007) showed how CWM for qualitative traits is less robust to missing trait data than for quantitative traits. They recommend acquiring trait values for those species that together account for at least 80% of the total abundance in each community. However, their results came from restricted conditions in terms of the indices used (CWM), the taxonomical scope (plants) and the particular types of communities (grasslands and woody sites). Other studies explored this question in more depth across a larger range of circumstances. For example, Majeková et al. (2016a) found that sensitivity of indices to the completeness of trait data differs substantially, and can even differ between taxonomic groups. It is actually not surprising that diversity indices that do not consider species' relative abundances, such as functional richness, are much more sensitive to missing data than indices that include the relative abundance of species, such as Rao's Q or functional dispersion (Pakeman 2014). This implies that if you are planning to use indices that are highly sensitive to missing trait data (such as FRic, FEve, FDiv; see Chapter 5; Pakeman 2014; Majeková et al. 2016a), you should measure more species for their traits. In general, it seems that indices that are sensitive to the influence of extreme trait values or outliers are also more sensitive to missing trait data. One solution to this problem might be to increase the normality of the trait distribution by data transformation, so that we enhance the resistance of these indices to missing data (Majeková et al. 2016a). Nonetheless, the *threshold of abundance* that should be considered differs between taxonomic groups (Majeková et al. 2016a) and also between datasets within a single taxonomic group (Pakeman & Quested 2007), which further complicates the provision of clear-cut instructions. This underscores the importance of deciding about this threshold during experimental design, where you should consider questions such as: what proportion of species do we need trait information for? Should we log-transform trait values? Should we transform species abundances? Which diversity indices should we use? The R package *traitor* (Majeková et al. 2016a) can be used for help in answering these questions (see 'R material Ch2'). It shows, for example, for what proportion of species trait data are available (for each plot and for the whole dataset) and helps to identify which species would be more effective to sample to increase such proportion (and thus increase the quality of estimations of CWM and FD). Unfortunately, this requires a knowledge of community abundance data (as in the trait game in Fig. 11.1), which results in a temporal separation of species sampling and trait acquisition, requiring a higher number of field visits.

An attractive solution for missing trait values is to impute these from available trait data. There are different procedures to perform data imputation. A very simple, yet rather useful, way is simply assigning the average trait value of species that are evolutionarily closely related (e.g. assign the average of the genus; Pakeman et al. 2011). For more sophisticated statistical procedures not within the scope of this chapter, we refer the reader to Taugourdeau et al. (2014) and Penone et al. (2014). While these computational approaches are clearly appealing, we also think they should be used carefully, because although trait values are often similar for phylogenetically related species (a case of niche conservatism), they can also be highly dissimilar due to adaptation (niche divergence).

The previous paragraphs condense two important lessons. First, you want to have trait information for as many species as possible. Second, if that is not possible, the most abundant species should be prioritized. You can combine these lessons and use the conclusions to your advantage, particularly for ecosystems with a few dominant species. Often it is simply not possible to have a very accurate characterization of the functional traits of all species present, but there is no reason to characterize all species to the same degree of accuracy. In general, even a relatively poor estimation of the trait values of species with low abundance would be preferable to no estimation at all. Accordingly,

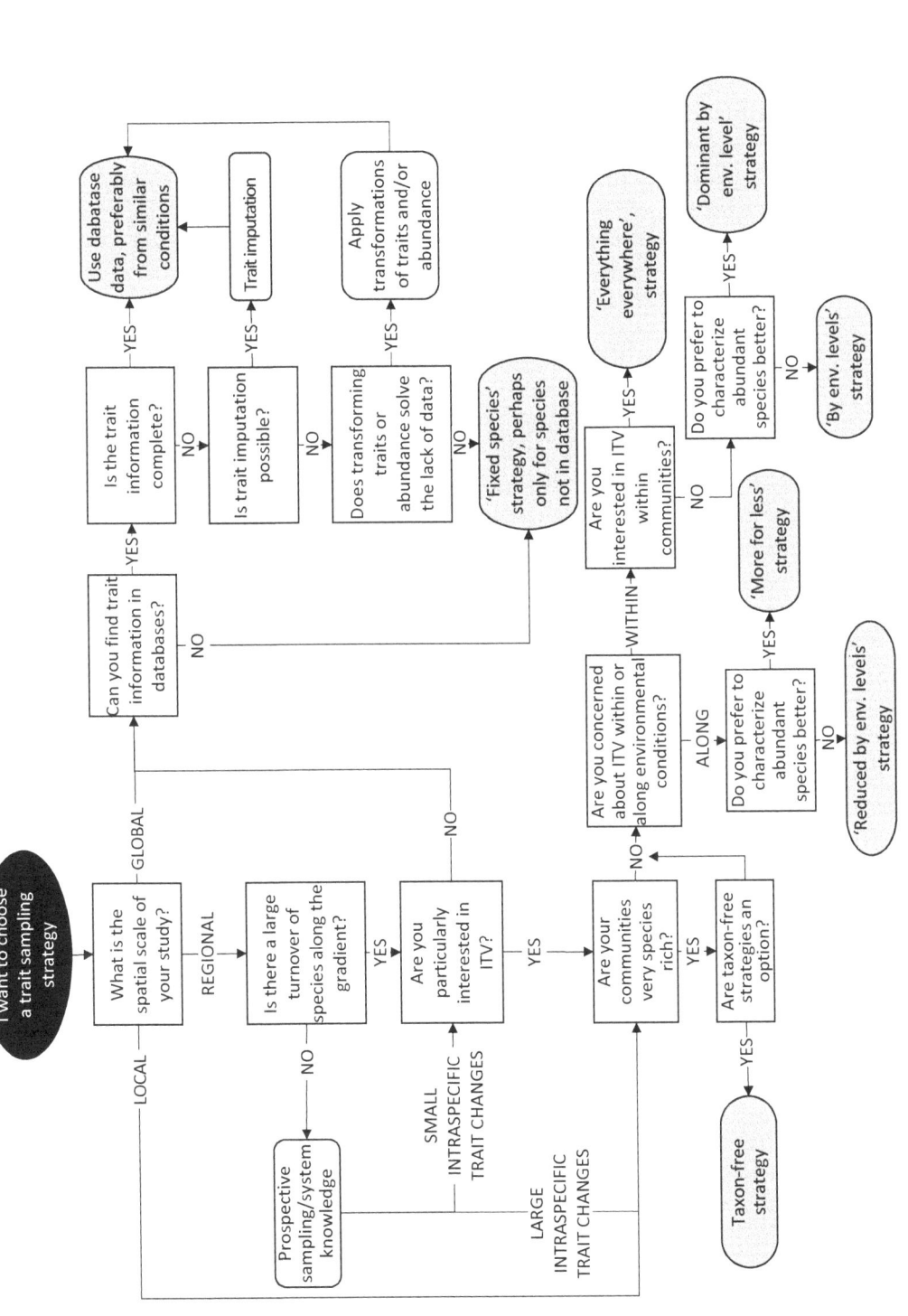

Figure 11.5 Decision flowchart for selecting a trait sampling strategy best suited for your needs. See the main text for a complete definition of each strategy.

Carmona et al. (2015a) showed that a sampling strategy consisting of measuring a higher number of individuals of the most abundant species in a plot is more efficient than sampling the same number of individuals across all species per plot. This approach is actually a *compromise* between the taxon-free strategy, where individuals are sampled randomly, independent of the species they belong to, and a sampling based on measuring an equal amount of individuals for all species. A good example of this approach can be found in Le Bagousse-Pinguet et al. (2014), where the authors selected 1–5 individuals per site and species, depending on the frequency of the species in a plot, and used these individuals to estimate the species' average trait values.

11.6 A Visual Guide to Choosing Sampling Strategies

While it might seem inefficient to collect too much trait data, it is generally preferable to err on the side of caution, since deficient trait information is likely to reduce the power of the analysis used to answer your ecological question. We have designed a flowchart that we hope might serve you when deciding about how best to acquire trait data (Fig. 11.5). However, we must reiterate, there is no failsafe strategy for this endeavour, so please do not rely on it blindly! If there is a key message in this chapter it is that you should rely on *common sense* when deciding which sampling strategy best fits your study and be really careful to tailor your sampling strategy to your question and resources.

Take-Home Messages

- Trait databases are useful. However, whenever possible, it is better to use traits measured in the specific system under study, particularly if traits in the databases were measured under very different conditions.
- We should strive for accurate and precise estimations of trait values with respect to our ecological question. However, we must also balance sampling effort, which could be complicated.
- It is virtually impossible to measure traits in the majority of individuals within a population and in all populations in a region. Similarly, it is very often unfeasible to measure all species in a region. We thus need to carefully plan a sampling scheme which sacrifices some aspects of trait variability.
- The choice of sampling schemes depends on a significant number of factors. For a given sampling effort, sampling schemes can be organized along a trade-off between two extreme cases. In one extreme we can sample several individuals from a single population for each species. In the other extreme, we can sample one individual per species in each environmental condition in which the species occur. Multiple combinations exist along this trade-off.
- In general, the larger the environmental variability and turnover of species across environmental conditions, the smaller the relative effect of intraspecific trait variability in the sampling scheme (with the consequence that sampling schemes could be adapted to this case).

12 Applied Trait-Based Ecology

We can easily agree that an important step for effectively solving any environmental problem is to understand how nature works. In the previous chapters we saw how trait-based approaches are used to address a variety of ecological questions, ranging from how individuals, populations and communities respond to environmental changes (Chapter 4), how the processes of community assembly shape biodiversity patterns (Chapter 7) and how traits affect ecosystem functioning and services (Chapters 9 and 10). The use of trait-based approaches is also rapidly increasing in applied research, helping us to gain insights into solutions for various environmental challenges. In fact, basic and applied research often feed each other with analytical tools, field and laboratory techniques, conceptual models and data. For instance, the study of the effects of plants on biogeochemistry started in agricultural systems (Liebig 1842; Silvertown et al. 2006b) and was later expanded to wild plants (Chapin 1980). Now, in turn, concepts linking plant traits and biogeochemistry, which were developed for wild plants, are used for understanding the effects of their domesticated relatives in agroecosystems (García-Palacios et al. 2013). This spiral of knowledge, as described by Cornwell and Cornelissen (2013), illustrates the mutual benefits of contributions from applied to basic ecology, and back (Lawton 1996). In this chapter we will show how trait-based approaches are helping to find solutions and broadening the scope of applied environmental science. Specifically, we will focus on how traits can be used to build a bridge between basic and applied ecology, employing ecological theory to solve environmental problems.

12.1 Biomonitoring: Biodiversity and Ecosystem Health

The unprecedented human-induced changes and losses in biodiversity have led to great uncertainties about the future of ecological systems. Rapid environmental changes are putting species at risk, especially those that cannot cope with new conditions such as increased disturbance regimes, increased nutrient inputs, decreases in areas of continuous habitat and so on (response perspective). In turn, changes in community species composition can have strong effects on ecosystem processes and service provision (effect perspective). Biodiversity monitoring is thus an essential task for the coming decades if we are to meet recently proposed political goals (e.g. the Aichi Targets for 2020, the United Nations Convention on Biological Diversity, the Paris Agreement, and the Sustainable Development Goals for 2030 of the United Nations), but also to ensure the provision of ecosystem services that impact human well-being (Díaz et al. 2015). Biodiversity monitoring is concerned with finding essential biodiversity variables (Branquinho et al. 2019) that enable us to assess and follow the state and health of ecosystems. Biodiversity monitoring is not a new concept, but as in other disciplines of ecology and conservation, it has increasingly seen a shift away from a more species-centred focus towards more functional approaches (Vandewalle et al. 2010; Pfestorf et al. 2013). But many questions remain open. How and when should we use trait-based metrics in place of taxonomic-based metrics? Should trait-based metrics focus on differences among species or within species? How do we define an essential set of traits that relate to species response to a given environemtal change? Can this set of traits be easily measured and potentially used across different taxa?

Traits can be used to implement a question-driven monitoring approach (Lindenmayer & Likens 2010), which uses the causal links between drivers and biodiversity (and resulting ecosystem processes) as its fundamental unit, offering a potentially powerful framework for generating and testing predictions. One of the first and most important considerations when choosing biodiversity indicators is the need for some additional information on non-biodiversity variables. Thus, for choosing effective *indicators of biodiversity change*, we must first clearly define which drivers of biodiversity change we are interested in investigating. This is the same reasoning one must take when selecting the right species response traits in relation to environmental conditions and their changes (as discussed in Chapter 4). The conceptual framework presented by Branquinho et al. (2019) shows that depending on the nature and intensity of the driver, different facets of diversity might be hierarchically affected, from individual organisms' performance to community functional, and taxonomical composition (Fig. 12.1). Low-intensity drivers are expected to affect individual organisms' performance without necessarily leaving detectable signatures in species abundances and species composition. The impact of low-intensity drivers on performance can, however, differ between individuals of the same species (intraspecific trait variation leading to tolerant vs sensitive individuals; see Chapter 6). Therefore, intraspecific trait-based metrics should be more responsive to this type of driver compared to species abundance or composition metrics. For example, a reduction of the ratio between leaf area and sapwood area measured along with time in the same individuals can indicate a decrease

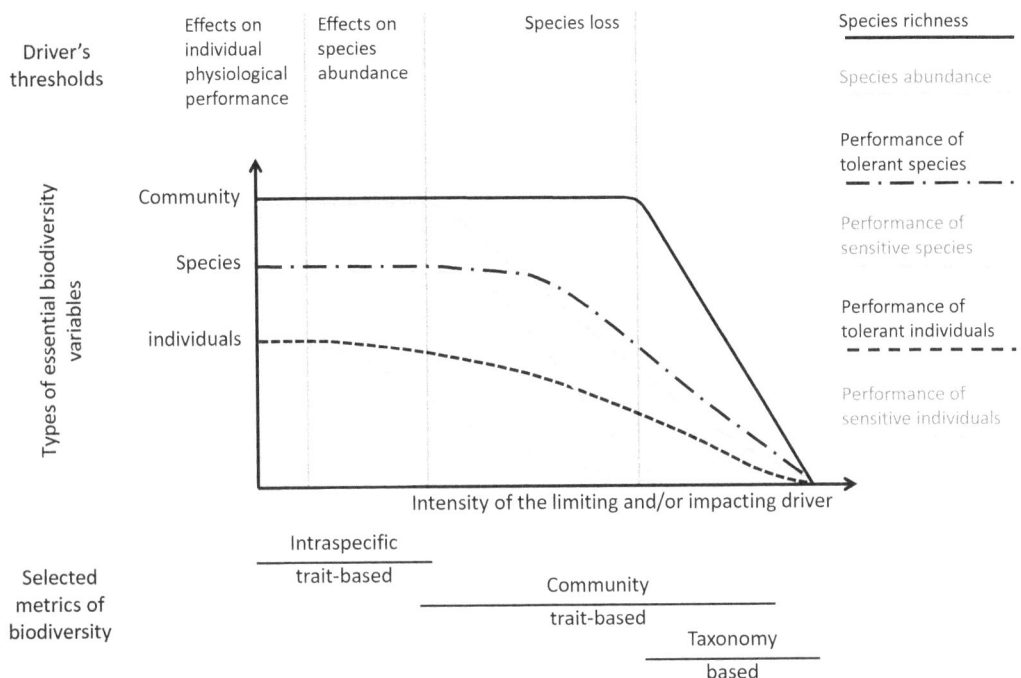

Figure 12.1 Conceptual figure showing the effects of driver intensity on Biodiversity-Change Indicators, from individual's performance to community species composition. A low-intensity driver is reflected by intraspecific trait-based metrics. Increasing the intensity of the driver should be reflected in community trait-based metrics, as sensitive species will reduce abundance before tolerant ones, with consequent changes in community trait abundance distribution. Under high-intensity drivers, species loss should be reflected in taxonomic-based metrics. Adapted from Branquinho et al. (2019), distributed under a Creative Commons Attribution License (CC BY) http://creativecommons.org/licenses/by/4.0/. © The Author(s) 2019; Open Access.

in water availability, as lower values of this trait can result in increased water economy. However, ecophysiological traits may also reflect changes in community functional composition. Increasing the intensity of the driver will affect sensitive species before tolerant species, resulting in changes of species' relative abundances with consequent changes in community functional composition reflected in community trait-based metrics. In this case, metrics such as species richness do not change. Following the previous example, an intensification of drought periods will ultimately lead to increased mortality of individuals from species that lack strategies (and associated traits) to either avoid or tolerate this stress. This will leave signatures in community trait-based metrics such as community weighted mean trait values (CWM) and functional diversity (FD). Finally, high-intensity drivers will ultimately lead to species loss, showing strong patterns in taxonomy-based metrics, as well as trait-based metrics.

The traditional paradigm of biodiversity monitoring is founded on the assumption that patterns of species abundance are often driven by some key environmental factors. Therefore, shifts in species composition are good indicators of shifts in environmental factors. Monitoring species (or a group of species) that are known to be sensitive to certain environmental factors are used to indicate changes in environmental conditions, which can be associated with anthropogenic disturbances like, for instance, pollution, eutrophication and land-use change. Such species or groups of species are generally referred to as 'bioindicators'. Prominent examples are lichens as indicators of air quality, or certain aquatic invertebrates (e.g. gilled snails, stoneflies) as indicators of water quality (Metcalfe 1989; Conti & Cecchetti 2001). This approach is especially useful when human-induced effects vary in short timescales. The discharge of pollutants in rivers (e.g. domestic sewage, leachate from agricultural fields or landfills) and in the air (e.g. fossil fuel combustion, fine particulate matter) is intermittent. Therefore, we would need to continuously measure abiotic conditions (e.g. pollutant concentration) to properly describe the impacts of human activities on ecosystems, while organisms integrate these short-term events experienced during their life cycle. The response of organisms to these scattered events, such as occasional discharge of pollutants, is often reflected in more long-lasting shifts in species composition patterns. This means that bioindicators can integrate environmental variation in more buffered variations of species composition. In this way, bioindicators can even be used to estimate average values of environmental conditions and their consequences for ecosystem processes. For instance, both plant species composition and CH_4 emissions in wetlands are determined by the same driver, i.e. shifts in water table level. Therefore, by assessing plant species composition, it is possible to estimate annual emissions of CH_4 from wetlands (Dias et al. 2010), the latter being much more expensive and time-consuming to measure. These taxonomic-based approaches can be very effective, but they lack generality beyond locations and regional floras or faunas. Therefore, in a new environment someone would need to again investigate which species are sensitive and which are tolerant, except when sensitivity to the driver is conserved in higher taxonomic levels (e.g. insect orders Trichoptera, Plecoptera and Ephemeroptera as indicators of water quality). Additionally, as mentioned by Branquinho et al. (2019), a taxonomic-based approach can be ineffective at detecting low-intensity drivers. So, how can the use of traits add to the traditional species-centric monitoring approaches?

One of the first explicit calls to incorporate trait metrics as a tool for biodiversity monitoring and ecosystem service assessments was put forward by Díaz et al. (2007). They presented a conceptual and methodological framework comprising models of increasing complexity, which test the link between drivers and ecosystem properties and services. As we saw in Chapter 9, these models start by considering abiotic factors, then adding community-weighted mean trait values (CWM), functional diversity and idiosyncratic species effects in order to find the best predictive model. Since then, a number of studies have proposed conceptual and methodological approaches that underline the importance of the functional aspects of species in the response of biodiversity to multiple stressors and drivers (among others Vandewalle et al. 2010; Pfestorf et al. 2013). These ideas were based on findings across different taxonomic groups that showed shifts in functional community components after some change, mostly degradation, in environmental conditions. For instance, rotifer functional group composition shows a stronger response to water quality than species composition (Oh et al. 2017). Similarly, functional groups of birds, rather than their taxonomical diversity, are used to indicate the recovery of bird communities in coastal dune forests in South Africa (Rolo et al. 2017). Also, Nunes et al. (2017) showed that both CWM traits and plant functional diversity respond to increasing aridity in Mediterranean vegetation, while no predictable shifts in species diversity were apparent. Even when using classical and well-established bioindicators, such as lichen species (Giordani et al. 2012) or testate amoebae (Fournier et al. 2012), the addition of information about their traits is advocated to increase predictability and generalizations beyond taxa and locations. In this way, trait-based metrics should be considered as complementary to the existing biodiversity indicators, such as species richness (Feld et al. 2009; Vandewalle et al. 2010; Matos et al. 2017).

This leads to the important question of how we should use trait-based metrics in combination with taxonomic-based metrics. Functional indicators might be good for predicting community responses to drivers but do not provide taxonomic information as to which species are at risk of being lost from the system. For example, taller vegetation could be an indicator of increasing nutrient load or decreasing disturbance. Thus, a simple indicator such as plant height could be useful for generalizing results beyond locations. However, this approach might be ineffective for determining the risk of losing a given threatened species or an endemic taxon. Therefore, it is generally believed that functional indicators could complement taxonomic indicators to achieve multiple monitoring goals (de Bello & Mudrák 2013; Fig. 12.2). Metrics based on taxonomic, functional and phylogenetic aspects of diversity often show a spatial mismatch for different taxa (Devictor et al. 2010; Monnet et al. 2014; Cadotte & Tucker 2018) suggesting that they provide complementary information on the effect of drivers. Using metrics based on only one aspect of diversity thus may result in poor decisions when managing (semi-)natural ecosystems or prioritizing areas for conservation (Cadotte & Tucker 2018). Therefore, the functional trait structure of communities and ecosystems is not set out to substitute taxonomic or other indicators, but rather to form a complementary addition that provides different information, often of causal underlying mechanisms, than that of more traditional indicators. This conclusion has resulted in different propositions on how to combine these complementary aspects of diversity for biodiversity monitoring to achieve more integrative conservation strategies (Devictor et al. 2010; Pardo et al. 2017; Cadotte & Tucker 2018).

One important criterion of a useful biodiversity indicator is its applicability across different habitats and its ease of use for practitioners. The conservationists who perform the actual monitoring are often not the same people who draw up and suggest indicators in a more academic setting. It is therefore crucial that indicator traits are also easy to measure or access, and have a known, standardized measurement protocol (see Chapter 2). In this context, it is important to consider whether the same traits can be used as indicators in different regions or biomes. One of the promises of functional ecology is to deliver general models that are relatively independent of species composition (Chapters 1 and 4), increasing their geographic domain of application. However, the response of a trait to a particular environmental change can differ between regions,

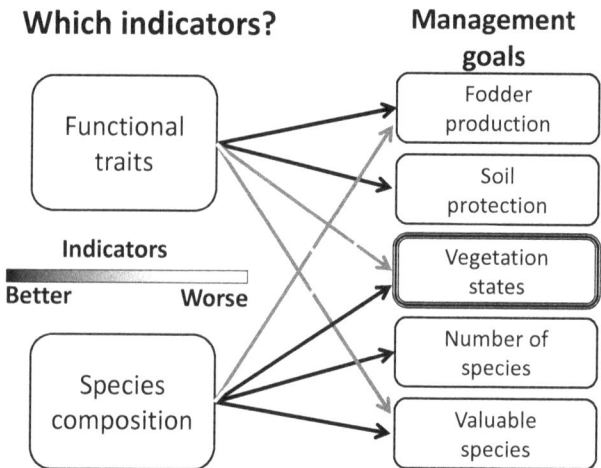

Figure 12.2 Hypothetical framework suggesting biodiversity indicators for different management goals. It shows that the suitability of functional and taxonomic metrics change with the management goal and should therefore be considered complementary. Taken from de Bello and Mudrák (2013), with permission from Wiley. © 2013 International Association for Vegetation Science.

possibly due to the mediation of different climatic conditions (Vesk & Westoby 2001; Pakeman 2004; Hatfield et al. 2018; Thornhill et al. 2018). Taking grazing as an example of an environmental pressure, inconsistent responses may be due to the presence of different trait strategies across sites, so that effects of land-use changes depend on the ecological strategies that are in fact available in the species pool (de Bello et al. 2005). For instance, if grazing promotes annual plant species under drier conditions, but relatively few of these short-lived annuals occur in wetter, temperate habitats, then longevity cannot be used as a general predictor. Within bioregions with more similar species pools, trait responses can indeed be consistent. For instance, Majeková et al. (2016b) showed that the functional responses of single species and entire communities to land-use intensification are consistent across meadows within a region, despite wide variation in productivity and soil moisture.

In fact, grassland monitoring provides a good example of how trait-based approaches can be used to increase generality of monitoring schemes, while also allowing for good communication with practitioners (Garnier et al. 2016). For instance, Herb'type© comprises a tool for monitoring grassland ecosystem services, in particular their productivity, developmental time and nutritive value (Duru et al. 2010). It is based on classifying grass species into five groups (see Chapter 3 and the 'R material Ch3' for more details on species classification approaches) according to their growth rate, phenology and six plant traits: SLA, LDMC, leaf life span, mechanical leaf resistance, flowering date and maximum plant height. This classification distinguishes species based on their strategies of resource acquisition or conservation, and timing of plant development. It has been successfully used to predict grassland productivity in Europe and South America (Duru et al. 2008; Cruz et al. 2010). To implement Herb'type©, a survey is needed to access the relative abundance of each grass functional group, plus legumes and other dicotyledons. This is done in a simplified way by scoring dominant species on small plots, which makes it easy for practitioners to implement. This tool can be used to assess the value of a grassland in terms of its ecosystem services, informing best management practices, and assisting the monitoring of biodiversity changes associated with service provision.

The examples provided in this section demonstrate how including traits in biodiversity monitoring can increase our ability to detect, and often understand, ecosystem responses to environmental and management changes. This can be especially important when biodiversity-change drivers are not strong enough to cause species loss, resulting in low signal within taxonomic-based metrics (Branquinho et al. 2019). But as we learned in Chapter 9, shifts in species' relative abundance can already result in strong changes in ecosystem processes due to changes in the distribution of effect trait values. In this way, traits can also be used to make a link between shifts in biodiversity and their consequent effect on ecosystem functioning and service provision via a response-and-effect framework (Suding et al. 2008).

12.2 Traits in Agricultural Systems

Modern agriculture faces an astonishing challenge imposed by the necessity to support a growing human population while simultaneously decreasing economic costs and environmental impacts. We can say that, in a broad sense, two different roads are being taken to achieve this. More traditionally, gains in monoculture productivity are achieved by increasing agricultural inputs (e.g. fertilizers and pesticides), improving management techniques, machinery and, more recently, by high-tech approaches like genetically modified organisms (GMO), the latter often followed by an increase in pesticide input with potentially strong environmental and health impacts. Alternatively, increasing diversity in crops in both space and time (culture rotation) is recognized as an important strategy for increasing yield and reducing risks in the face of environmental variation. While traditional agricultural research heavily focuses on crop yield, trait-based approaches can be used for placing agricultural systems in a broader perspective by searching the causes and consequences of shifts in agroecosystem structure and functioning. This includes the effects of agroecosystems dynamics on non-crop biodiversity maintenance, and how, in turn, non-crop biodiversity influences agroecosystem yield and stability, and the contribution of agroecosystems to global primary productivity and other biogeochemical cycles (Martin & Isaac 2015).

To apply trait-based approaches to agroecosystems, we need to consider different sources of trait variability. One of the most important sources of variability is the consequence of phenotypic changes resulting from domestication. Naturalists and ecologists have long been interested in the consequences of domestication on phenotypes (e.g. Darwin 1859). Applying a response-to-effect trait framework (see Chapters 9 and 10) to phenotypic changes caused by domestication can help us to understand and predict the functioning of agricultural systems (Garnier & Navas 2012; Damour et al. 2018). Plant species have been selected for many characteristics. While several traits can be selected (e.g. larger or smaller seeds, larger fruits, growth rate) for different purposes (e.g. tolerance to salinization, attractiveness, productivity), it is important to note that traits do not vary independently, and selection is constrained by different trade-offs. Selection for one trait might select against another trait that someone would also like to optimize. This leads to the notion of the domestication syndrome (Milla et al. 2015), in which selection operating on a particular trait results in a change of values of associated traits. For instance, the selection of cultivated varieties with shorter life cycles has also unintentionally selected for traits that favour higher rates of resource acquisition (Milla et al. 2015; Roucou et al. 2018). This has important consequences for ecosystem functioning as demonstrated by García-Palacios et al. (2013) with cultivars showing higher rates of decomposition compared to their undomesticated relatives, which could ultimately result in a more open and leaky nutrient cycle in agricultural systems.

The original response-to-effect trait framework (Lavorel & Garnier 2002) describes the biophysical functioning of the ecosystem based on response and effect traits ('Biophysical module'). Recently, Damour et al. (2018) expanded this framework for agroecological systems to include a biophysical and a decision module. In agroecological systems, the 'Biophysical module' should account for both planted and spontaneous species (planned and associated biodiversity, respectively; Altieri 1999), which are subjected to different degrees of control (Fig. 12.3). While planted species are directly selected and managed by farmers, spontaneous species generally receive less attention, and management practices can either increase or decrease environmental suitability for them. The 'Decision module' consists of farmers' choices that follow a set of rules, which result from production objectives, biophysical constraints, market demand, social and cultural preferences and constraints related to farmers' technical knowledge (Damour et al. 2018). Decisions include choosing species, based on their effect traits, which can deliver given ecosystem services (e.g. crop production, nitrogen fixation) and their response traits, allowing them to cope with some set of environmental conditions (Martin & Isaac 2015). While the traditional agronomic trait approach focuses strictly on crop yield, the functional approach can help broaden the scope of farmers' decisions based on species traits. For instance, Isaac et al. (2018) found evidence that farmers diagnose major management concerns and adjust their practice, at least in part, according to intraspecific variation in traits related to the plant economic spectrum (Wright et al. 2004), which ranks plants according to their light environment and soil fertility preferences. Selecting the right traits can not only help to identify the species that will result in a desired ecosystem service, but also help to design crop mixtures that improve productivity or enhance multifunctionality in agroecosystems as predicted by the niche complementarity hypothesis (Malézieux et al. 2009; Blesh 2018).

In this context, crop weeds are a main component of the associated diversity in agricultural systems (Altieri 1999). Although weeds are mainly known for their negative effects on crops, leading to a substantial reduction in productivity, this purely negative view of weeds is being challenged by the rediscovery of their positive effects

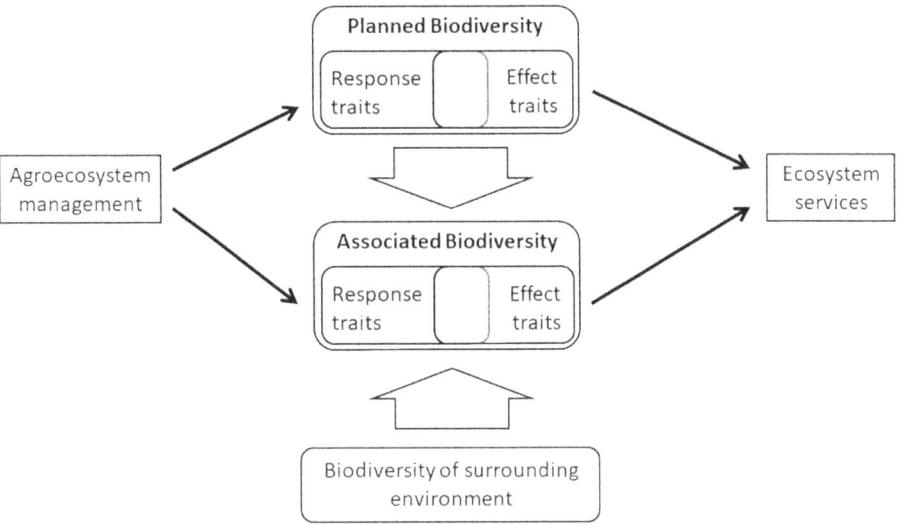

Figure 12.3 Scheme showing drivers of biodiversity in agricultural ecosystems. Both planted (i.e. 'planned biodiversity') and spontaneous (i.e. 'associated biodiversity') species respond to management practices through their response traits. Associated biodiversity is also affected by the biodiversity of the surrounding environment, which acts as a source of species, and by interactions with planted species, which are the main biotic component in agricultural systems, often representing a strong biotic filter. Effect traits of both planned and associated biodiversity determine the provision of ecosystem services in agroecosystems. Modified from Altieri (1999), with permission from Elsevier. Copyright © 1999 Elsevier Science B.V. All rights reserved; adapted version, permission via RightsLink and additional email communication.

(Gunton et al. 2011). This change in how we perceive the role of weeds in agricultural systems highlights the importance of developing integral management practices, where weeds are considered as an integral (and sometimes desirable) part of the system, as has been elegantly explained by Garnier et al. (2016). The application of the trait-based approach to crop weed needs to account for hierarchical filters. Climatic and edaphic conditions establish a backdrop, which can still be substantially modified by management practices. Additionally, crops themselves, which are by far the dominant biotic component in agricultural systems, can strongly affect environmental suitability for weeds. Studies show that weeds that are considered problematic often show a similar suite of traits (Storkey 2006; Gunton et al. 2011). Weeds with a high growth rate, annual life cycle and ruderal strategy (following the C-S-R classification; see Chapter 3 for more details) are often able to escape herbicide applications, thus contributing to their 'weediness'. The interaction between weeds and crops will ultimately determine productivity and economic losses. Studies demonstrate that differences in leaf area and plant height between weeds and cultivated plants, measured at early stages of crop development, are good indicators of the intensity of their competitive interactions and the consequent losses in yield (Booth & Swanton 2002; McDonald et al. 2010). Leaf area and plant height are traits related to resource acquisition and in this way can indicate competitive hierarchies. However, an important service crop weeds can provide is their support (via both providing resource and habitat) of overall biodiversity. For instance, flower traits (e.g. corolla length, type of reward) and flowering phenology can have important impacts on resource provision for pollinators and also, perhaps less intuitively, for natural enemies of herbivores, which often rely on supplementary plant resources (Perović et al. 2018). This can translate into the provision of important ecosystem services, including pollination and biological control in agroecological systems, that potentially overrule negative effects due to resource competition between weeds and crops.

12.3 Ecosystem-Based Solutions

The field of ecosystem management finds itself in the intersection between science, policy and practice and is regularly confronted with new ideas and terminologies trying to connect these different fields of knowledge. The recently formulated concept of nature-based solutions seeks to promote nature as a means of providing solutions to environmental problems (Nesshöver et al. 2017). Nature-based solutions (hereafter NbS) can be considered as an umbrella term, covering many ecosystem-based approaches for connecting nature to human well-being, e.g. ecosystem-based adaptation, green (and blue) infrastructure, ecosystem services and natural capital (Table 12.1). Here we use the term nature-based solutions to refer to initiatives that use natural elements (species and ecosystems) as a strategy to provide benefits to people (Díaz et al. 2018). Below we discuss how the use of trait-based approaches can aid the management of natural elements within the landscape so that we maximize the services they provide.

12.3.1 Green Urban Infrastructure

Cities are often perceived as the antithesis of nature. While urbanization strongly shapes the landscape, converting vast areas of natural systems to an anthropic environment, the relevant ecological processes are still performed by the organisms found in urban landscapes. The recent recognition of the importance of natural elements for ecosystem service provision, which contributes to human well-being in urban societies, has boosted research in urban ecology (Luederitz et al. 2015; Ziter 2016). By providing ecosystem services, green urban infrastructures (GUI; i.e. trees, parks, gardens, lawn, green roofs etc.) are considered an essential element for making cities more sustainable, helping cities to adapt to climate change and reducing risks of disasters, e.g. floods and landslides (Brink et al. 2016; Nesshöver et al. 2017). GUI often provide multiple ecosystem services (see Fig. 12.4), contrary to grey infrastructures, which are traditionally designed for a single purpose. For instance, drainage and stormwater storage systems provide flood risk reduction only. On the other hand, green roofs and urban trees reduce flooding risk through temporary water storage and transpiration, while also providing thermal comfort, aesthetic values, air purification and the promotion of biodiversity. This example also illustrates that different GUI can contribute to the same

Table 12.1 Non-exhaustive overview of different concepts related to nature-based solutions. Modified from Nesshöver et al. (2017), distributed under a Creative Commons Attribution License (CC BY). Copyright © 2017, Elsevier.

	Concept			
	Green/blue infrastructure	Ecosystem-based adaptation/mitigation	Ecosystem services	Natural capital
Definition	Planned and managed, spatially interconnected network of multifunctional natural, semi-natural and man-made green (land-based) and blue (water related) features. This includes agricultural land, green corridors, green roofs, urban trees, urban parks, forest reserves, wetlands, rivers, coastal sand, artificial channels, ponds, urban waste water networks and other aquatic ecosystems.	Policies and measures that take into account the role of ecosystem services in adapting, mitigating or reducing the vulnerability of society to climate change. Such policies can involve national and regional governments, local communities, private companies and NGOs in addressing the different pressures on ecosystem services, which can be relevant at distinct scales.	Human benefits derived directly or indirectly from ecosystem structure and processes. This includes provisioning services (e.g. food, water and building materials), cultural services (e.g. recreation, tourism, sense of place) and regulatory and supporting services (e.g. protection against flood or erosion, climate regulation, soil formation and nutrient cycling). We can understand the natural capital as the stock of assets, and ecosystem services as the flow of benefits derived from those assets.	The stock of living and non-living components of ecosystems that directly or indirectly yield benefits (i.e. ecosystem services) to humans.
Aim of employing concept	To highlight the ecological, social and economic benefits derived from natural solutions. It helps us to understand the value of the benefits natural elements provide to human society, encouraging investments to protect and enhance them. It aims at regulating storm flows, flood risk, air temperatures, greenhouse gases, and enhancing overall environmental quality.	To recognize the importance of natural and semi-natural features in reducing societal vulnerability to disasters and climate changes. This should help to avoid relying only on infrastructure that is expensive to build, when nature can often provide cheaper and more durable solutions.	To understand and describe how nature benefits humans, helping to inform and improve social and political processes to improve the management and governance of ecosystems.	By enabling natural systems to be valued and managed equally to other forms of capital (financial, human, social, manufactured), it is expected that considering natural capital can help to improve decision-making for various sectors.

Table 12.1 (*cont.*)

	Concept			
	Green/blue infrastructure	Ecosystem-based adaptation/ mitigation	Ecosystem services	Natural capital
Potential relation to NbS	Similar to NbS in some areas and can sometimes be synonymous. Or green/blue infrastructures can be seen as means or tools through which we can achieve NbS.	Ecosystem-based adaptation should be part of NbS, to ensure solutions are climate-adapted.	Ecosystem services concept can be an excellent way to consider solutions during NbS design and appraisal; however, their use should not be restricted to single or few ecosystem services and their beneficiaries.	The natural capital concept can help demonstrate the role of nature in meeting human needs, and hence the value of considering NbS versus other types of interventions.

12.3 Ecosystem-Based Solutions

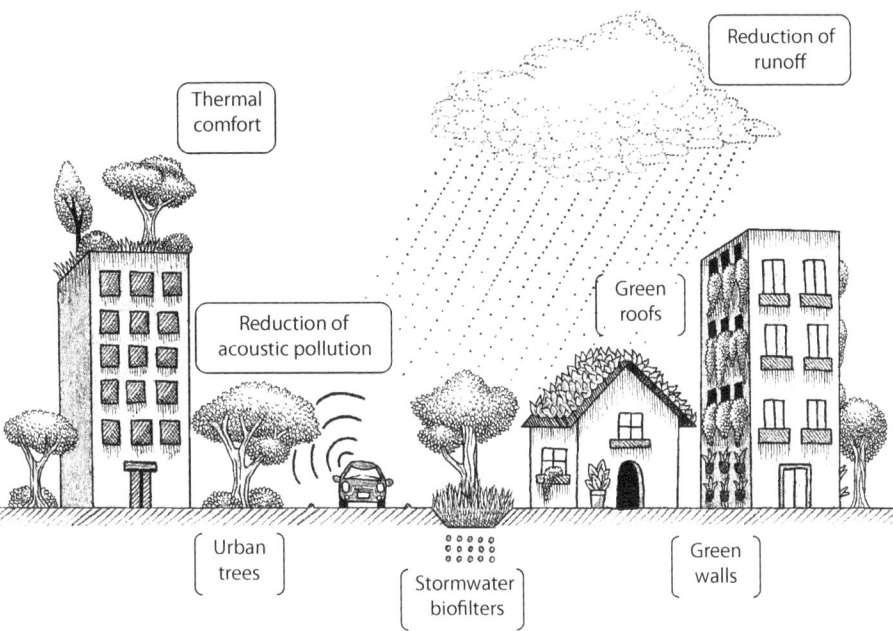

Figure 12.4 Co-benefits provided by different types of green urban infrastructure. Figure shows different green urban infrastructures (between brackets) providing the same ecosystem service (grey boxes). For instance, green roofs, stormwater biofilters and urban trees can reduce runoff through complementary mechanisms (e.g. stormwater storage, enhanced infiltration and transpiration). Nevertheless, green urban infrastructures also provide multiple ecosystem services. For instance, urban trees reduce runoff, provide thermal comfort and reduce acoustic pollution. Figure created by Luís Gustavo Barretto.

ecosystem service and can be planned jointly as a strategy for maximizing its provision (Demuzere et al. 2014).

While the benefits of GUIs are well characterized as a function of their presence and cover in urban landscapes (Demuzere et al. 2014), some studies show that species strongly differ in their ability to provide ecosystem services (Read et al. 2010; Grote et al. 2016). This opens questions regarding which traits maximize the provision of different services in these systems, and how community functional composition can mediate the provision of ecosystem services in urban environments (Livesley et al. 2016; Schwarz et al. 2017). Identifying traits involved in ecological processes responsible for ecosystem services and disservices can considerably improve our ability to plan more efficient GUIs, while also informing us of their interaction with sociocultural aspects, such as preferences for particular tree species planted in cities (Conway 2016).

Urban trees are an important GUI, providing many services but also disservices to urban inhabitants (von Döhren & Haase 2015). Grote et al. (2016) showed that tree species most commonly found in Western European cities strongly differ in their traits and, consequently, in their provision of services and disservices. The tree traits considered most important for mitigation of air pollution are canopy density, foliage longevity and water-use strategy, while reproductive traits are responsible for species allergenicity through pollen production. However, studies addressing residents' preferences for tree species show that the preferred trees do not necessarily possess the traits that enhance provision and regulatory services (Pataki et al. 2013; Conway 2016). For instance, aesthetic reasons, which can be quite dependent on the cultural context, rank high in people's motivations for selecting species. This cultural service can be linked to showy flowers and, in some places, to autumn leaf colours. Some desired traits, like slow growth and low debris production, can oppose the provision of other services like thermal comfort and carbon sequestration. This means that adding the socio-cultural dimension to the selection of urban tree traits might bring additional trade-offs between different ecosystem services.

Stormwater biofilters are another type of multifunctional GUI. They are designed to work as temporary ponds, reducing flooding risks during peak rain discharges, filtering pollutants from runoff water and adding aesthetic value to urban settings (Levin & Mehring 2015). Read et al. (2010) show that some root traits (i.e. length of the longest root, rooting depth, total root length and root mass), especially in combination with high growth rates, give species the capacity to effectively remove N and P from stormwater. Winfrey et al. (2018) showed that stormwater biofilters often decrease species richness with time without reducing functional diversity. This suggests that the initial species composition planted in biofilters provides functional redundancy, permitting functional stability in the face of environmental and species compositional changes. Increasing functional diversity is expected to increase niche complementarity, leading to enhanced service provision, for instance the retention of different nutrients and pollutants such as copper (Levin & Mehring 2015). Also, the introduction of engineering species, i.e. species with traits showing strong effects on the physical environment (Jones et al. 1994), can be used to design biofilters with enhanced functions. For example, earthworms are able to increase soil porosity, promoting higher rates of water infiltration and, when combined with plants that have the right root traits, might result in optimal service provisioning.

12.3.2 Restoration with Functional Targets

Restoration is broadly defined as the practice of assisting the recovery of degraded ecosystems. Traditionally, restorationists have adopted the main goal of hastening the return of a system to a pre-disturbance stable state. Restoration success is most often measured in relation to a static compositional target that is assumed to reflect proper functioning based on a reference system (i.e. similar undisturbed system) (Perring et al. 2015). Taking a more functional perspective, instead of targeting a given species composition, the target is the provision of a preferred ecosystem process or service delivery. To this end, different species compositions could provide the same results, thus species selection should be determined by their traits. This trait-based selection of species with similar functions to some reference species might be particularly important when assembling communities that can cope with present and future environmental stress.

With this in mind, several researchers have recently questioned the use of historical ecosystem conditions as a reference and target for restoration due to the pervasive and unprecedented rates of human-induced environmental changes (Harris et al. 2006). For instance, atmospheric CO_2 concentration is an important factor determining the balance between herbaceous and woody components in savanna ecosystems (Bond et al. 2003). Also, climate change can shift the balance of carbon fluxes between soil and the atmosphere (Dorrepaal et al. 2009), leading to a reduction of soil carbon concentration (Bellamy et al. 2005). In a rapidly changing world, the use of historical references, like the ecosystem properties mentioned above, might have limited value for setting restoration targets. This led researchers to think about future-oriented restoration targets, which can be classified into two different modes: responsive and proactive targets (Harris et al. 2006). Responsive targets seek to restore or build ecosystems that are resilient to future changes. This includes considering functional and phylogenetic facets of diversity as a way to enhance the chance that a community will be able to cope with environmental changes (Elmqvist et al. 2003). Also, if we take a landscape perspective to restoration projects, connectivity becomes a key property to be enhanced or restored (Perring et al. 2015). This means that we should focus not only on traits that directly relate to drivers of environmental change, and traits that define ecosystem services, but also traits related to dispersal ability. Proactive targets seek to mitigate or reverse the causes of environmental changes: for instance, designing restoration practices to sequester carbon or to influence local climate (Harris et al. 2006).

This paradigm shift in restoration ecology has led restorationists to embrace multiple goals, and to aim at functional targets in restoration programmes (Laughlin 2014a; Perring et al. 2015). In this sense, restoration projects with functional targets can also be framed under the umbrella of NbS, as they can be designed to enhance the provision of specific ecosystem services (e.g. carbon sequestration, erosion control, flood prevention and water purification; Keesstra et al. 2018). Functional ecology can be especially important as a conceptual and methodological framework for designing ecosystems that enhance specific functional properties (e.g. resilience or a given service provision). If we know which traits are responsible for a given functional property, it is possible to translate restoration goals into functional trait targets, i.e. selection of species with the right trait values for the restoration goal. In this sense, traits can be especially important for bringing ecological theory to the design of restoration projects. For instance, when restoring an ecosystem under high biological invasion pressure, managers can increase the abundance of native species that outcompete the invading species (Funk et al. 2008). This can be done by choosing native species that are functionally similar to the invader, as predicted by the limiting similarity theory (Conti et al. 2018), or by increasing the abundance of native species with traits that provide a higher competitive ability, as predicted by the competitive hierarchy theory (Keddy 1992a). Additionally, in a scenario of fast environmental change, restoration should seek to build resilient communities. The insurance hypothesis predicts that this can be accomplished by enhancing the response of diversity (i.e. diversity of response traits *sensu* Lavorel & Garnier 2002; see also Chapter 9) of species providing an important function for the community (Elmqvist et al. 2003), such as, for instance, providing resources (e.g. fruits or flowers) during winter for non-migrating bird species. When focusing on the effect of species on ecosystem processes, both functional identity (mass-ratio hypothesis; Grime 1998) and functional diversity (diversity hypothesis; Díaz et al. 2007) can be important. The call to use ecological theory to improve restoration practices is not new (Palmer et al. 2016), and the use of functional approaches is now helping to bring theory and practice together (Perring et al. 2015).

But how can we put this into practice when designing communities with a given functional composition? It can be quite straightforward to select species within a specific range of trait values. For instance, we can choose species with low SLA to maximize survival in habitats with stressful conditions. However, if practitioners want to select species based on several traits (e.g. wood density, leaf area, life form), or to manipulate different components of community functional structure (e.g. community weighted trait means (CWM) and functional diversity), this can turn it into a much more complex task. This is because traits often show complex interrelationships (Messier et al. 2017; Kleyer et al. 2019) and components of functional structure do not vary independently (Dias et al. 2013b). Recently, Laughlin (2014a,b) proposed that this task could be accomplished by using trait-based models of community assembly for translating trait targets into ranges of species abundances. The maximum entropy model (Maxent), or community assembly by trait selection model (CATS; Shipley et al. 2006), uses entropy maximization for predicting a distribution of relative abundances of species based on species traits and community functional structure. The trait space model (Laughlin et al. 2012) is proposed as an additional modelling approach, incorporating intraspecific trait variation. These models are flexible and can be adjusted for a variety of purposes. For instance, Rosenfield and Müller (2017) used the trait space model to monitor whether trajectories of restoration projects were heading towards or away from functional targets based on reference areas. These models can even be used to maximize the similarity between the CWM of several traits within the planned community, and the traits values of an invader species (Laughlin 2014a), with the aim of increasing invasion resistance (Funk et al. 2008).

12.4 Exotic Species' Impacts on Ecosystems

Invasive species are considered a significant threat to biodiversity, while also being responsible for major environmental and economic problems (Vilà et al. 2011). Many studies suggest that a trait-based approach may help to implement management practices for invasive species control. Communities should be more resistant to invasion by exotic species that are functionally similar to native species and, therefore, are expected to compete for the same resources; or by exotic species which have traits that result in lower competitive ability as compared to native species (Price & Pärtel 2013; Conti et al. 2018). As mentioned above, this knowledge of species traits can be used on restoration or enrichment projects, aiming at increasing the resistance of local communities to invasion. Invasion by exotic species can be detrimental for nature conservation purposes, leading to biotic homogenization and possibly losing the identity of natural communities as we know them. Another important question, and one still highly debated in the literature, is: what impact do exotic species have on ecosystem functioning? Of course, there are exemplary cases showing strong effects of exotic species on ecosystem processes and properties (Vilà et al. 2011). However, when these effects are not so apparent at first sight, exotic species impacts are often assumed to exist rather than being quantified. This was well illustrated by Didham et al. (2005) with a provocative question on whether exotic species are the drivers of ecological changes or merely 'passengers' of such changes. While exotic species can in some cases promote strong environmental changes, in other cases they enter into the system after a disturbance, and thus are actually 'profiting' from such changes.

If we want to predict the effects of exotic species on ecosystem processes, then we must consider effect traits of both the exotic species and the community which might be invaded (Finerty et al. 2016). Traditional frameworks for understanding the ecological impact of invaders consider only the abundance and the effect, per individual (or biomass), of the invader (Parker et al. 1999). However, this approach is not helpful if we wish to understand whether the invader species is bringing more of the same traits as are already present in the native community, or, on the contrary, if it is capable of changing community functional composition of traits related to ecosystem functioning (effect traits). Finerty et al. (2016) showed that species effects on decomposition can be predicted by their traits rather than from their exotic or native status. More importantly, exotic species only affected decomposition in litter mixtures when they were abundant enough to shift community functional composition (CWM, functional diversity or both). This shows that simply comparing the potential effects (per biomass basis) of exotics and natives on ecosystem processes provides limited information as to whether a given exotic species is expected to promote relevant effects within the system (Liao et al. 2008; Jo et al. 2016).

This reinforces the idea that we should, for the most part, avoid judging the impact of exotic species based on where they come from, but instead base it on who they are and what they do (Davis et al. 2011). For this, we need to consider the trait composition of the community where they are present. However, the importance of the biogeographical origin of invading species should not be flagrantly dismissed (Buckley & Catford 2016; Prescott & Zukswert 2016). In fact, the distinct biogeographical origin of an exotic species can have significant consequences for its interactions with its new neighbouring native species (see Buckley & Catford 2016 for a review). This can be particularly important for species interactions with natural enemies. For instance, studies found that exotic plant species show lower negative plant–soil feedbacks when growing in conditioned soil from a new location when compared with conditioned soil from its original range (Callaway et al. 2004; Diez et al. 2010). Also, negative plant–soil feedback tends to increase with time since colonization, demonstrating that exotic species tend to accumulate natural enemies which will feed on them given enough time (Diez et al. 2010). This shows how evolutionary processes can mediate species interactions and modulate exotic species invasiveness.

Evolutionary changes in traits related to dispersal can also play an important role in species invasion and range expansion. This was formalized in the theory of 'runaway' evolution and spatial selection (Phillips et al. 2010). According to this theory, populations that are at the edge of a species' range should comprise individuals with high dispersal capacity. If such dispersal traits are heritable, breeding between these individuals will result in phenotypes with even higher dispersal abilities. This will be especially true when other factors, like climatic conditions, do not restrict range expansion. This scenario will occur for many introduced species which encounter new expanses of area with suitable climatic conditions ripe for colonization. For instance, Phillips et al. (2006) showed that individuals of the invasive cane toad (*Bufo marinus*) showed higher values of relative leg length, a trait related to dispersal capacity, at the expanding front of their range. Also, Monty and Mahy (2010) showed increased relative pappus diameter in diaspores of the invading Asteraceae *Senecio inaequidens*. On the other hand, a similar study could not find evidence of runaway evolution for the related species *Senecio madagascariensis* which is currently invading Australia (Bartle et al. 2013). In order to understand the factors that can limit runaway evolution at range edges, such as the Allee effect (i.e. positive correlation between population density and the mean individual fitness; for instance, a low pollination success resulting from low population density) or environmental filtering, we must consider additional traits other than those related to dispersal alone.

12.5 Ecological Literacy

The concept of *nature* is changing among ecologists. In a world of rapid change and ubiquitous human influence, the separation of Nature and Culture is becoming meaningless (Cronon 1996). Preserving 'untouched' nature reserves is without doubt important for species conservation and the regulation of important climatic and biogeochemical processes. However, when designing sustainable socio-economic systems, it is also crucial that we consider the role nature plays in service provision within urban, rural and natural systems. This idea is well illustrated by the restorative continuum proposed by McDonald et al. (2016), idealizing a continuum of practices that can contribute to sustainability, ranging from the choice of renewable energy sources and building cities with enhanced biodiversity support, to restoration and preservation of entire natural ecosystems.

From this perspective, 'what do species do in ecosystems?' (Lawton 1994) is becoming a central question in both basic and applied ecology. Trait-based approaches have contributed considerably to our understanding of how species both respond to environmental changes and affect ecosystem functioning (Chapters 9 and 10), demonstrating their potential for supporting effective biodiversity monitoring, conservation management and provision of services, as we have discussed in this chapter. Despite this potential, using traits to elaborate policies for solving, preventing or mitigating environmental problems is still a challenge. Recent advances in conceptual frameworks (Andrés et al. 2012; Branquinho et al. 2019) and data standardization and management (Pérez-Harguindeguy et al. 2013; Moretti et al. 2017; Kissling et al. 2018) are paving the way for a wider use of trait-based approaches in applied sciences. However, the concept of the trait, and its relation to function, is not widely disseminated among either the general population or professionals. This lack of literacy in functional ecology might represent a strong barrier to communicating with decision makers, and incorporating trait-based approaches in environmental policies. Even more specialized professionals, like biologists, agronomists and environmental scientists, might not show a full comprehension of the concepts related to trait-based approaches. Functional ecology chapters are still missing from ecology textbooks, which might explain why many practitioners are still unfamiliar with trait-based ecology concepts and methods. Nowadays, there is still a shortage of textbooks that can be used for teaching and disseminating general concepts, as well as the rationale and methodology related to trait-based approaches (e.g. Swenson 2014a; Garnier et al. 2016). We hope the present book will also increase literacy in trait-based ecology, which will ultimately help the adoption of this approach for solving environmental problems. Additionally, actions promoting ecological literacy for the general population (Jordan et al. 2009) should also include basic principles on the relationship between traits and function (or functioning).

Take-Home Messages

- Trait-based metrics can be used as a tool for biodiversity monitoring, often showing more sensibility to environmental changes as compared to taxonomic-based metrics. This can be especially important for low-intensity environmental drivers, because they are expected to affect individual organisms' performance without necessarily leaving detectable signatures in species abundances and species composition.
- Trait-based approaches can help to broaden the scope of applied environmental sciences, helping to harness ecological theory to solving environmental problems.
- By considering response and effect traits it is possible to restore or create new ecological communities that are more resilient to environmental changes, or that enhance desirable ecosystem services.
- Poor literacy on functional ecology might represent a strong barrier to communicating with decision makers, and incorporating trait-based approaches in environmental policies.

References

Abrams, P. (1983) The theory of limiting similarity. *Annual Review of Ecology and Systematics*, 14, 359–376.

Ackerly, D. D. & Monson, R. K. (2003) Waking the sleeping giant: the evolutionary foundations of plant function. *International Journal of Plant Sciences*, 164, S1–S6.

Ackerly, D. D., Knight, C. A., Weiss, S. B., Barton, K. & Starmer, K. P. (2002) Leaf size, specific leaf area and microhabitat distribution of chaparral woody plants: contrasting patterns in species level and community level analyses. *Oecologia*, 130, 449–457.

Adler, P. B., HilleRisLambers, J. & Levine, J. M. (2007) A niche for neutrality. *Ecology Letters*, 10, 95–104.

Adler, P. B., Salguero-Gómez, R., Compagnoni, A., Hsu, J. S., Ray-Mukherjee, J., Mbeau-Ache, C. & Franco, M. (2014) Functional traits explain variation in plant life history strategies. *Proceedings of the National Academy of Sciences of the United States of America*, 111, 740–745.

Albert, C. H. (2015) Intraspecific trait variability matters. *Journal of Vegetation Science*, 26, 7–8.

Albert, C. H., de Bello, F., Boulangeat, I., Pellet, G., Lavorel, S. & Thuiller, W. (2012) On the importance of intraspecific variability for the quantification of functional diversity. *Oikos*, 121, 116–126.

Albert, C. H., Grassein, F., Schurr, F. M., Vieilledent, G. & Violle, C. (2011) When and how should intraspecific variability be considered in trait-based plant ecology? *Perspectives in Plant Ecology, Evolution and Systematics*, 13, 217–225.

Albert, C. H., Thuiller, W., Yoccoz, N. G., Douzet, R., Aubert, S. & Lavorel, S. (2010) A multi-trait approach reveals the structure and the relative importance of intra- vs. interspecific variability in plant traits. *Functional Ecology*, 24, 1192–1201.

Albouy, C., Leprieur, F., Le Loc'h, F., Mouquet, N., Meynard, C. N., Douzery, E. J. P. & Mouillot, D. (2014) Projected impacts of climate warming on the functional and phylogenetic components of coastal Mediterranean fish biodiversity. *Ecography*, 38, 681–689.

Ali, A. & Yan, E. R. (2017) Relationships between biodiversity and carbon stocks in forest ecosystems: a systematic literature review. *Tropical Ecology*, 58, 1–14.

Alonso, C. (2005) Pollination success across an elevation and sex ratio gradient in gynodioecious *Daphne laureola*. *American Journal of Botany*, 92, 1264–1269

Altermatt, F., Fronhofer, E. A., Garnier, A., Giometto, A., Hammes, F., Klecka, J., Legrand, D., Maechler, E., Massie, T. M., Pennekamp, F., Plebani, M., Pontarp, M., Schtickzelle, N., Thuillier, V. & Petchey, O. L. (2015) Big answers from small worlds: a user's guide for protist microcosms as a model system in ecology and evolution. *Methods in Ecology and Evolution*, 6, 218–231.

Altieri, M. A. (1999) The ecological role of biodiversity in agroecosystems. *Agriculture, Ecosystems and Environment*, 74, 19–31.

Andrés, S. M., Calvet Mir, L., van den Bergh, J. C. J. M., Ring, I. & Verburg, P. H. (2012) Ineffective biodiversity policy due to five rebound effects. *Ecosystem Services*, 1, 101–110.

Aqueel, M. A. & Leather, S. R. (2011) Effect of nitrogen fertilizer on the growth and survival of *Rhopalosiphum padi* (L.) and *Sitobion avenae* (F.) (Homoptera: Aphididae) on different wheat cultivars. *Crop Protection*, 30, 216–221.

Armbruster, P., Hutchinson, R. A. & Cotgreave, P. (2002) Factors influencing community structure in a South American tank bromeliad fauna. *Oikos*, 96, 225–234.

Arnold, S. J. (1983) Morphology, performance and fitness. *American Zoologist*, 23, 347–361.

Aubin, I., Messier, C., Gachet, S., Lawrence, K., McKenney, D., Arseneault, A., Bell, W., De Grandpré, L., Shipley, B., Ricard, J. P. & Munson, A. D. (2012) *TOPIC – Traits of Plants in Canada*. Sault Ste Marie, Ontario: Natural Resources Canada, Canadian Forest Service.

Auger, S. & Shipley, B. (2013) Inter-specific and intra-specific trait variation along short environmental gradients in an old-growth temperate forest. *Journal of Vegetation Science*, 24, 419–428.

Austin, M. P. & Smith, T. M. (1989) A new model for the continuum concept. *Vegetatio*, 83, 35–47.

Baraloto, C., Paine, C. E. T., Patino, S., Bonal, D., Herault, B. & Chave, J. (2010) Functional trait variation and sampling strategies in species-rich plant communities. *Functional Ecology*, 24, 208–216.

Bardgett, D. R. & Wardle, A. D. (2010) *Aboveground–belowground linkages: biotic interactions, ecosystem processes, and global change*. Oxford University Press. Oxford, UK.

Bardgett, R. D., Mommer, L. & De Vries, F. T. (2014) Going underground: root traits as drivers of ecosystem processes. *Trends in Ecology & Evolution*, 29, 692–699.

Barluenga, M., Stolting, K. N., Salzburger, W., Muschick, M. & Meyer, A. (2006) Sympatric speciation in Nicaraguan crater lake cichlid fish. *Nature*, 439, 719–723.

Bartle, K., Moles, A. T. & Bonser, S. P. (2013) No evidence for rapid evolution of seed dispersal ability in range edge populations of the invasive species *Senecio madagascariensis*. *Austral Ecology*, 38, 915–920.

Bartomeus, I., Gravel, D., Tylianakis, J. M., Aizen, M. A., Dickie, I. A. & Bernard-Verdier, M. (2016) A common framework for identifying linkage rules across different types of interactions. *Functional Ecology*, 30, 1894–1903.

Bascompte, J. & Jordano, P. (2007) Plant–animal mutualistic networks: the architecture of biodiversity. *Annual Review of Ecology, Evolution, and Systematics*, 38, 567–593.

Bastazini, V. A. G., Ferreira, P. M. A., Azambuja, B. O., Casas, G., Debastiani, V. J., Guimarães, P. R. & Pillar, V. D. (2017) Untangling the tangled bank: a novel method for partitioning the effects of phylogenies and traits on ecological networks. *Evolutionary Biology*, 44, 312–324.

Beier, C. M., Patterson, T. M. & Chapin, F. S. (2008) Ecosystem services and emergent vulnerability in managed ecosystems: a geospatial decision-support tool. *Ecosystems*, 11, 923–938.

Bellamy, P. H., Loveland, P. J., Bradley, R. I., Lark, R. M. & Kirk, G. J. D. (2005) Carbon losses from all soils across England and Wales 1978–2003. *Nature*, 437, 245–248.

Benard, M. F. (2006) Survival trade-offs between two predator-induced phenotypes in Pacific treefrogs (*Pseudacris regilla*). *Ecology*, 87, 340–346.

Bennett, J. A. & Pärtel, M. (2017) Predicting species establishment using absent species and functional neighborhoods. *Ecology and Evolution*, 7, 2223–2237.

Bennett, J. A., Lamb, E. G., Hall, J. C., Cardinal-McTeague, W. M. & Cahill, J. F. (2013) Increased competition does not lead to increased phylogenetic overdispersion in a native grassland. *Ecology Letters*, 16, 1168–1176.

Berg, M. P. & Bengtsson, J. (2007) Temporal and spatial variability in soil food web structure. *Oikos*, 116, 1789–1804.

Berg, M. P. & Ellers, J. (2010) Trait plasticity in species interactions: a driving force of community dynamics. *Evolutionary Ecology*, 24, 617–629.

Berg, M. P., Kiers, E. T., Driessen, G., van der Heijden, M., Kooi, B. W., Kuenen, F., Liefting, M., Verhoef, H. A. & Ellers, J. (2010) Adapt or disperse: understanding species persistence in a changing world. *Global Change Biology*, 16, 587–598.

Bernard-Verdier, M., Navas, M. L., Vellend, M., Violle, C., Fayolle, A. & Garnier, E. (2012) Community assembly along a soil depth gradient: contrasting patterns of plant trait convergence and divergence in a Mediterranean rangeland. *Journal of Ecology*, 100, 1422–1433.

Bertelsmeier, C., Luque, G. M., Confais, A. & Courchamp, F. (2013) Ant Profiler – a database of ecological characteristics of ants (Hymenoptera: Formicidae). *Myrmecological News*, 18, 73–76.

Bezzel, E. (1985) *Kompendium der Vögel mitteleuropas: Nonpasseriformes-Nichtsingvögel*. Wiesbaden: Aula-Verlag.

Bijlsma, R. & Loeschcke, V. (2005) Environmental stress, adaptation and evolution: an overview. *Journal of Evolutionary Biology*, 18, 744–749.

Bílá, K., Moretti, M., Bello, F., Dias, A. T. C., Pezzatti, G. B., Van Oosten, A. R. & Berg, M. P. (2014) Disentangling community functional components in a litter – macrodetritivore model system reveals the predominance of the mass ratio hypothesis. *Ecology and Evolution*, 4, 408–416.

Bimler, M. D., Stouffer, D. B., Lai, H. R. & Mayfield, M. M. (2018) Accurate predictions of coexistence in natural systems require the inclusion of facilitative interactions and environmental dependency. *Journal of Ecology*, 106, 1839–1852.

Blanck, A. & Lamouroux, N. (2007) Large-scale intraspecific variation in life-history traits of European freshwater fish. *Journal of Biogeography*, 34, 862–875.

Blanquart, F., Kaltz, O., Nuismer, S. L. & Gandon, S. (2013) A practical guide to measuring local adaptation. *Ecology Letters*, 16, 1195–1205.

Blesh, J. (2018) Functional traits in cover crop mixtures: biological nitrogen fixation and multifunctionality. *Journal of Applied Ecology*, 55, 38–48.

Blomberg, S. P., Garland, T. & Ives, A. R. (2003) Testing for phylogenetic signal in comparative data: behavioral traits are more labile. *Evolution*, 57, 717–745.

Blonder, B. (2016) Do hypervolumes have holes? *American Naturalist*, 187, E93–E105.

Blonder, B., Lamanna, C., Violle, C. & Enquist, B. J. (2014) The n-dimensional hypervolume. *Global Ecology and Biogeography*, 23, 595–609.

Blonder, B., Morrow, C. B., Maitner, B., Harris, D. J., Lamanna, C., Violle, C., Enquist, B. J. & Kerkhoff, A. J. (2018) New approaches for delineating n-dimensional hypervolumes. *Methods in Ecology and Evolution*, 9, 305–319.

Blüthgen, N. (2010) Why network analysis is often disconnected from community ecology: a critique and an ecologist's guide. *Basic and Applied Ecology*, 11, 185–195.

Bolnick, D. I. & Fitzpatrick, B. M. (2007) Sympatric speciation: models and empirical evidence. *Annual Review of Ecology Evolution and Systematics*, 38, 459–487.

Bolnick, D. I., Amarasekare, P., Araujo, M. S., Burger, R., Levine, J. M., Novak, M., Rudolf, V. H. W., Schreiber, S. J., Urban, M. C. & Vasseur, D. A. (2011) Why intraspecific trait variation matters in community ecology. *Trends in Ecology & Evolution*, 26, 183–192.

Bommarco, R., Biesmeijer, J. C., Meyer, B., Potts, S. G., Poyry, J., Roberts, S. P. M., Steffan-Dewenter, I. & Ockinger, E. (2010) Dispersal capacity and diet breadth modify the response of wild bees to habitat loss. *Proceedings of the Royal Society B – Biological Sciences*, 277, 2075–2082.

Bond, W. J., Midgley, G. F. & Woodward, F. I. (2003) The importance of low atmospheric CO_2 and fire in promoting the spread of grasslands and savannas. *Global Change Biology*, 9, 973–982.

Booth, B. D. & Swanton, C. J. (2002) Assembly theory applied to weed communities. *Weed Science*, 50, 2–13.

Borgy, B., Violle, C., Choler, P., Garnier, E., Kattge, J., Loranger, J., Amiaud, B., Cellier, P., Debarros, G., Denelle, P., Diquelou, S., Gachet, S., Jolivet, C., Lavorel, S., Lemauviel-Lavenant, S., Mikolajczak, A., Munoz, F., Olivier, J. & Viovy, N. (2017) Sensitivity of community-level trait–environment relationships to data representativeness: a test for functional biogeography. *Global Ecology and Biogeography*, 26, 729–739.

Bossdorf, O., Richards, C. L. & Pigliucci, M. (2008) Epigenetics for ecologists. *Ecology Letters*, 11, 106–115.

Botta-Dukat, Z. (2005) Rao's quadratic entropy as a measure of functional diversity based on multiple traits. *Journal of Vegetation Science*, 16, 533–540.

Botta-Dukat, Z. (2018) The generalized replication principle and the partitioning of functional diversity into independent alpha and beta components. *Ecography*, 41, 40–50.

Botta-Dukat, Z. & Czucz, B. (2016) Testing the ability of functional diversity indices to detect trait convergence and divergence using individual-based simulation. *Methods in Ecology and Evolution*, 7, 114–126.

Bouget, C., Brustel, H. & Zagatti, P. (2008) The French information system on saproxylic beetle ecology (frisbee): an ecological and taxonomical database to help with the assessment of forest conservation status. *Revue d'Ecologie (Terre Vie)*, 10, 33–36.

Bowman, R. (1961) Morphological differentiation and adaptation in the Galápagos finches. Diferenciación morfológica y adaptación en los pinzones de las Galápagos. *University of California Publications in Zoology*, 58, 1–302.

Branquinho, C., Serrano, H. C., Nunes, A., Pinho, P. & Matos, P. (2019) *Essential biodiversity change indicators for evaluating the effects of Anthropocene in ecosystems at a global scale. From assessing to conserving biodiversity* (eds. Casetta, E., da Silva, J. M., Vecchi, D., Casetta, E., da Silva, J. M. & Vecchi, D.), pp. 137–166. Cham: Springer.

Brathen, K. A., Ims, R. A., Yoccoz, N. G., Fauchald, P., Tveraa, T. & Hausner, V. H. (2007) Induced shift in ecosystem productivity? Extensive scale effects of abundant large herbivores. *Ecosystems*, 10, 773–789.

Brink, E., Aalders, T., Ádám, D., Feller, R., Henselek, Y., Hoffmann, A., Ibe, K., Matthey-Doret, A., Meyer, M., Negrut, N. L., Rau, A. L., Riewerts, B., von Schuckmann, L., Törnros, S., von Wehrden, H., Abson, D. J. & Wamsler, C. (2016) Cascades of green: a review of ecosystem-based adaptation in urban areas. *Global Environmental Change*, 36, 111–123.

Brooker, R. W., Maestre, F. T., Callaway, R. M., Lortie, C. L., Cavieres, L. A., Kunstler, G., Liancourt, P., Tielbörger, K., Travis, J. M. J., Anthelme, F., Armas, C., Coll, L., Corcket, E., Delzon, S., Forey, E., Kikvidze, Z., Olofsson, J., Pugnaire, F., Quiroz, C. L., Saccone, P., Schiffers, K., Seifan, M., Touzard, B. & Michalet, R. (2008) Facilitation in plant communities: the past, the present, and the future. *Journal of Ecology*, 96, 18–34.

Brousseau, P. M., Gravel, D. & Handa, I. T. (2018) On the development of a predictive functional trait approach for studying terrestrial arthropods. *Journal of Animal Ecology*, 87, 1209–1220.

Brown, J. J., Mennicken, S., Massante, J. C., Dijoux, S., Telea, A., Benedek, A. M., Götzenberger, L., Majeková, M., Lepš, J., Smilauer, P., Hrcek, J. & de Bello, F. (2019) A novel method to predict dark diversity using unconstrained ordination analysis. *Journal of Vegetation Science*, 30, 610–619.

Brown, R. (1828) A brief account of microscopical observations made on the particles contained in the pollen of plants. *Philosophical Magazine*, 4, 161–173.

Brun, P., Payne, M. R. & Kiørboe, T. (2016) *A trait database for marine copepods*. Pangea https://doi.pangaea.de/10.1594/PANGAEA.862968.

Bruno, J. F., Stachowicz, J. J. & Bertness, M. D. (2003) Inclusion of facilitation into ecological theory. *Trends in Ecology & Evolution*, 18, 119–125.

Bu, W., Huang, J., Xu, H., Zang, R., Ding, Y., Li, Y., Lin, M., Wang, J. & Zhang, C. (2019) Plant functional traits are the mediators in regulating effects of abiotic site conditions on aboveground carbon stock – evidence from a 30 ha tropical forest plot. *Frontiers in Plant Science*, 9, 1958.

Buckley, Y. M. & Catford, J. (2016) Does the biogeographic origin of species matter? Ecological effects of native and non-native species and the use of origin to guide management. *Journal of Ecology*, 104, 4–17.

Budrys, E., Budriene, A. & Orlovskyte, S. (2014) Cavity-nesting wasps and bees database. http://scales.ckff.si/scaletool/?menu=6&submenu=3.

Buisson, L., Grenouillet, G., Villéger, S., Canal, J. & Laffaille, P. (2013) Toward a loss of functional diversity in stream fish assemblages under climate change. *Global Change Biology*, 19, 387–400.

Bulleri, F., Bruno, J. F., Silliman, B. R. & Stachowicz, J. J. (2016) Facilitation and the niche: implications for coexistence, range shifts and ecosystem functioning. *Functional Ecology*, 30, 70–78.

Burns, M., Hedin, M. & Tsurusaki, N. (2018) Population genomics and geographical parthenogenesis in Japanese harvestmen (Opiliones, Sclerosomatidae, *Leiobunum*). *Ecology and Evolution*, 8, 36–52.

Bush, G. L. (1969) Sympatric host race formation and speciation in frugivorous flies of genus *Rhagoletis* (Diptera, Tephritidae). *Evolution*, 23, 237–251.

Butterfield, B. J. & Suding, K. N. (2013) Single-trait functional indices outperform multi-trait indices in linking environmental gradients and ecosystem services in a complex landscape. *Journal of Ecology*, 101, 9–17.

Cadotte, M. W. (2017) Functional traits explain ecosystem function through opposing mechanisms. *Ecology Letters*, 20, 989–996.

Cadotte, M. W. & Tucker, C. M. (2017) Should environmental filtering be abandoned? *Trends in Ecology & Evolution*, 32, 429–437.

Cadotte, M. W. & Tucker, C. M. (2018) Difficult decisions: strategies for conservation prioritization when taxonomic, phylogenetic and functional diversity are not spatially congruent. *Biological Conservation*, 225, 128–133.

Cadotte, M. W., Albert, C. H. & Walker, S. C. (2013) The ecology of differences: assessing community assembly with trait and evolutionary distances. *Ecology Letters*, 16, 1234–1244.

Cadotte, M. W., Carboni, M., Si, X. & Tatsumi, S. (2019) Do traits and phylogeny support congruent community diversity patterns and assembly inferences? *Journal of Ecology*, 107, 2065–2077.

Cadotte, M. W., Carscadden, K. & Mirotchnick, N. (2011) Beyond species: functional diversity and the maintenance of ecological processes and services. *Journal of Applied Ecology*, 48, 1079–1087.

Cadotte, M. W., Cavender-Bares, J., Tilman, D. & Oakley, T. H. (2009) Using phylogenetic, functional and trait diversity to understand patterns of plant community productivity. *PLOS ONE*, 4, e5695.

Cadotte, M. W., Mai, D. V., Jantz, S., Collins, M. D., Keele, M. & Drake, J. A. (2006) On testing the competition–colonization trade-off in a multispecies assemblage. *American Naturalist*, 168, 704–709.

Callaway, R. M., Thelen, G. C., Rodriguez, A. & Holben, W. E. (2004) Soil biota and exotic plant invasion. *Nature*, 427, 731–733.

Carboni, M., Münkemüller, T., Lagergne, S., Choler, P., Borge, B., Violle, C., Essl, F., Roquet, C., Munoz, F., DivGrass Consortium & Thuiller, W. (2016) What it takes to invade grassland ecosystems: traits, introduction history and filtering processes. *Ecology Letters*, 19, 219–229.

Cardinale, B. J., Duffy, J. E., Gonzalez, A., Hooper, D. U., Perrings, C., Venail, P., Narwani, A., Mace, G. M., Tilman, D., Wardle, D. A., Kinzig, A. P., Daily, G. C., Loreau, M., Grace, J. B., Larigauderie, A., Srivastava, D. S. & Naeem, S. (2012) Biodiversity loss and its impact on humanity. *Nature*, 486, 59–67.

Carmona, C. P., Azcarate, F. M., de Bello, F., Ollero, H. S., Lepš, J. & Peco, B. (2012) Taxonomical and functional diversity turnover in Mediterranean grasslands: interactions between grazing, habitat type and rainfall. *Journal of Applied Ecology*, 49, 1084–1093.

Carmona, C. P., de Bello, F., Mason, N. W. H. & Lepš, J. (2016) Traits without borders: integrating functional diversity across scales. *Trends in Ecology & Evolution*, 31, 382–394.

Carmona, C. P., de Bello, F., Mason, N. W. H. & Lepš, J. (2019) Trait probability density (TPD): measuring functional diversity across scales based on TPD with R. *Ecology*, 100, e08276.

Carmona, C. P., de Bello, F., Sasaki, T., Uchida, K. & Partel, M. (2017a) Towards a common toolbox for rarity: a response to Violle et al. *Trends in Ecology & Evolution*, 32, 889–891.

Carmona, C. P., Guerrero, I., Morales, M. B., Onate, J. J. & Peco, B. (2017b) Assessing vulnerability of functional diversity to species loss: a case study in Mediterranean agricultural systems. *Functional Ecology*, 31, 427–435.

Carmona, C. P., Mason, N. W. H., Azcarate, F. M. & Peco, B. (2015b) Inter-annual fluctuations in rainfall shift the functional structure of Mediterranean grasslands across gradients of productivity and disturbance. *Journal of Vegetation Science*, 26, 538–551.

Carmona, C. P., Rota, C., Azcarate, F. M. & Peco, B. (2015a) More for less: sampling strategies of plant functional traits across local environmental gradients. *Functional Ecology*, 29, 579–588.

Carpenter, S. R., Mooney, H. A., Agard, J., Capistrano, D., DeFries, R. S., Díaz, S., Dietz, T., Duraiappah, A. K., Oteng-Yeboah, A., Pereira, H. M., Perrings, C., Reid, W. V., Sarukhan, J., Scholes, R. J. & Whyte, A. (2009) Science for managing ecosystem services: beyond the Millennium Ecosystem Assessment. *Proceedings of the National Academy of Sciences of the United States of America*, 106, 1305–1312.

Cernansky, R. (2017) The biodiversity revolution. *Nature*, 546, 22–24.

Chalmandrier, L., Münkemüller, T., Colace, M. P., Renaud, J., Aubert, S., Carlson, B. Z., Clement, J. C., Legay, N., Pellet, G., Saillard, A., Lavergne, S. & Thuiller, W. (2017) Spatial scale and intraspecific trait variability mediate assembly rules in alpine grasslands. *Journal of Ecology*, 105, 277–287.

Champely, S. & Chessel, D. (2002) Measuring biological diversity using Euclidean metrics. *Environmental and Ecological Statistics*, 9, 167–177.

Chapin, F. S. (1980) The mineral nutrition of wild plants. *Annual Review of Ecology and Systematics*, 11, 233–260.

Chase, J. M. & Leibold, M. A. (2003) *Ecological niches: linking classical and contemporary approaches*. University of Chicago Press. Chicago, IL.

Chesson, P. (2000) Mechanisms of maintenance of species diversity. *Annual Review of Ecology and Systematics*, 31, 343–366.

Chesson, P. & Huntly, N. (1997) The roles of harsh and fluctuating conditions in the dynamics of ecological communities. *American Naturalist*, 150, 519–553.

Chiu, C.-H. & Chao, A. (2014) Distance-based functional diversity measures and their decomposition: a framework based on Hill numbers. *PLOS ONE*, 9, e113561.

Chollet, S., Rambal, S., Fayolle, A., Hubert, D., Foulquie, D. & Garnier, E. (2014) Combined effects of climate, resource availability, and plant traits on biomass produced in a Mediterranean rangeland. *Ecology*, 95, 737–748.

Chown, S. L. (2012) Trait-based approaches to conservation physiology: forecasting environmental change risks from the bottom up. *Philosophical Transactions of the Royal Society B – Biological Sciences*, 367, 1615–1627.

Chuang, A. & Peterson, C. R. (2016) Expanding population edges: theories, traits, and trade-offs. *Global Change Biology*, 22, 494–512.

Cianciaruso, M. V., Batalha, M. A., Gaston, K. J. & Petchey, O. L. (2009) Including intraspecific variability in functional diversity. *Ecology*, 90, 81–89.

Cingolani, A. M., Cabido, M., Gurvich, D. E., Renison, D. & Díaz, S. (2007) Filtering processes in the assembly of plant communities: are species presence and abundance driven by the same traits? *Journal of Vegetation Science*, 18, 911–920.

Clausen, J., Keck, D. D. & Hiesey, W. M. (1948) *Experimental studies on the nature of species. III. Environmental responses of climatic races of Achillea*. Washington, DC: Carnegie Institution.

Concepción, E. D., Götzenberger, L., Nobis, M. P., de Bello, F., Obrist, M. K. & Moretti, M. (2017) Contrasting trait assembly patterns in plant and bird communities along environmental and human-induced land-use gradients. *Ecography*, 40, 753–763.

Connor, E. F. & Simberloff, D. (1979) The assembly of species communities: chance or competition? *Ecology*, 60, 1132–1140.

Conti, G. & Díaz, S. (2013) Plant functional diversity and carbon storage – an empirical test in semi-arid forest ecosystems. *Journal of Ecology*, 101, 18–28.

Conti, L., Block, S., Parepa, M., Münkemüller, T., Thuiller, W., Acosta, A. T. R., van Kleunen, M., Dullinger, S., Essl, F., Dullinger, I., Moser, D., Klonner, G., Bossdorf, O. & Carboni, M. (2018) Functional trait differences and trait plasticity mediate biotic resistance to potential plant invaders. *Journal of Ecology*, 106, 1607–1620.

Conti, M. E. & Cecchetti, G. (2001) Biological monitoring: lichens as bioindicators of air pollution assessment – a review. *Environmental Pollution*, 114, 471–492.

Conway, T. M. (2016) Tending their urban forest: residents' motivations for tree planting and removal. *Urban Forestry and Urban Greening*, 17, 23–32.

Cordlandwehr, V., Meredith, R. L., Ozinga, W. A., Bekker, R. M., van Groenendael, J. M. & Bakker, J. P. (2013) Do plant traits retrieved from a database accurately predict on-site measurements? *Journal of Ecology*, 101, 662–670.

Cornelissen, J. H. C., Lang, S. I., Soudzilovskaia, N. A. & During, H. J. (2007) Comparative cryptogam ecology: a review of bryophyte and lichen traits that drive biogeochemistry. *Annals of Botany*, 99, 987–1001.

Cornelissen, J. H. C., Lavorel, S., Garnier, E., Díaz, S., Buchmann, N., Gurvich, D. E., Reich, P. B., ter Steege, H., Morgan, H. D., van der Heijden, M. G. A., Pausas, J. G. & Poorter, H. (2003)

A handbook of protocols for standardised and easy measurement of plant functional traits worldwide. *Australian Journal of Botany*, 51, 335–380.

Cornelissen, J. H. C., Pérez-Harguindeguy, N., Díaz, S., Grime, J. P., Marzano, B., Cabido, M., Vendramini, F. & Cerabolini, B. (1999) Leaf structure and defence control litter decomposition rate across species and life forms in regional floras on two continents. *New Phytologist*, 143, 191–200.

Cornell, H. V. & Harrison, S. P. (2014) What are species pools and when are they important? *Annual Review of Ecology, Evolution, and Systematics*, 45, 45–67.

Cornwell, W. K. & Ackerly, D. D. (2009) Community assembly and shifts in plant trait distributions across an environmental gradient in coastal California. *Ecological Monographs*, 79, 109–126.

Cornwell, W. K. & Cornelissen, J. H. C. (2013) A broader perspective on plant domestication and nutrient and carbon cycling. *New Phytologist*, 198, 331–333.

Cornwell, W. K. & Habacuc (2018, 11 April) traitecoevo/fungaltraits v0.0.3 (Version v0.0.3). Zenodo. http://doi.org/10.5281/zenodo.1216257.

Cornwell, W. K., Cornelissen, J. H. C., Amatangelo, K., Dorrepaal, E., Eviner, V. T., Godoy, O., Hobbie, S. E., Hoorens, B., Kurokawa, H., Pérez-Harguindeguy, N., Quested, H. M., Santiago, L. S., Wardle, D. A., Wright, I. J., Aerts, R., Allison, S. D., van Bodegom, P., Brovkin, V., Chatain, A., Callaghan, V. T., Díaz, S., Garnier, E., Gurvich, D. E., Kazakou, E., Klein, J. A., Read, J., Reich, P. B., Soudzilovskaia, N. A., Vaieretti, M. V. & Westoby, M. (2008) Plant species traits are the predominant control on litter decomposition rates within biomes worldwide. *Ecology Letters*, 11, 1065–1071.

Cornwell, W. K., Schwilk, D. W. & Ackerly, D. D. (2006) A trait-based test for habitat filtering: convex hull volume. *Ecology*, 87, 1465–1471.

Correa, S. B., Arujo, J. K., Penha, J., Nunes da Cunha, C., Bobier, K. E. & Anderson, J. T. (2016) Stability and generalization in seed dispersal networks: a case study of frugivorous fish in Neotropical wetlands. *Proceedings of the Royal Society B – Biological Sciences*, 283, 20161267.

Cosendai, A. C., Wagner, J., Ladinig, U., Rosche, C. & Horandl, E. (2013) Geographical parthenogenesis and population genetic structure in the alpine species *Ranunculus kuepferi* (Ranunculaceae). *Heredity*, 110, 560–569.

Craine, J. M. (2005) Reconciling plant strategy theories of Grime and Tilman. *Journal of Ecology*, 93, 1041–1052.

Craven, D., Eisenhauer, N., Pearse, W. D., Hautier, Y., Isbell, F., Roscher, C., Bahn, M., Beierkuhnlein, C., Bönisch, G., Buchmann, N., Byun, C., Catford, J. A., Cerabolini, B. E. L., Cornelissen, J. H. C., Craine, J. M., De Luca, E., Ebeling, A., Griffin, J. N., Hector, A., Hines, J., Jentsch, A., Kattge, J., Kreyling, J., Lanta, V., Lemoine, N., Meyer, S. T., Minden, V., Onipchenko, V., Polley, H. W., Reich, P. B., van Ruijven, J., Schamp, B., Smith, M. D., Soudzilovskaia, N. A., Tilman, D., Weigelt, A., Wilsey, B. & Manning, P. (2018) Multiple facets of biodiversity drive the diversity–stability relationship. *Nature Ecology & Evolution*, 2, 1579–1587.

Crisp, M. D. & Cook, L. G. (2012) Phylogenetic niche conservatism: what are the underlying evolutionary and ecological causes? *New Phytologist*, 196, 681–694.

Cronon, W. (1996) *Uncommon ground: rethinking the human place in nature*. New York: Norton & Company.

Cruz, P., De Quadros, F. L. F., Theau, J. P., Frizzo, A., Jouany, C., Duru, M. & Carvalho, P. C. F. (2010) Leaf traits as functional descriptors of the intensity of continuous grazing in native grasslands in the South of Brazil. *Rangeland Ecology and Management*, 63, 350–358.

Damour, G., Navas, M. L. & Garnier, E. (2018) A revised trait-based framework for agroecosystems including decision rules. *Journal of Applied Ecology*, 55, 12–24.

Darwin, C. (1859) *On the origin of species by means of natural selection, or the preservation of favoured races in the struggle for life*. London: John Murray.

David, J. F. (2014) The role of litter-feeding macroarthropods in decomposition processes: a reappraisal of common views. *Soil Biology and Biochemistry*, 76, 109–118.

Davis, M. A., Chew, M. K., Hobbs, R. J., Lugo, A. E., Ewel, J. J., Vermeij, G. J., . . . Briggs, J. C. (2011) Don't judge species on their origins. *Nature*, 474, 153–154.

Dawson, S. K., Boddy, L., Halbwachs, H., Bassler, C., Andrew, C., Crowther, T. W., Heilmann-Clausen, J., Norden, J., Ovaskainen, O. & Jonsson, M. (2019) Handbook for the measurement of macrofungal functional traits: a start with basidiomycete wood fungi. *Functional Ecology*, 33, 372–387.

de Bello, F. (2012) The quest for trait convergence and divergence in community assembly: are null-models the magic wand? *Global Ecology and Biogeography*, 21, 312–317.

de Bello, F. & Mudrák, O. (2013) Plant traits as indicators: loss or gain of information? *Applied Vegetation Science*, 16, 353–354.

de Bello, F., Berg, M. P., Dias, A. T. C., Diniz-Filho, J. A. F., Götzenberger, L., Hortal, J., Ladle, R. J. & Lepš, J. (2015) On the need for phylogenetic 'corrections' in functional trait-based approaches. *Folia Geobotanica*, 50, 349–357.

de Bello, F., Carmona, C. P., Lepš, J., Szava-Kovats, R. & Partel, M. (2016a) Functional diversity through the mean trait dissimilarity: resolving shortcomings with existing paradigms and algorithms. *Oecologia*, 180, 933–940.

de Bello, F., Carmona, C. P., Mason, N. W. H., Sebastià, M. T. & Lepš, J. (2013a) Which trait dissimilarity for functional diversity: trait means or trait overlap? *Journal of Vegetation Science*, 24, 807–819.

de Bello, F., Fibich, P., Zelený, D., Kopecky, M., Mudrak, O., Chytry, M., Pysek, P., Wild, J., Michalcova, D., Sadlo, J., Smilauer, P., Lepš, J. & Partel, M. (2016b) Measuring size and composition of species pools: a comparison of dark diversity estimates. *Ecology and Evolution*, 6, 4088–4101.

de Bello, F., Lavergne, S., Meynard, C. N., Lepš, J. & Thuiller, W. (2010b) The partitioning of diversity: showing Theseus a way out of the labyrinth. *Journal of Vegetation Science*, 21, 992–1000.

de Bello, F., Lavorel, S., Albert, C. H., Thuiller, W., Grigulis, K., Dolezal, J., Janecek, S. & Lepš, J. (2011) Quantifying the relevance of intraspecific trait variability for functional diversity. *Methods in Ecology and Evolution*, 2, 163–174.

de Bello, F., Lavorel, S., Díaz, S., Harrington, R., Cornelissen, J. H. C., Bardgett, R. D., Berg, M. P., Cipriotti, P., Feld, C. K., Hering, D., Martins da Silva, P., Potts, S. G., Sandin, L., Sousa, J. P., Storkey, J., Wardle, D. A. & Harrison, P. A. (2010a) Towards an assessment of multiple ecosystem processes and services via functional traits. *Biodiversity and Conservation*, 19, 2873–2893.

de Bello, F., Lavorel, S., Lavergne, S., Albert, C. H., Boulangeat, I., Mazel, F. & Thuiller, W. (2013b) Hierarchical effects of environmental filters on the functional structure of plant communities: a case study in the French Alps. *Ecography*, 36, 393–402.

de Bello, F., Lepš, J., Lavorel, S. & Moretti, M. (2007) Importance of species abundance for assessment of trait composition: an example based on pollinator communities. *Community Ecology*, 8, 163–170.

de Bello, F., Lepš, J. & Sebastià, M. T. (2005) Predictive value of plant traits to grazing along a climatic gradient in the Mediterranean. *Journal of Applied Ecology*, 42, 824–833.

de Bello, F., Price, J. N., Münkemüller, T., Liira, J., Zobel, M., Thuiller, W., Gerhold, P., Götzenberger, L., Lavergne, S., Lepš, J., Zobel, K. & Pärtel, M. (2012) Functional species pool framework to test for biotic effects on community assembly. *Ecology*, 93, 2263–2273.

de Bello, F., Šmilauer, P., Diniz-Filho, J. A. F., Carmona, C. P., Lososová, Z., Herben, T. & Götzenberger, L. (2017) Decoupling phylogenetic and functional diversity to reveal hidden signals in community assembly. *Methods in Ecology and Evolution*, 8, 1200–1211.

de Bello, F., Thuiller, W., Lepš, J., Choler, P., Clement, J. C., Macek, P., Sebastià, M. T. & Lavorel, S. (2009) Partitioning of functional diversity reveals the scale and extent of trait convergence and divergence. *Journal of Vegetation Science*, 20, 475–486.

de Bello, F., Vandewalle, M., Reitalu, T., Lepš, J., Prentice, H. C., Lavorel, S. & Sykes, M. T. (2013c) Evidence for scale- and disturbance-dependent trait assembly patterns in dry semi-natural grasslands. *Journal of Ecology*, 101, 1237–1244.

De Groot, R. S., Wilson, M. A. & Boumans, R. M. (2002) A typology for the classification, description and valuation of ecosystem functions, goods and services. *Ecological Economics*, 41, 393–408.

De Oliveira, T., Hättenschwiler, S. & Handa, I. T. (2010) Snail and millipede complementarity in decomposing Mediterranean forest leaf litter mixtures. *Functional Ecology*, 24, 937–946.

de Ruiter, P., Neutel, A.-M. & Moore, J. (1998) Biodiversity in soil ecosystems: the role of energy flow and community stability. *Applied Soil Ecology*, 10, 217–228.

Defossez, E., Pellissier, L. & Rasmann, S. (2018) The unfolding of plant growth form-defence syndromes along elevation gradients. *Ecology Letters*, 21, 609–618.

Dell, A. I., Pawar, S. & Savage, V. M. (2013) The thermal dependence of biological traits. *Ecology*, 94, 1205–1206.

Demuzere, M., Orru, K., Heidrich, O., Olazabal, E., Geneletti, D., Orru, H., Bhave, A. G., Mittal, N., Feliu, E. & Faehnle, M. (2014) Mitigating and adapting to climate change: multi-functional and multi-scale assessment of green urban infrastructure. *Journal of Environmental Management*, 146, 107–115.

Deraison, H., Badenhausser, I., Loeuille, N., Scherber, C. & Gross, N. (2015) Functional trait diversity across trophic levels determines herbivore impact on plant community biomass. *Ecology Letters*, 18, 1346–1355.

Desdevises, Y., Legendre, P., Azouzi, L. & Morand, S. (2003) Quantifying phylogenetically structured environmental variation. *Evolution*, 57, 2647–2652.

Devictor, V., Julliard, R., Couvet, D. & Jiguet, F. (2008) Birds are tracking climate warming, but not fast enough. *Proceedings of the Royal Society B – Biological Sciences*, 275, 2743–2748.

Devictor, V., Mouillot, D., Meynard, C., Jiguet, F., Thuiller, W. & Mouquet, N. (2010) Spatial mismatch and congruence between taxonomic, phylogenetic and functional diversity: the need for integrative conservation strategies in a changing world. *Ecology Letters*, 13, 1030–1040.

DeWitt, T. J. & Scheiner, S. M. (eds.) (2004) *Phenotypic plasticity: functional and conceptual approaches*. Oxford University Press. Oxford, UK.

Diamond, J. M. (1975) Assembly of species communities. In: *Ecology and Evolution of Communities* (eds. Cody, M. L. & Diamond, J. M.), pp. 342–444. Cambridge, MA: Harvard University Press.

Dias, A. T. C., Berg, M. P., de Bello, F., Van Oosten, A. R., Bila, K. & Moretti, M. (2013b) An experimental framework to identify community functional components driving ecosystem processes and services delivery. *Journal of Ecology*, 101, 29–37.

Dias, A. T. C., Hoorens, B., Logtestijn, R. S. P., Vermaat, J. E. & Aerts, R. (2010) Plant species composition can be used as a proxy to predict methane emissions in peatland ecosystems after land-use changes. *Ecosystems*, 13, 526–538.

Dias, A. T. C., Krab, E. J., Marien, J., Zimmer, M., Cornelissen, J. H., Ellers, J., Wardle, D. A. & Berg, M. P. (2013a) Traits underpinning desiccation resistance explain distribution patterns of terrestrial isopods. *Oecologia*, 172, 667–77.

Dias, A. T. C., Rosado, B. H., de Bello, F., Pistón, N. & de Mattos, E. A. (2020) Alternative plant designs: consequences for community assembly and ecosystem functioning. *Annals of Botany*, 125, 391–398.

Díaz, S. & Cabido, M. (2001) Vive la difference: plant functional diversity matters to ecosystem processes. *Trends in Ecology & Evolution*, 16, 646–655.

Díaz, S., Cabido, M. & Casanoves, F. (1998) Plant functional traits and environmental filters at a regional scale. *Journal of Vegetation Science*, 9, 113–122.

Díaz, S., Demissew, S., Joly, C., Lonsdale, W. M. & Larigauderie, A. (2015) A Rosetta Stone for nature's benefits to people. *PLOS Biology*, 13, 1–8.

Díaz, S., Fargione, J., Chapin, F. S., III & Tilman, D. (2006) Biodiversity loss threatens human well-being. *PLOS Biology*, 4, e277.

Díaz, S., Hodgson, J. G., Thompson, K., Cabido, M., Cornelissen, J. H. C., Jalili, A., Montserrat-Martí, G., Grime, J. P., Zarrinkamar, F., Asri, Y., Band, S. R., Basconcelo, S., Castro-Díez, P., Funes, G., Hamzehee, B., Khoshnevi, M., Pérez-Harguindeguy, N., Pérez-Rontomé, M. C., Shirvany, F. A., Vendramini, F., Yazdani, S., Abbas-Azimi, R., Bogaard, A., Boustani, S., Charles, M., Dehghan, M., de Torres-Espuny, L., Falczuk, V., Guerrero-Campo, J., Hynd, A., Jones, G., Kowsary, E., Kazemi-Saeed, F., Maestro-Martínez, M., Romo-Díez, A., Shaw, S., Siavash, B., Villar-Salvador, P. & Zak, M. R. (2004) The plant traits that drive ecosystems: evidence from three continents. *Journal of Vegetation Science*, 15, 295–304.

Díaz, S., Kattge, J., Cornelissen, J. H. C., Wright, I. J., Lavorel, S., Dray, S., Reu, B., Kleyer, M., Wirth, C., Prentice, I. C., Garnier, E., Bonisch, G., Westoby, M., Poorter, H., Reich, P. B., Moles, A. T., Dickie, J., Gillison, A. N., Zanne, A. E., Chave, J., Wright, S. J., Sheremet'ev, S. N., Jactel, H., Baraloto, C., Cerabolini, B., Pierce, S., Shipley, B., Kirkup, D., Casanoves, F., Joswig, J. S., Gunther, A., Falczuk, V., Ruger, N., Mahecha, M. D. & Gorne, L. D. (2016) The global spectrum of plant form and function. *Nature*, 529, 167–171.

Díaz, S., Lavorel, S., de Bello, F., Quetier, F., Grigulis, K. & Robson, M. (2007) Incorporating plant functional diversity effects in ecosystem service assessments. *Proceedings of the National Academy of Sciences of the United States of America*, 104, 20684–20689.

Díaz, S., Noy-Meir, I. & Cabido, M. (2001) Can grazing response of herbaceous plants be predicted from simple vegetative traits? *Journal of Applied Ecology*, 38, 497–508.

Díaz, S., Pascual, U., Stenseke, M., Martín-López, B., Watson, R. T., Molnár, Z., Hill, R., Chan, K. M. A., Baste, I. A., Brauman, K. A., Polasky, S., Church, A., Lonsdale, M., Larigauderie, A., Leadley, P. W., van Oudenhoven, A. P. E., van der Plaat, F., Schröter, M., Lavorel, S., Aumeeruddy-Thomas, Y., Bukvareva, E., Davies, K., Demissew, S., Erpul, G., Failler, P., Guerra, C. A., Hewitt, C. L., Keune, H., Lindley, S. & Shirayama, Y. (2018) Assessing nature's contributions to people. *Science*, 359, 270–272.

Díaz, S., Purvis, A., Cornelissen, J. H. C., Mace, G. M., Donoghue, M. J., Ewers, R. M., Jordano, P. & Pearse, W. D. (2013) Functional traits, the phylogeny of function, and ecosystem service vulnerability. *Ecology and Evolution*, 3, 2958–2975.

Didham, R. K., Tylianakis, J. M., Hutchison, M. A., Ewers, R. M. & Gemmell, N. J. (2005) Are invasive species the drivers of ecological change? *Trends in Ecology and Evolution*, 20, 470–474.

Diez, J. M., Dickie, I., Edwards, G., Hulme, P. E., Sullivan, J. J. & Duncan, R. P. (2010) Negative soil feedbacks accumulate over time for non-native plant species. *Ecology Letters*, 13, 803–809.

Diniz-Filho, J. A. F., Bini, L. M., Rangel, T. F. L. V. B., Morales-Castilla, I., Olalla-Tárraga, M. Á., Rodríguez, M. Á. & Hawkins, B. A. (2012) On the selection of phylogenetic eigenvectors for ecological analyses. *Ecography*, 35, 239–249.

Donatti, C. I., Guimarães, P. R., Galetti, M., Pizo, M. A., Marquitti, F. M. D. & Dirzo, R. (2011) Analysis of a hyper-diverse seed dispersal network: modularity and underlying mechanisms. *Ecology Letters*, 14, 773–781.

Dorrepaal, E., Toet, S., Van Logtestijn, R. S. P., Swart, E., Van De Weg, M. J., Callaghan, V. T. & Aerts, R. (2009) Carbon respiration from subsurface peat accelerated by climate warming in the subarctic. *Nature*, 460, 616–619.

Doughty, C. E., Wolf, A. & Malhi, Y. (2013) The legacy of the Pleistocene megafauna extinctions on nutrient availability in Amazonia. *Nature Geoscience*, 6, 761–764.

Dray, S. & Legendre, P. (2008) Testing the species traits–environment relationships: the fourth-corner problem revisited. *Ecology*, 89, 3400–3412.

Duru, M., Cruz, P., Jouany, C. & Theau, J. P. (2010) Herb'type©: un nouvel outil pour évaluer les services de production fournis par les prairies permanentes. *Productions Animales*, 23, 319–332.

Duru, M., Cruz P., Raouda, A. H. K., Ducourtieux, C. & Theau, J. P. (2008) Relevance of plant functional types based on leaf dry matter content for assessing digestibility of native grass species and species-rich grassland communities in spring. *Agronomy Journal*, 100, 1622–1630.

Dwyer, J. M. & Laughlin, D. C. (2017) Constraints on trait combinations explain climatic drivers of biodiversity: the importance of trait covariance in community assembly. *Ecology Letters*, 20, 872–882.

Eggenberger, H., Frey, D., Pellissier, L., Ghazoul, J., Fontana, S. & Moretti, M. (2019) Urban bumblebees are smaller and more phenotypically diverse than their rural counterparts. *Journal of Animal Ecology*, 88, 1522–1533.

Eisenhauer, N., Sabais, A. C. W. & Scheu, S. (2011) Collembola species composition and diversity effects on ecosystem functioning vary with plant functional group identity. *Soil Biology & Biochemistry*, 43, 1697–1704.

Ellenberg, H. (1974) Zeigerwerte der Gefäßpflanzen mitteleuropas. *Scripta Geobotanica*, 9, 3–122

Ellers, J., Berg, M. P., Dias, A. T., Fontana, S., Ooms, A. & Moretti, M. (2018) Diversity in form and function: vertical distribution of soil fauna mediates multidimensional trait variation. *Journal of Animal Ecology*, 87, 933–944.

Ellers, J., Rog, S., Braam, C. & Berg, M. P. (2011) Genotypic richness and phenotypic dissimilarity enhance population performance. *Ecology*, 92, 1605–1615.

Ellison, C. E., Hall, C., Kowbel, D., Welch, J., Brem, R. B., Glass, N. L. & Taylor, J. W. (2011) Population genomics and local adaptation in wild isolates of a model microbial eukaryote. *Proceedings of the National Academy of Sciences of the United States of America*, 108, 2831–2836.

Elmqvist, T., Folke, C., Nystrom, M., Peterson, G., Bengtsson, J., Walker, B. & Norberg, J. (2003) Response diversity, ecosystem change, and resilience. *Frontiers in Ecology and the Environment*, 1, 488–494.

Elser, J. J., Fagan, W. F., Kerkhoff, A. J., Swenson, N. G. & Enquist, B. J. (2010) Biological stoichiometry of plant production: metabolism, scaling and ecological response to global change. *New Phytologist*, 186, 593–608.

Elton, C. S. (1946) Competition and the structure of ecological communities. *Journal of Animal Ecology*, 15, 54–68.

Elzinga, J. A., Atlan, A., Biere, A., Gigord, L., Weis, A. E. & Bernasconi, G. (2007) Time after time: flowering phenology and biotic interactions. *Trends in Ecology & Evolution*, 22, 432–439.

Falkner, G., Obrdlik, P., Castella, E. & Speight, M. C. D. (2001) *Shelled Gastropoda of Western Europe*. Munich: Friedrich-Held-Gesellschaft.

Farwig, N., Schabo, D. G. & Albrecht, J. (2017) Trait-associated loss of frugivores in fragmented forest does not affect seed removal rates. *Journal of Ecology*, 105, 20–28.

Feld, C. K., Da Silva, P. M., Sousa, J. P., De Bello, F., Bugter, R., Grandin, U., Hering, D., Lavorel, S., Mountford, O., Pardo, I., Pärtel, M., Römbke, J., Sandin, L., Bruce Jones, K. & Harrison, P. (2009) Indicators of biodiversity and ecosystem services: a synthesis across ecosystems and spatial scales. *Oikos*, 118, 1862–1871.

Felsenstein, J. (1985) Phylogenies and the comparative method. *American Naturalist*, 125, 1–15.

Felten, D. & Emmerling, C. (2009) Earthworm burrowing behaviour in 2D terraria with single- and multi-species assemblages. *Biology and Fertility of Soils*, 45, 789–797.

Finegan, B., Pena-Claros, M., de Oliveira, A., Ascarrunz, N., Bret-Harte, M. S., Carreno-Rocabado, G., Casanoves, F., Díaz, S., Velepucha, P. E., Fernandez, F., Licona, J. C., Lorenzo, L., Negret, B. S., Vaz, M. & Poorter, L. (2015) Does functional trait diversity predict above-ground biomass and productivity of tropical forests? Testing three alternative hypotheses. *Journal of Ecology*, 103, 191–201.

Finerty, G. E., de Bello, F., Bílá, K., Berg, M. P., Dias, A. T. C., Pezzatti, G. B. & Moretti, M. (2016) Exotic or not, leaf trait dissimilarity modulates the effect of dominant species on mixed litter decomposition. *Journal of Ecology*, 104, 1400–1409.

Fontana, S., Berg, M. P. & Moretti, M. (2019) Intraspecific niche partitioning in macrodetritivores enhances mixed leaf litter decomposition. *Functional Ecology*, 33, 2391–2401.

Fontana, S., Petchey, O. L. & Pomati, F. (2015) Individual-level trait diversity concepts and indices to comprehensively describe community change in multidimensional trait space. *Functional Ecology*, 30, 808–818.

Forsman, A. (2014) Effects of genotypic and phenotypic variation on establishment are important for conservation, invasion, and infection biology. *Proceedings of the National Academy of Sciences of the United States of America*, 111, 302–307.

Forsman, A. & Hagman, M. (2009) Association of coloration mode with population declines and endangerment in Australian frogs. *Conservation Biology*, 23, 1535–1543.

Forsman, A. & Wennersten, L. (2016) Inter-individual variation promotes ecological success of populations and species: evidence from experimental and comparative studies. *Ecography*, 39, 630–648.

Fournier, B., Malysheva, E., Mazei, Y., Moretti, M. & Mitchell, E. A. D. (2012) Toward the use of testate amoeba functional traits as indicator of floodplain restoration success. *European Journal of Soil Biology*, 49, 85–91.

Franken, O., Huizinga, M., Ellers, J. & Berg, M. P. (2018) Heated communities: large inter- and intraspecific variation in heat tolerance across trophic levels of a soil arthropod community. *Oecologia*, 186, 311–322.

Franzen, M. & Betzholtz, P. E. (2012) Species traits predict island occupancy in noctuid moths. *Journal of Insect Conservation*, 16, 155–163.

Freckleton, R. P. (2009) The seven deadly sins of comparative analysis. *Journal of Evolutionary Biology*, 22, 1367–1375.

Freschet, G. T., Aerts, R. & Cornelissen, J. H. C. (2012) A plant economics spectrum of litter decomposability. *Functional Ecology*, 26, 56–65.

Froese, R. & Pauly, D. (2018) *FishBase*. World Wide Web electronic publication.

Fukami, T., Bezemer, T. M., Mortimer, S. R. & van der Putten, W. H. (2005) Species divergence and trait convergence in experimental plant community assembly. *Ecology Letters*, 8, 1283–1290.

Funk, J. L. & Wolf, A. A. (2016) Testing the trait-based community framework: do functional traits predict competitive outcomes? *Ecology*, 97, 2206–2211.

Funk, J. L., Cleland, E. E., Suding, K. N. & Zavaleta, E. S. (2008) Restoration through reassembly: plant traits and invasion resistance. *Trends in Ecology and Evolution*, 23, 695–703.

Galetti, M., Guevara, R., Côrtes, M. C., Fadini, R., Von Matter, S., Leite, A. B., Labecca, F., Ribeiro, T., Carvalho, C. S., Collevatti, R. G., Pires, M. M. G., Brancalion, P. H. S., Ribeiro, M. C. & Jordano, P. (2013) Functional extinction of birds drives rapid evolutionary changes in seed size. *Science*, 340, 1086–1090.

Galland, T., Adeux, G., Dvořáková, H., E-Vojtkó, A., Orbán, I., Lussu, M., Puy, J., Blažek, P., Lanta, V., Lepš, J., de Bello, F., Pérez Carmona, C., Valencia, E. & Götzenberger, L. (2019) Colonization resistance and establishment success along gradients of functional and phylogenetic diversity in experimental plant communities. *Journal of Ecology*, 107, 2090–2104.

Garamszegi, L. Z. (ed.) (2014) *Modern phylogenetic comparative methods and their application in evolutionary biology*. Berlin: Springer.

García-Palacios, P., Milla, R., Delgado-Baquerizo, M., Martín-Robles, N., Álvaro-Sánchez, M. & Wall, D. H. (2013) Side-effects of plant domestication: ecosystem impacts of changes in litter quality. *New Phytologist*, 198, 504–513.

Garnier, E. & Navas, M. L. (2012) A trait-based approach to comparative functional plant ecology: concepts, methods and applications for agroecology. A review. *Agronomy for Sustainable Development*, 32, 365–399.

Garnier, E., Cortez, J., Billes, G., Navas, M. L., Roumet, C., Debussche, M., Laurent, G., Blanchard, A., Aubry, D., Bellmann, A., Neill, C. & Toussaint, J. P. (2004) Plant functional markers capture ecosystem properties during secondary succession. *Ecology*, 85, 2630–2637.

Garnier, E., Laurent, G., Bellmann, A., Debain, S., Berthelier, P., Ducout, B., Roumet, C. & Navas, M. L. (2001) Consistency of species ranking based on functional leaf traits. *New Phytologist*, 152, 69–83.

Garnier, E., Lavorel, S., Ansquer, P., Castro, H., Cruz, P., Dolezal, J., Eriksson, O., Fortunel, C., Freitas, H., Golodets, C., Grigulis, K., Jouany, C., Kazakou, E., Kigel, J., Kleyer, M., Lehsten, V., Lepš, J., Meier, T., Pakeman, R., Papadimitriou, M., Papanastasis, V. P., Quested, H., Quetier, F., Robson, M., Roumet, C., Rusch, G., Skarpe, C., Sternberg, M., Theau, J. P., Thebault, A., Vile, D. & Zarovali, M. P. (2007) Assessing the effects of land-use change on plant traits, communities and ecosystem functioning in grasslands: a standardized methodology and lessons from an application to 11 European sites. *Annals of Botany*, 99, 967–985.

Garnier, E., Navas, M. L. & Grigulis, K. (2016) *Plant functional diversity: organism traits, community structure, and ecosystem properties*. Oxford University Press. Oxford, UK.

Garnier, E., Stahl, U., Laporte, M. A., Kattge, J., Mougenot, I., Kühn, I., Laporte, B., Amiaud, B., Ahrestani, F. S., Bönisch, G., Bunker, D. E., Cornelissen, J. H. C., Díaz, S., Enquist, B. J., Gachet, S., Jaureguiberry, P., Kleyer, M., Lavorel, S., Maicher, L., Pérez-Harguindeguy, N., Poorter, H., Schildhauer, M., Shipley, B., Violle, C., Weiher, E., Wirth, C., Wright, I. J. & Klotz, S. (2017) Towards a thesaurus of plant characteristics: an ecological contribution. *Journal of Ecology*, 105, 298–309.

Gaston, K. J. (1996) Biodiversity – congruence. *Progress in Physical Geography*, 20, 105–112.

Gause, G. F. (1934) *The struggle for existence*. Baltimore: The Williams & Wilkins Company.

Geange, S. W., Pledger, S., Burns, K. C. & Shima, J. S. (2011) A unified analysis of niche overlap incorporating data of different types. *Methods in Ecology and Evolution*, 2, 175–184.

Geber, M. A. & Griffen, L. R. (2003) Inheritance and natural selection on functional traits. *International Journal of Plant Sciences*, 164, S21–S42.

Gerhold, P., Cahill, J. F., Winter, M., Bartish, I. V. & Prinzing, A. (2015) Phylogenetic patterns are not proxies of community assembly mechanisms (they are far better). *Functional Ecology*, 29, 600–614.

Germain, R. M., Mayfield, M. M. & Gilbert, B. (2018a) The 'filtering' metaphor revisited: competition and environment jointly structure invasibility and coexistence. *Biology Letters*, 14, 20180460.

Germain, R. M., Williams, J. L., Schluter, D. & Angert, A. L. (2018b) Moving character displacement beyond characters using contemporary coexistence theory. *Trends in Ecology & Evolution*, 33, 74–84.

Gerz, M., Bueno, C. G., Ozinga, W. A., Zobel, M. & Moora, M. (2018) Niche differentiation and expansion of plant species are associated with mycorrhizal symbiosis. *Journal of Ecology*, 106, 254–264.

Gilchrist, G. W., Huey, R. B. & Serra, L. (2001) Rapid evolution of wing size clines in Drosophila subobscura. *Genetica*, 112–113, 273–286.

Giordani, P., Brunialti, G., Bacaro, G. & Nascimbene, J. (2012) Functional traits of epiphytic lichens as potential indicators of environmental conditions in forest ecosystems. *Ecological Indicators*, 18, 413–420.

Goberna, M. & Verdú, M. (2015) Predicting microbial traits with phylogenies. *ISME J*, 10, 959–967.

Godoy, O., Stouffer, D. B., Kraft, N. J. B. & Levine, J. M. (2017) Intransitivity is infrequent and fails to promote annual plant coexistence without pairwise niche differences. *Ecology*, 98, 1193–1200.

Gohli, J. & Voje, K. L. (2016) An interspecific assessment of Bergmann's rule in 22 mammalian families. *BMC Evolutionary Biology*, 16, 222.

Goncalves, F., Bovendorp, R. S., Beca, G., Bello, C., Costa-Pereira, R., Muylaert, R. L., Rodarte, R. R., Villar, N., Souza, R., Graipel, M. E., Cherem, J. J., Faria, D., Baumgarten, J., Alvarez, M. R., Vieira, E. M., Caceres, N., Pardini, R., Leite, Y. L. R., Costa, L. P., Mello, M. A. R., Fischer, E., Passos, F. C., Varzinczak, L. H., Prevedello, J. A., Cruz-Neto, A. P., Carvalho, F., Percequillo, A. R., Paviolo, A., Nava, A., Duarte, J. M. B., de la Sancha, N. U., Bernard, E., Morato, R. G., Ribeiro, J. F., Becker, R. G., Paise, G., Tomasi, P. S., Velez-Garcia, F., Melo, G. L., Sponchiado, J., Cerezer, F., Barros, M. A. S., de Souza, A. Q. S., dos Santos, C. C., Gine, G. A. F., Kerches-Rogeri, P., Weber, M. M., Ambar, G., Cabrera-Martinez, L. V., Eriksson, A., Silveira, M., Santos, C. F., Alves, L., Barbier, E., Rezende, G. C., Garbino, G. S. T., Rios, E. O., Silva, A., Nascimento, A. T. A., de Carvalho, R. S., Feijo, A., Arrabal, J., Agostini, I., Lamattina, D., Costa, S., Vanderhoeven, E., de Melo, F. R., Laroque, P. D., Jerusalinsky, L., Valenca-Montenegro, M. M., Martins, A. B., Ludwig, G., de Azevedo, R. B., Anzoategui, A., da Silva, M. X., Moraes, M. F. D., Vogliotti, A., Gatti, A., Puttker, T., Barros, C. S., Martins, T. K., Keuroghlian, A., Eaton, D. P., Neves, C. L., Nardi, M. S., Braga, C., Goncalves, P. R., Srbek-Araujo, A. C., Mendes, P., de Oliveira, J. A., Soares, F. A. M., Rocha, P. A., Crawshaw, P., Ribeiro, M. C. & Galetti, M. (2018) Atlantic mammal traits: a data set of morphological traits of mammals in the Atlantic Forest of South America. *Ecology*, 99, 498.

Gonzalez-Suarez, M. & Revilla, E. (2013) Variability in life-history and ecological traits is a buffer against extinction in mammals. *Ecology Letters*, 16, 242–251.

Gossner, M. M., Simons, N. K., Achtziger, R., Blick, T., Dorow, W. H. O., Dziock, F., Kohler, F., Rabitsch, W. & Weisser, W. W. (2015) A summary of eight traits of Coleoptera, Hemiptera, Orthoptera and Araneae, occurring in grasslands in Germany. *Scientific Data*, 2, 150013.

Gotelli, N. J. & Graves, G. R. (1996) *Null models in ecology*. Washington, DC: Smithsonian Institution Press.

Gotelli, N. J. & McCabe, D. J. (2002) Species co-occurrence: a meta-analysis of J. M. Diamond's assembly rules model. *Ecology*, 83, 2091–2096.

Götzenberger, L., Botta-Dukat, Z., Lepš, J., Pärtel, M., Zobel, M. & de Bello, F. (2016) Which randomizations detect convergence and divergence in trait-based community assembly? A test of commonly used null models. *Journal of Vegetation Science*, 27, 1275–1287.

Götzenberger, L., de Bello, F., Bråthen, K. A., Davison, J., Dubuis, A., Guisan, A., Lepš, J., Lindborg, R., Moora, M., Pärtel, M., Pellissier, L., Pottier, J., Vittoz, P., Zobel, K. & Zobel, M. (2012) Ecological assembly rules in plant communities – approaches, patterns and prospects. *Biological Reviews of the Cambridge Philosophical Society*, 87, 111–27.

Gower, J. C. (1971) A general coefficient of similarity and some of its properties. *Biometrics*, 27, 623–637.

Grandcolas, P., Nattier, R., Legendre, F. & Pellens, R. (2011) Mapping extrinsic traits such as extinction risks or modelled bioclimatic niches on phylogenies: does it make sense at all? *Cladistics*, 27, 181–185.

Grant, P. R. & Grant, B. R. (2006) Evolution of character displacement in Darwin's finches. *Science*, 313, 224–226.

Greenleaf, S. S., Williams, N. M., Winfree, R. & Kremen, C. (2007) Bee foraging ranges and their relationship to body size. *Oecologia*, 153, 589–596.

Grigulis, K., Lavorel, S., Krainer, U., Legay, N., Baxendale, C., Dumont, M., Kastl, E., Arnoldi, C., Bardgett, R. D., Poly, F., Pommier, T., Schloter, M., Tappeiner, U., Bahn, M. & Clément, J. C. (2013) Relative contributions of plant traits and soil microbial properties to mountain grassland ecosystem services. *Journal of Ecology*, 101, 47–57.

Grime, J. P. (1979) *Plant strategies and vegetation processes*. Chichester: John Wiley & Sons.

Grime, J. P. (1998) Benefits of plant diversity to ecosystems: immediate, filter and founder effects. *Journal of Ecology*, 86, 902–910.

Grime, J. P. (2006) Trait convergence and trait divergence in herbaceous plant communities: mechanisms and consequences. *Journal of Vegetation Science*, 17, 255–260.

Grime, J. P. & Pierce, S. (2012) *The evolutionary strategies that shape ecosystems*. Chichester: Wiley-Blackwell.

Grime, J. P., Hodgson, J. G. & Hunt, R. (1988) *Comparative plant ecology: a functional approach to common British species*. Dordrecht: Springer.

Grimm, A., Prieto Ramírez, A. M., Moulherat, S., Reynaud, J. & Henle, K. (2014) Life-history trait database of European reptile species. *Nature Conservation*, 9, 45–67.

Gross, N., Borger, L., Soriano-Morales, S. I., Le Bagousse-Pinguet, Y., Quero, J. L., Garcia-Gomez, M., Valencia-Gomez, E. & Maestre, F. T. (2013) Uncovering multiscale effects of aridity and biotic interactions on the functional structure of Mediterranean shrublands. *Journal of Ecology*, 101, 637–649.

Gross, N., Le Bagousse-Pinguet, Y., Liancourt, P., Berdugo, M., Gotelli, N. J. & Maestre, F. T. (2017) Functional trait diversity maximizes ecosystem multifunctionality. *Nature Ecology & Evolution*, 1, 0132.

Gross, N., Robson, T. M., Lavorel, S., Albert, C., LeBagusse-Pinguet, Y. & Guillemin, R. (2008) Plant response traits mediate the effects of subalpine grasslands on soil moisture. *New Phytologist*, 180, 652–662.

Grote, R., Samson, R., Alonso, R., Amorim, J. H., Cariñanos, P., Churkina, G., Fares, S., Thiec, L. D., Niinemets, Ü., Mikkelsen, T. N., Paoletti, E., Tiwary, A. & Calfapietra, C. (2016) Functional traits of urban trees: air pollution mitigation potential. *Frontiers in Ecology and the Environment*, 14, 543–550.

Guisan, A. & Zimmermann, N. E. (2000) Predictive habitat distribution models in ecology. *Ecological Modelling*, 135, 147–186.

Guisan, A., Thuiller, W. & Zimmermann, N. E. (2017) *Habitat suitability and distribution models: with applications in R.* Cambridge University Press. Cambridge, UK.

Gunton, R. M., Petit, S. & Gaba, S. (2011) Functional traits relating arable weed communities to crop characteristics. *Journal of Vegetation Science*, 22, 541–550.

Hadfield, J. D. & Nakagawa, S. (2010) General quantitative genetic methods for comparative biology: phylogenies, taxonomies and multi-trait models for continuous and categorical characters. *Journal of Evolutionary Biology*, 23, 494–508.

Handa, I. T., Aerts, R., Berendse, F., Berg, M. P., Bruder, A., Butenschoen, O., Chauvet, E., Gessner, M. O., Jabiol, J., Makkonen, M., McKie, B. G., Malmqvist, B., Peeters, E. T. H. M., Scheu, S., Schmid, B., van Ruijven, J., Vos, V. C. A. & Hättenschwiler, S. (2014) Consequences of biodiversity loss for litter decomposition across biomes. *Nature*, 509, 218–221.

Handa, T., Raymond-Léonard, L., Boisvert-Marsh, L., Dupuch, A. & Aubin, I. (2017) *CRITTER: Canadian repository of invertebrate traits and trait-like ecological records.* Sault Ste Marie, Ontario: Natural Resources Canada, Canadian Forest Service

Hanisch M., Schweiger O., Cord A. F., Volk M. & Knapp S. (2020) Plant functional traits shape multiple ecosystem services, their trade-offs and synergies in grasslands. *Journal of Applied Ecology*. https://doi.org/10.1111/1365-2664.13644.

Hanski, I. & Gaggiotti, O. (2004) Metapopulation biology: past, present, and future. In *Ecology, genetics and evolution of metapopulations* (eds. Hanski, I. & Gaggiotti, O.), pp. 3–22. New York: Academic Press.

Hardy, O. J. (2008) Testing the spatial phylogenetic structure of local communities: statistical performances of different null models and test statistics on a locally neutral community. *Journal of Ecology*, 96, 914–926.

Harris, J. A., Hobbs, R. J., Higgs, E. & Aronson, J. (2006) Ecological restoration and global climate change. *Restoration Ecology*, 14, 170–176.

Harrison, P. A., Berry, P. M., Simpson, G., Haslett, J. R., Blicharska, M., Bucur, M., Dunford, R., Egoh, B., Garcia-Llorente, M., Geamănă, N., Geertsema, W., Lommelen, E., Meiresonne, L. & Turkelboom, F. (2014) Linkages between biodiversity attributes and ecosystem services: a systematic review. *Ecosystem Services*, 9, 191–203.

Harvey, P. H. & Pagel, M. D. (1991) *The comparative method in evolutionary ecology.* Oxford University Press. Oxford, UK.

Harvey, P. H., Read, A. F. & Nee, S. (1995) Why ecologists need to be phylogenetically challenged. *Journal of Ecology*, 83, 535–536.

Hatfield, J. H., Orme, C. D. L., Tobias, J. A. & Banks-Leite, C. (2018) Trait-based indicators of bird species sensitivity to habitat loss are effective within but not across data sets. *Ecological Applications*, 28, 28–34.

Hättenschwiler, S. & Gasser, P. (2005) Soil animals alter plant litter diversity effects on decomposition. *Proceedings of the National Academy of Sciences of the United States of America*, 102, 1519–1524.

Hättenschwiler, S., Tiunov, A. V. & Scheu, S. (2005) Biodiversity and litter decomposition in terrestrial ecosystems. *Annual Review of Ecology Evolution and Systematics*, 36, 191–218.

Hawkins, B. A., Leroy, B., Rodríguez, M. Á., Singer, A., Vilela, B., Villalobos, F., Wang, X. & Zelený, D. (2017) Structural bias in aggregated species-level variables driven by repeated species co-occurrences: a pervasive problem in community and assemblage data. *Journal of Biogeography*, 44, 1199–1211.

He, Q., Bertness, M. D. & Altieri, A. H. (2013) Global shifts towards positive species interactions with increasing environmental stress. *Ecology Letters*, 16, 695–706.

Heemsbergen, D. A., Berg, M. P., Loreau, M., van Haj, J. R., Faber, J. H. & Verhoef, H. A. (2004) Biodiversity effects on soil processes explained by interspecific functional dissimilarity. *Science*, 306, 1019–1020.

Herman, J. J. & Sultan, S. E. (2016) DNA methylation mediates genetic variation for adaptive transgenerational plasticity. *Proceedings of the Royal Society B – Biological Sciences*, 283, 20160988.

Hevia, V., Martín-López, B., Palomo, S., García-Llorente, M., de Bello, F. & González, J. A. (2017) Trait-based approaches to analyze links between the drivers of change and ecosystem services: synthesizing existing evidence and future challenges. *Ecology and Evolution*, 7, 831–844.

Hijmans, R. J. & Graham, C. H. (2006) The ability of climate envelope models to predict the effect of climate change on species distributions. *Global Change Biology*, 12, 2272–2281.

HilleRisLambers, J., Adler, P. B., Harpole, W. S., Levine, J. M. & Mayfield, M. M. (2012) Rethinking community assembly through the lens of coexistence theory. *Annual Review of Ecology, Evolution, and Systematics*, 43, 227–248.

Hintze, C., Heydel, F., Hoppe, C., Cunze, S., König, A. & Tackenberg, O. (2013) D3: The Dispersal and Diaspore Database – baseline data and statistics on seed dispersal. *Perspectives in Plant Ecology, Evolution and Systematics*, 15, 180–192.

Hochkirch, A., Deppermann, J. & Groning, J. (2008) Phenotypic plasticity in insects: the effects of substrate color on the coloration of two ground-hopper species. *Evolution & Development*, 10, 350–359.

Holyoak, M., Leibold, M. A. & Holt, R. D. (2005) *Metacommunities: spatial dynamics and ecological communities*. University of Chicago Press. Chicago, IL.

Homburg, K., Homburg, N., Schäfer, F., Schuldt, A., Assmann, T., Dytham, C. & Ewers, R. (2014) Carabids.org – a dynamic online database of ground beetle species traits (Coleoptera, Carabidae). *Insect Conservation and Diversity*, 7, 195–205.

Hooper, D. U., Chapin, F. S., Ewel, J. J., Hector, A., Inchausti, P., Lavorel, S., Lawton, J. H., Lodge, D. M., Loreau, M., Naeem, S., Schmid, B., Setala, H., Symstad, A. J., Vandermeer, J. & Wardle, D. A. (2005) Effects of biodiversity on ecosystem functioning: a consensus of current knowledge. *Ecological Monographs*, 75, 3–35.

Hortal, J., de Bello, F., Diniz-Filho, J. A. F., Lewinsohn, T. M., Lobo, J. M. & Ladle, R. J. (2015) Seven shortfalls that beset large-scale knowledge of biodiversity. *Annual Review of Ecology, Evolution, and Systematics*, 46, 523–549.

Hubbell, S. P. (2001) *The unified neutral theory of biodiversity and biogeography*. Princeton University Press. Princeton, NJ.

Huber, S. K., De Leon, L. F., Hendry, A. P., Bermingham, E. & Podos, J. (2007) Reproductive isolation of sympatric morphs in a population of Darwin's finches. *Proceedings of the Royal Society B – Biological Sciences*, 274, 1709–1714.

Hughes, A. R., Inouye, B. D., Johnson, M. T. J., Underwood, N. & Vellend, M. (2008) Ecological consequences of genetic diversity. *Ecology Letters*, 11, 609–623.

Hulshof, C. M., Violle, C., Spasojevic, M. J., McGill, B., Damschen, E., Harrison, S. & Enquist, B. J. (2013) Intra-specific and inter-specific variation in specific leaf area reveal the importance of abiotic and biotic drivers of species diversity across elevation and latitude. *Journal of Vegetation Science*, 24, 921–931.

Hutchinson, G. E. (1959) Homage to Santa Rosalia or why are there so many kinds of animals? *American Naturalist*, 93, 149–159.

Hutchinson, G. E. (1961) The paradox of the plankton. *American Naturalis*, 95, 137–145.

Ibanez, S. (2012) Optimizing size thresholds in a plant–pollinator interaction web: towards a mechanistic understanding of ecological networks. *Oecologia*, 170, 233–242.

Ibanez, S., Lavorel, S., Puijalon, S. & Moretti, M. (2013) Herbivory mediated by coupling between biomechanical traits of plants and grasshoppers. *Functional Ecology*, 27, 479–489.

IPBES (2019) Summary for policymakers of the global assessment report on biodiversity and ecosystem services of the Intergovernmental Science-Policy Platform on Biodiversity and Ecosystem Services. IPBES secretariat.

Isaac, M. E., Cerda, R., Rapidel, B., Martin, A. R., Dickinson, A. K. & Sibelet, N. (2018) Farmer perception and utilization of leaf functional traits in managing agroecosystems. *Journal of Applied Ecology*, 55, 69–80.

Iversen, C. M., McCormack, M. L., Powell, A. S., Blackwood, C. B., Freschet, G. T., Kattge, J., Roumet, C., Stover, D. B., Soudzilovskaia, N. A., Valverde-Barrantes, O. J., Bodegom, P. M. & Violle, C. (2017) A global Fine-Root Ecology Database to address below-ground challenges in plant ecology. *New Phytologist*, 215, 15–26.

Jablonski, N. G. & Chaplin, G. (2010) Human skin pigmentation as an adaptation to UV radiation. *Proceedings of the National Academy of Sciences of the United States of America*, 107, 8962–8968.

Jakobsson, A. & Eriksson, O. (2000) A comparative study of seed number, seed size, seedling size and recruitment in grassland plants. *Oikos*, 88, 494–502.

Jamil, T., Ozinga, W. A., Kleyer, M. & ter Braak, C. J. F. (2013) Selecting traits that explain species–environment relationships: a generalized linear mixed model approach. *Journal of Vegetation Science*, 24, 988–1000.

Jo, I., Fridley, J. D. & Frank, D. A. (2016) More of the same? In situ leaf and root decomposition rates do not vary between 80 native and nonnative deciduous forest species. *New Phytologist*, 209, 115–122.

Johnson, S. D. & Steiner, K. E. (1997) Long-tongued fly pollination and evolution of floral spur length in the *Disa draconis* Complex (Orchidaceae). *Evolution*, 51, 45–53.

Jonas, C. S. & Geber, M. A. (1999) Variation among populations of *Clarkia unguiculata* (Onagraceae) along altitudinal and latitudinal gradients. *American Journal of Botany*, 86, 333–343.

Jones, C. G., Lawton, J. H. & Shachak, M. (1994) Organisms as ecosystem engineers. *Oikos*, 69, 373–386.

Jordan, R., Singer, F., Vaughan, J. & Berkowitz, A. (2009) What should every citizen know about ecology? *Frontiers in Ecology and the Environment*, 7, 495–500.

Jost, L. (2007) Partitioning diversity into independent alpha and beta components. *Ecology*, 88, 2427–2439.

Jung, V., Violle, C., Mondy, C., Hoffmann, L. & Muller, S. (2010) Intraspecific variability and trait-based community assembly. *Journal of Ecology*, 98, 1134–1140.

Junker, R. R., Kuppler, J., Bathke, A. C., Schreyer, M. L. & Trutschnig, W. (2016) Dynamic range boxes – a robust nonparametric approach to quantify size and overlap of n-dimensional hypervolumes. *Methods in Ecology and Evolution*, 7, 1503–1513.

Kaldhusdal, A., Brandl, R., Müller, J., Möst, L. & Hothorn, T. (2014) Spatio-phylogenetic multispecies distribution models. *Methods in Ecology and Evolution*, 6, 187–197.

Karger, D. N., Cord, A. F., Kessler, M., Kreft, H., Kuehn, I., Pompe, S., Sandel, B., Cabral, J. S., Smith, A. B., Svenning, J. C., Tuomisto, H., Weigelt, P. & Wesche, K. (2016) Delineating probabilistic species pools in ecology and biogeography. *Global Ecology and Biogeography*, 25, 489–501.

Kattge, J., Díaz, S., Lavorel, S., Prentice, C., Leadley, P., Bonisch, G., Garnier, E., Westoby, M., Reich, P. B., Wright, I. J., Cornelissen, J. H. C., Violle, C., Harrison, S. P., van Bodegom, P. M., Reichstein, M., Enquist, B. J., Soudzilovskaia, N. A., Ackerly, D. D., Anand, M., Atkin, O., Bahn, M., Baker, T. R., Baldocchi, D., Bekker, R., Blanco, C. C., Blonder, B., Bond, W. J., Bradstock, R., Bunker, D. E., Casanoves, F., Cavender-Bares, J., Chambers, J. Q., Chapin, F. S., Chave, J., Coomes, D., Cornwell, W. K., Craine, J. M., Dobrin, B. H., Duarte, L., Durka, W., Elser, J., Esser, G., Estiarte, M., Fagan, W. F., Fang, J., Fernandez-Mendez, F., Fidelis, A., Finegan, B., Flores, O., Ford, H., Frank, D., Freschet, G. T., Fyllas, N. M., Gallagher, R. V., Green, W. A., Gutierrez, A. G., Hickler, T., Higgins, S. I., Hodgson, J. G., Jalili, A., Jansen, S., Joly, C. A., Kerkhoff, A. J., Kirkup, D., Kitajima, K., Kleyer, M., Klotz, S., Knops, J. M. H., Kramer, K., Kuhn, I., Kurokawa, H., Laughlin, D., Lee, T. D., Leishman, M., Lens, F., Lenz, T., Lewis, S. L., Lloyd, J., Llusia, J., Louault, F., Ma, S., Mahecha, M. D., Manning, P., Massad, T., Medlyn, B. E., Messier, J., Moles, A. T., Muller, S. C., Nadrowski, K., Naeem, S., Niinemets, U., Nollert, S., Nuske, A., Ogaya, R., Oleksyn, J., Onipchenko, V. G., Onoda, Y., Ordonez, J., Overbeck, G., Ozinga, W. A., et al. (2011) TRY – a global database of plant traits. *Global Change Biology*, 17, 2905–2935.

Kazakou, E., Violle, C., Roumet, C., Navas, M. L., Vile, D., Kattge, J. & Garnier, E. (2014) Are trait-based species rankings consistent across data sets and spatial scales? *Journal of Vegetation Science*, 25, 235–247.

Keane, R. M. & Crawley, M. J. (2002) Exotic plant invasions and the enemy release hypothesis. *Trends in Ecology & Evolution*, 17, 164–170.

Keck, F., Rimet, F., Bouchez, A. & Franc, A. (2016) phylosignal: an R package to measure, test, and explore the phylogenetic signal. *Ecology and Evolution*, 6, 2774–2780.

Keddy, P. A. (1992a) A pragmatic approach to functional ecology. *Functional Ecology*, 6, 621–626.

Keddy, P. A. (1992b) Assembly and response rules: two goals for predictive community ecology. *Journal of Vegetation Science*, 3, 157–164.

Keesstra, S., Nunes, J., Novara, A., Finger, D., Avelar, D., Kalantari, Z. & Cerdà, A. (2018) The superior effect of nature based solutions in land management for enhancing ecosystem services. *Science of the Total Environment*, 610–611, 997–1009.

Kichenin, E., Wardle, D. A., Peltzer, D. A., Morse, C. W. & Freschet, G. F. (2013) Contrasting effects of plant inter- and intraspecific variation on community-level trait measures along an environmental gradient. *Functional Ecology*, 27, 1254–1261.

Kingston, T. & Rossiter, S. J. (2004) Harmonic-hopping in Wallacea's bats. *Nature*, 429, 654–657.

Kissling, W. D., Walls, R., Bowser, A., Jones, M. O., Kattge, J., Agosti, D., Amengual, J., Basset, A., van Bodegom, P. M., Cornelissen, J. H. C., Denny, E. G., Deudero, S., Egloff, W., Elmendorf, S. C., Alonso García, E., Jones, K. D., Jones, O. R., Lavorel, S., Lear, D., Navarro, L. M., Pawar, S., Pirzl, R., Rüger, N., Sal, S., Salguero-Gómez, R., Schigel, D., Schulz, K. S., Skidmore, A. & Guralnick, R. P. (2018) Towards global data products of Essential Biodiversity Variables on species traits. *Nature Ecology and Evolution*, 2, 1531–1540.

Klaiber, J., Altermatt, F., Birrer, S., Chittaro, Y., Dziock, F., Gonseth, Y., Hoess, R., Keller, D., Köchler, H., Luka, H., Manzke, U., Müller, A., Pfeifer, M. A., Roesti, C., Schlegel, J., Schneider, K., Sonderegger, P., Walter, T., Holderegger, R. & Bergamini, A. (2017) *Fauna Indicativa*. Birmensdorf: Eidgenössische Forschungsanstalt für Wald, Schnee und Landschaft WSL.

Klaus, V. H., Kleinebecker, T., Boch, S., Muller, J., Socher, S. A., Prati, D., Fischer, M. & Holzel, N. (2012) NIRS meets Ellenberg's indicator values: prediction of moisture and nitrogen values of agricultural grassland vegetation by means of near-infrared spectral characteristics. *Ecological Indicators*, 14, 82–86.

Kleyer, M., Bekker, R. M., Knevel, I. V., Bakker, J. P., Thompson, K., Sonnenschein, M., Poschlod, P., van Groenendael, J. M., Klimes, L., Klimešová, J., Klotz, S., Rusch, G. M. et al. (2008) The LEDA Traitbase: a database of life-history traits of the Northwest European flora. *Journal of Ecology*, 96, 1266–1274.

Kleyer, M., Dray, S., Bello, F., Lepš, J., Pakeman, R. J., Strauss, B., Thuiller, W. & Lavorel, S. (2012) Assessing species and community functional responses to environmental gradients: which multivariate methods? *Journal of Vegetation Science*, 23, 805–821.

Kleyer, M., Trinogga, J., Cebrián-Piqueras, M. A., Trenkamp, A., Fløjgaard, C., Ejrnæs, R., Bouma, T. J., Minden, V., Maier, M., Mantilla-Contreras, J., Albach, D. C. & Blasius, B. (2019) Trait correlation network analysis identifies biomass allocation traits and stem specific length as hub traits in herbaceous perennial plants. *Journal of Ecology*, 107, 829–842.

Klimešová, J., Danihelka, J., Chrtek, J., de Bello, F. & Herben, T. (2017) CLO-PLA: a database of clonal and bud-bank traits of the Central European flora. *Ecology*, 98, 1179.

Klotz, S., Kühn, I., Durka, W. & Briemle, G. (2002) *BIOLFLOR: eine Datenbank mit biologisch-ökologischen Merkmalen zur Flora von Deutschland (vol. 38)*. Bundesamt für naturschutz Bonn.

Klumpp, K. & Soussana, J. F. (2009) Using functional traits to predict grassland ecosystem change: a mathematical test of the response-and-effect trait approach. *Global Change Biology*, 15, 2921–2934.

Knight, T. M., McCoy, M. W., Chase, J. M., McCoy, K. A. & Holt, R. D. (2005) Trophic cascades across ecosystems. *Nature*, 437, 880–883.

Koide, R. T., Fernandez, C. & Malcolm, G. (2014) Determining place and process: functional traits of ectomycorrhizal fungi that affect both community structure and ecosystem function. *New Phytologist*, 201, 433–439.

Kostikova, A., Litsios, G., Salamin, N. & Pearman, P. B. (2013) Linking life-history traits, ecology, and niche breadth evolution in North American eriogonoids (Polygonaceae). *American Naturalist*, 182, 760–774.

Kotowska, A. M., Cahill, J. F. & Keddie, B. A. (2010) Plant genetic diversity yields increased plant productivity and herbivore performance. *Journal of Ecology*, 98, 237–245.

Kraft, N. J. B. & Ackerly, D. D. (2010) Functional trait and phylogenetic tests of community assembly across spatial scales in an Amazonian forest. *Ecological Monographs*, 80, 401–422.

Kraft, N. J. B., Adler, P. B., Godoy, O., James, E. C., Fuller, S., Levine, J. M. & Fox, J. (2015) Community assembly, coexistence and the environmental filtering metaphor. *Functional Ecology*, 29, 592–599.

Kraft, N. J. B., Cornwell, W. K., Webb, C. O. & Ackerly, D. D. (2007) Trait evolution, community assembly, and the phylogenetic structure of ecological communities. *The American Naturalist*, 170, 271–283.

Krasnov, B. R., Shenbrot, G. I., Khokhlova, I. S. & Degen, A. A. (2016) Trait-based and phylogenetic associations between parasites and their hosts: a case study with small mammals and fleas in the Palearctic. *Oikos*, 125, 29–38.

Kremen, C., Williams, N. M., Aizen, M. A., Gemmill-Herren, B., LeBuhn, G., Minckley, R., Packer, L., Potts, S. G., Roulston, T. A., Steffan-Dewenter, I., Vázquez, D. P., Winfree, R., Adams, L., Crone, E. E., Greenleaf, S. S., Keitt, T. H., Klein, A. M., Regetz, J. & Ricketts, T. H. (2007) Pollination and other ecosystem services produced by mobile organisms: a conceptual framework for the effects of land-use change. *Ecology Letters*, 10, 299–314.

Krishna, A., Guimarães, P. R., Jordano, P. & Bascompte, J. (2008) A neutral-niche theory of nestedness in mutualistic networks. *Oikos*, 117, 1609–1618.

Kurokawa, H., Peltzer, D. A. & Wardle, D. A. (2010) Plant traits, leaf palatability and litter decomposability for co-occurring woody species differing in invasion status and nitrogen fixation ability. *Functional Ecology*, 24, 513–523.

Laliberté, E. & Legendre, P. (2010) A distance-based framework for measuring functional diversity from multiple traits. *Ecology*, 91, 299–305.

Laliberté, E., Wells, J. A., DeClerck, F., Metcalfe, D. J., Catterall, C. P., Queiroz, C., Aubin, I., Bonser, S. P., Ding, Y., Fraterrigo, J. M., McNamara, S., Morgan, J. W., Merlos, D. S., Vesk, P. A. & Mayfield, M. M. (2010) Land-use intensification reduces functional redundancy and response diversity in plant communities. *Ecology Letters*, 13, 76–86.

Lamanna, C., Blonder, B., Violle, C., Kraft, N. J. B., Sandel, B., Simova, I., Donoghue, J. C., Svenning, J. C., McGill, B. J., Boyle, B., Buzzard, V., Dolins, S., Jørgensen, P. M., Marcuse-Kubitza, A., Morueta-Holme, N., Peet, R. K., Piel, W. H., Regetz, J., Schildhauer, M., Spencer, N., Thiers, B., Wiser, S. K. & Enquist, B. J. (2014) Functional trait space and the latitudinal diversity gradient. *Proceedings of the National Academy of Sciences of the United States of America*, 111, 13745–13750.

Lanta, V. & Lepš, J. (2008) Effect of plant species richness on invasibility of experimental plant communities. *Plant Ecology*, 198, 253–263.

Larsen, T. H., Williams, N. M. & Kremen, C. (2005) Extinction order and altered community structure rapidly disrupt ecosystem functioning. *Ecology Letters*, 8, 538–547.

Laughlin, D. C. (2011) Nitrification is linked to dominant leaf traits rather than functional diversity. *Journal of Ecology*, 99, 1091–1099.

Laughlin, D. C. (2014a) Applying trait-based models to achieve functional targets for theory-driven ecological restoration. *Ecology Letters*, 17, 771–784.

Laughlin, D. C. (2014b) The intrinsic dimensionality of plant traits and its relevance to community assembly. *Journal of Ecology*, 102, 186–193.

Laughlin, D. C. (2018) Rugged fitness landscapes and Darwinian demons in trait-based ecology. *New Phytologist*, 217, 501–503.

Laughlin, D. C. & Laughlin, D. E. (2013) Advances in modeling trait-based plant community assembly. *Trends in Plant Science*, 18, 584–593.

Laughlin, D. C. & Messier, J. (2015) Fitness of multidimensional phenotypes in dynamic adaptive landscapes. *Trends in Ecology & Evolution*, 30, 487–496.

Laughlin, D. C., Joshi, C., Richardson, S. J., Peltzer, D. A., Mason, N. W. H. & Wardle, D. A. (2015) Quantifying multimodal trait distributions improves trait-based predictions of species abundances and functional diversity. *Journal of Vegetation Science*, 26, 46–57.

Laughlin, D. C., Joshi, C., van Bodegom, P. M., Bastow, Z. A. & Fulé, P. Z. (2012) A predictive model of community assembly that incorporates intraspecific trait variation. *Ecology Letters*, 15, 1291–1299.

Lavania, U. C., Srivastava, S., Lavania, S., Basu,' S., Misra, N. K. & Mukai, Y. (2012) Autopolyploidy differentially influences body size in plants, but facilitates enhanced accumulation of secondary metabolites, causing increased cytosine methylation. *Plant Journal*, 71, 539–549.

Lavorel, S. (2013) Plant functional effects on ecosystem services. *Journal of Ecology*, 101, 4–8.

Lavorel, S. & Garnier, E. (2002) Predicting changes in community composition and ecosystem functioning from plant traits: revisiting the Holy Grail. *Functional Ecology*, 16, 545–556.

Lavorel, S. & Grigulis, K. (2012) How fundamental plant functional trait relationships scale-up to trade-offs and synergies in ecosystem services. *Journal of Ecology*, 100, 128–140.

Lavorel, S., Grigulis, K., Lamarque, P., Colace, M.-P., Garden, D., Girel, J., Pellet, G. & Douzet, R. (2011) Using plant functional traits to understand the landscape distribution of multiple ecosystem services. *Journal of Ecology*, 99, 135–147.

Lavorel, S., Grigulis, K., McIntyre, S., Williams, N. S. G., Garden, D., Dorrough, J., Berman, S., Quetier, F., Thebault, A. & Bonis, A. (2008) Assessing functional diversity in the field – methodology matters! *Functional Ecology*, 22, 134–147.

Lavorel, S., McIntyre, S., Landsberg, J. & Forbes, T. D. A. (1997) Plant functional classifications: from general groups to specific groups based on response to disturbance. *Trends in Ecology & Evolution*, 12, 474–478.

Lavorel, S., Storkey, J., Bardgett, R. D., de Bello, F., Berg, M. P., Le Roux, X., Moretti, M., Mulder, C., Pakeman, R. J., Díaz, S. & Harrington, R. (2013) A novel framework for linking functional diversity of plants with other trophic levels for the quantification of ecosystem services. *Journal of Vegetation Science*, 24, 942–948.

Lawton, J. H. (1994) What do species do in ecosystems? *Oikos*, 71, 367–374.

Lawton, J. H. (1996) Corncrake pie and prediction in ecology. *Oikos*, 76, 3–4.

Le Bagousse-Pinguet, Y., Borger, L., Quero, J. L., Garcia-Gomez, M., Soriano, S., Maestre, F. T. & Gross, N. (2015) Traits of neighbouring plants and space limitation determine intraspecific trait variability in semi-arid shrublands. *Journal of Ecology*, 103, 1647–1657.

Le Bagousse-Pinguet, Y., de Bello, F., Vandewalle, M., Lepš, J. & Sykes, M. T. (2014) Species richness of limestone grasslands increases with trait overlap: evidence from within- and between-species functional diversity partitioning. *Journal of Ecology*, 102, 466–474.

Le Bagousse-Pinguet, Y., Gross, N., Maestre, F. T., Maire, V., de Bello, F., Fonseca, C. R., Kattge, J., Valencia, E., Lepš, J. & Liancourt, P. (2017) Testing the environmental filtering concept in global drylands. *Journal of Ecology*, 105, 1058–1069.

Le Bagousse-Pinguet, Y., Soliveres, S., Gross, N., Torices, R., Berdugo, M. & Maestre, F. T. (2019) Phylogenetic, functional, and taxonomic richness have both positive and negative effects on ecosystem multifunctionality. *Proceedings of the National Academy of Sciences of the United States of America*, 116, 8419–8424.

Le Lann, C., Visser, B., Meriaux, M., Moiroux, J., van Baaren, J., van Alphen, J. J. M. & Ellers, J. (2014) Rising temperature reduces divergence in resource use strategies in coexisting parasitoid species. *Oecologia*, 174, 967–977.

Lecerf, A. & Chauvet, E. (2008) Intraspecific variability in leaf traits strongly affects alder leaf decomposition in a stream. *Basic and Applied Ecology*, 9, 598–605.

Lefcheck, J. S. & Duffy, J. E. (2015) Multitrophic functional diversity predicts ecosystem functioning in experimental assemblages of estuarine consumers. *Ecology*, 96, 2973–2983.

Lefcheck, J. S., Byrnes, J. E. K., Isbell, F., Gamfeldt, L., Griffin, J. N., Eisenhauer, N., Hensel, M. J. S., Hector, A., Cardinale, B. J. & Duffy, J. E. (2015) Biodiversity enhances ecosystem multifunctionality across trophic levels and habitats. *Nature Communications*, 6, 7.

Lennon, J. T. & Lehmkuhl, B. K. (2016) A trait-based approach to bacterial biofilms in soil. *Environmental Microbiology*, 18, 2732–2742.

Lepš, J., de Bello, F., Lavorel, S. & Berman, S. (2006) Quantifying and interpreting functional diversity of natural communities: practical considerations matter. *Preslia*, 78, 481–501.

Lepš, J., de Bello, F., Šmilauer, P. & Dolezal, J. (2011) Community trait response to environment: disentangling species turnover vs. intraspecific trait variability effects. *Ecography*, 34, 856–863.

Letten, A. D., Keith, D. A. & Tozer, M. G. (2014) Phylogenetic and functional dissimilarity does not increase during temporal heathland succession. *Proceedings of the Royal Society B – Biological Sciences*, 281, 20142102.

Levin, L. A. & Mehring, A. S. (2015) Optimization of bioretention systems through application of ecological theory. *Wiley Interdisciplinary Reviews: Water*, 2, 259–270.

Levine, J. M., Adler, P. B. & Yelenik, S. G. (2004) A meta-analysis of biotic resistance to exotic plant invasions. *Ecology Letters*, 7, 975–989.

Levine, J. M., Bascompte, J., Adler, P. B. & Allesina, S. (2017) Beyond pairwise mechanisms of species coexistence in complex communities. *Nature*, 546, 56–64.

Lewinsohn, T. M., Prado, P. I., Jordano, P., Bascompte, J. & Olesen, J. M. (2006) Structure in plant–animal interaction assemblages. *Oikos*, 113, 174–184.

Liancourt, P., Callaway, R. M. & Michalet, R. (2005) Stress tolerance and competitive-response ability determine the outcome of biotic interactions. *Ecology*, 86, 1611–1618.

Liao, C., Peng, R., Luo, Y., Zhou, X., Wu, X., Fang, C., Chen, J. & Li, B. (2008) Altered ecosystem carbon and nitrogen cycles by plant invasion: a meta-analysis. *New Phytologist*, 177, 706–714.

Liebig, J. (1842) *Chemistry in its application to agriculture and physiology*. 2nd ed. London: Taylor and Walton.

Liefting, M., Weerenbeck, M., van Dooremalen, C. & Ellers, J. (2010) Temperature-induced plasticity in egg size and resistance of eggs to temperature stress in a soil arthropod. *Functional Ecology*, 24, 1291–1298.

Lindenmayer, D. B. & Likens, G. E. (2010) The science and application of ecological monitoring. *Biological Conservation*, 143, 1317–1328.

Liow, L. H. (2007) Does versatility as measured by geographic range, bathymetric range and morphological variability contribute to taxon longevity? *Global Ecology and Biogeography*, 16, 117–128.

Lislevand, T., Figuerola, J. & Székely, T. (2007) Avian body sizes in relation to fecundity, mating system, display behavior, and resource sharing. *Ecology*, 88, 1605.

Liu, G. F., Freschet, G. T., Pan, X., Cornelissen, J. H. C., Li, Y. & Dong, M. (2010) Coordinated variation in leaf and root traits across multiple spatial scales in Chinese semi-arid and arid ecosystems. *New Phytologist*, 188, 543–553.

Livesley, S. J., McPherson, G. M. & Calfapietra, C. (2016) The urban forest and ecosystem services: impacts on urban water, heat, and pollution cycles at the tree, street, and city scale. *Journal of Environmental Quality*, 45, 119–124.

Livingston, G., Matias, M., Calcagno, V., Barbera, C., Combe, M., Leibold, M. A. & Mouquet, N. (2012) Competition–colonization dynamics in experimental bacterial metacommunities. *Nature Communications*, 3, 1234.

Loiola, P. P., de Bello, F., Chytry, M., Götzenberger, L., Carmona, C. P., Pysek, P. & Lososova, Z. (2018) Invaders among locals: alien species decrease phylogenetic and functional diversity while increasing dissimilarity among native community members. *Journal of Ecology*, 106, 2230–2241.

Loreau, M. & Hector, A. (2001) Partitioning selection and complementarity in biodiversity experiments. *Nature*, 412, 72–76.

Loreau, M. & Hector, A. (2019) Not even wrong: comment by Loreau and Hector. *Ecology*, 100, e02794.

Loreau, M., Mouquet, N. & Holt, R. D. (2003) Meta-ecosystems: a theoretical framework for a spatial ecosystem ecology. *Ecology Letters*, 6, 673–679.

Loring, P. A., Chapin, F. S. & Gerlach, S. C. (2008) The services-oriented architecture: ecosystem services as a framework for diagnosing change in social ecological systems. *Ecosystems*, 11, 478–489.

Lortie, C. J., Brooker, R. W., Choler, P., Kikvidze, Z., Michalet, R., Pugnaire, F. I. & Callaway, R. M. (2004) Rethinking plant community theory. *Oikos*, 107, 433–438.

Losos, J. B. (2008) Phylogenetic niche conservatism, phylogenetic signal and the relationship between phylogenetic relatedness and ecological similarity among species. *Ecology Letters*, 11, 995–1003.

Lososová, Z., de Bello, F., Chytrý, M., Kühn, I., Pyšek, P., Sádlo, J., Winter, M. & Zelený, D. (2015). Alien plants invade more phylogenetically clustered community types and cause even stronger clustering. *Global Ecology and Biogeography*, 24, 786–794.

Lotka, A. J. (1925) *Elements of physical biology*. Baltimore: Williams and Wilkins.

Lu, Z.-X., Yu, X.-P., Heong, K.-L. & Hu, C. (2007) Effect of nitrogen fertilizer on herbivores and its stimulation to major insect pests in rice. *Rice Science*, 14, 56–66.

Luck, G. W., Harrington, R., Harrison, P. A., Kremen, C., Berry, P. M., Bugter, R., Dawson, T. P., de Bello, F., Díaz, S., Feld, C. K., Haslett, J. R., Hering, D., Kontogianni, A., Lavorel, S., Rounsevell, M., Samways, M. J., Sandin, L., Settele, J., Sykes, M. T., van den Hove, S., Vandewalle, M. & Zobel, M. (2009) Quantifying the contribution of organisms to the provision of ecosystem services. *Bioscience*, 59, 223–235.

Luck, G. W., Lavorel, S., McIntyre, S. & Lumb, K. (2012) Improving the application of vertebrate trait-based frameworks to the study of ecosystem services. *Journal of Animal Ecology*, 81, 1065–1076.

Luederitz, C., Brink, E., Gralla, F., Hermelingmeier, V., Meyer, M., Niven, L., Panzer, L., Partelow, S., Rau, A. L., Sasaki, R., Abson, D. J., Lang, D. J., Wamsler, C. & von Wehrden, H. (2015) A review of urban ecosystem services: six key challenges for future research. *Ecosystem Services*, 14, 98–112.

MacArthur, R. & Levins, R. (1967) Limiting similarity convergence and divergence of coexisting species. *American Naturalist*, 101, 377–385.

MacArthur, R. H. & Wilson, E. O. (1967) *The theory of island biogeography*. Princeton University Press. Princeton, NJ.

Madin, J. S., Anderson, K. D., Andreasen, M. H., Bridge, T. C. L., Cairns, S. D., Connolly, S. R., Darling, E. S., Díaz, M., Falster, D. S., Franklin, E. C., Gates, R. D., Harmer, A. M. T., Hoogenboom, M. O., Huang, D., Keith, S. A., Kosnik, M. A., Kuo, C.-Y., Lough, J. M., Lovelock, C. E., Luiz, O., Martinelli, J., Mizerek, T., Pandolfi, J. M., Pochon, X., Pratchett, M. S., Putnam, H. M., Roberts, T. E., Stat, M., Wallace, C. C., Widman, E. & Baird, A. H. (2016) The Coral Trait Database, a curated database of trait information for coral species from the global oceans. *Scientific Data*, 3, 160017.

Maestre, F. T., Callaway, R. M., Valladares, F. & Lortie, C. J. (2009) Refining the stress-gradient hypothesis for competition and facilitation in plant communities. *Journal of Ecology*, 97, 199–205.

Majeková, M., Janeček, Š, Mudrák, O., Horník, J., Janečková, P., Bartoš, M., Fajmon, K., Jiráská, Š., Götzenberger, L., Šmilauer, P., Lepš, J. & de Bello, F. (2016b) Consistent functional response of meadow species and communities to land-use changes across productivity and soil moisture gradients. *Applied Vegetation Science*, 19, 196–205.

Majeková, M., Paal, T., Plowman, N. S., Bryndova, M., Kasari, L., Norberg, A., Weiss, M., Bishop, T. R., Luke, S. H., Sam, K., Le Bagousse–Pinguet, Y., Lepš, J., Götzenberger, L. & de Bello, F. (2016a) Evaluating functional diversity: missing trait data and the importance of species abundance structure and data transformation. *PLOS ONE*, 11., e0149270

Makkonen, M., Berg, M. P., van Hal, J. R., Callaghan, V. T., Press, M. C. & Aerts, R. (2011) Traits explain the responses of a sub-arctic Collembola community to climate manipulation. *Soil Biology and Biochemistry*, 43, 377–384.

Malézieux, E., Crozat, Y., Duparz, C., Laurans, M., Makowski, D., Ozier-Lafontaine, H., Rapidel, B., de Tourdonnet, S. & Valantin-Morison, M. (2009) Mixing plant species in cropping systems: concepts, tools and models. A review. *Agronomy for Sustainable Development*, 29, 43–62.

Manly, B. F. J. (1995) A note on the analysis of species cooccurrences. *Ecology*, 76, 1109–1115.

Margalef, R. (1963) On certain unifying principles in ecology. *American Naturalist*, 97, 357–374.

Marichal, R., Praxedes, C., Decaens, T., Grimaldi, M., Oszwald, J., Brown, G. G., Desjardins, T., da Silva, M. L., Martinez, A. F., Oliveira, M. N. D., Velasquez, E. & Lavelle, P. (2017) Earthworm functional traits, landscape degradation and ecosystem services in the Brazilian Amazon deforestation arc. *European Journal of Soil Biology*, 83, 43–51.

Martello, F., de Bello, F., Morini, M. S. D., Silva, R. R., de Souza-Campana, D. R., Ribeiro, M. C. & Carmona, C. P. (2018) Homogenization and impoverishment of taxonomic and functional diversity of ants in Eucalyptus plantations. *Scientific Reports*, 8, 3266

Martin, A. R. & Isaac, M. E. (2015) Plant functional traits in agroecosystems: a blueprint for research. *Journal of Applied Ecology*, 52, 1425–1435.

Mason, N. W. H., de Bello, F., Dolezal, J. & Lepš, J. (2011) Niche overlap reveals the effects of competition, disturbance and contrasting assembly processes in experimental grassland communities. *Journal of Ecology*, 99, 788–796.

Mason, N. W. H., de Bello, F., Mouillot, D., Pavoine, S. & Dray, S. (2013) A guide for using functional diversity indices to reveal changes in assembly processes along ecological gradients. *Journal of Vegetation Science*, 24, 794–806.

Mason, N. W. H., Lanoiselee, C., Mouillot, D., Wilson, J. B. & Argillier, C. (2008) Does niche overlap control relative abundance in French lacustrine fish communities? A new method incorporating functional traits. *Journal of Animal Ecology*, 77, 661–669.

Mason, N. W. H., Mouillot, D., Lee, W. G. & Wilson, J. B. (2005) Functional richness, functional evenness and functional divergence: the primary components of functional diversity. *Oikos*, 111, 112–118.

Matos, P., Geiser, L., Hardman, A., Glavich, D., Pinho, P., Nunes, A., Soares, A. M. V. M. & Branquinho, C. (2017) Tracking global change using lichen diversity: towards a global-scale ecological indicator. *Methods in Ecology and Evolution*, 8, 788–798.

Mayfield, M. M. & Levine, J. M. (2010) Opposing effects of competitive exclusion on the phylogenetic structure of communities. *Ecology Letters*, 13, 1085–1093.

Mayfield, M. M. & Stouffer, D. B. (2017) Higher-order interactions capture unexplained complexity in diverse communities. *Nature Ecology & Evolution*, 1, 0062.

McCann, K. S. (2000) The diversity–stability debate. *Nature*, 405, 228–233.

McDonald, A. J., Riha, S. J. & Ditommaso, A. (2010) Early season height differences as robust predictors of weed growth potential in maize: new avenues for adaptive management? *Weed Research*, 50, 110–119.

McDonald, T., Gann, G. D., Jonson, J. & Dixon, K. W. (2016) *International standards for the practice of ecological restoration – including principles and key concepts*. Washington, DC: Society for Ecological Restoration.

McGill, B. J., Enquist, B. J., Weiher, E. & Westoby, M. (2006) Rebuilding community ecology from functional traits. *Trends in Ecology & Evolution*, 21, 178–185.

McGuire, K. L. (2007) Common ectomycorrhizal networks may maintain monodominance in a tropical rain forest. *Ecology*, 88, 567–574.

Meier, C. L. & Bowman, W. D. (2008) Links between plant litter chemistry, species diversity, and below-ground ecosystem function. *Proceedings of the National Academy of Sciences of the United States of America*, 105, 19780–19785.

Memmott, J., Waser, N. M. & Price, V. M. (2004) Tolerance of pollination networks to species extinctions. *Proceedings of the Royal Society B: Biological Sciences*, 271, 2605–2611.

Messier, J., Lechowicz, M. J., McGill, B. J., Violle, C. & Enquist, B. J. (2017) Interspecific integration of trait dimensions at local scales: the plant phenotype as an integrated network. *Journal of Ecology*, 105, 1775–1790.

Messier, J., McGill, B. J. & Lechowicz, M. J. (2010) How do traits vary across ecological scales? A case for trait-based ecology. *Ecology Letters*, 13, 838–848.

Metcalfe, J. L. (1989) Biological water quality assessment of running waters based on macroinvertebrate communities: history and present status in Europe. *Environmental Pollution*, 60, 101–139.

Metz, J., Liancourt, P., Kigel, J., Harel, D., Sternberg, M. & Tielborger, K. (2010) Plant survival in relation to seed size along environmental gradients: a long-term study from semi-arid and Mediterranean annual plant communities. *Journal of Ecology*, 98, 697–704.

Micó, E., Ramilo, P., Thorn, S., Müller, J., Galante, E. & Carmona, C. P. (2020) Contrasting functional structure of saproxylic beetle assemblages associated to different microhabitats. *Scientific Reports*, 10, 1520.

Milla, R., Osborne, C. P., Turcotte, M. M. & Violle, C. (2015) Plant domestication through an ecological lens. *Trends in Ecology and Evolution*, 30, 463–469.

Minden, V. & Kleyer, M. (2011) Testing the effect–response framework: key response and effect traits determining above-ground biomass of salt marshes. *Journal of Vegetation Science*, 22, 387–401.

Miner, B. G., Sultan, S. E., Morgan, S. G., Padilla, D. K. & Relyea, R. A. (2005) Ecological consequences of phenotypic plasticity. *Trends in Ecology and Evolution*, 20, 685–692.

Mitchell, N., Carlson, J. E. & Holsinger, K. E. (2018) Correlated evolution between climate and suites of traits along a fast–slow continuum in the radiation of *Protea*. *Ecology and Evolution*, 8, 1853–1866.

Möbius, K. A. (1877) *Die Auster und die Austernwirthschaft*. Berlin: Verlag von Wiegandt, Hemple & Parey.

Mody, K., Unsicker, S. B. & Linsenmair, K. E. (2007) Fitness related diet-mixing by intraspecific host-plant-switching of specialist insect herbivores. *Ecology*, 88, 1012–1020.

Mokany, K., Ash, J. & Roxburgh, S. (2008) Functional identity is more important than diversity in influencing ecosystem processes in a temperate native grassland. *Journal of Ecology*, 96, 884–893.

Moles, A. T. & Westoby, M. (2004) Seedling survival and seed size: a synthesis of the literature. *Journal of Ecology*, 92, 372–383.

Moles, A. T., Gruber, M. A. M. & Bonser, S. P. (2008) A new framework for predicting invasive plant species. *Journal of Ecology*, 96, 13–17.

Mols, C. M. M. & Visser, M. E. (2002) Great tits can reduce caterpillar damage in apple orchards. *Journal of Applied Ecology*, 39, 888–899.

Monnet, A. C., Jiguet, F., Meynard, C. N., Mouillot, D., Mouquet, N., Thuiller, W. & Devictor, V. (2014) Asynchrony of taxonomic, functional and phylogenetic diversity in birds. *Global Ecology and Biogeography*, 23, 780–788.

Monty, A. & Mahy, G. (2010) Evolution of dispersal traits along an invasion route in the wind-dispersed *Senecio inaequidens* (Asteraceae). *Oikos*, 119, 1563–1570.

Moore, B. D., Andrew, R. L., Kulheim, C. & Foley, W. J. (2014) Explaining intraspecific diversity in plant secondary metabolites in an ecological context. *New Phytologist*, 201, 733–750.

Morales-Castilla, I., Davies, T. J., Pearse, W. D. & Peres-Neto, P. (2017) Combining phylogeny and co-occurrence to improve single species distribution models. *Global Ecology and Biogeography*, 26, 740–752

Moretti, M., de Bello, F., Ibanez, S., Fontana, S., Pezzatti, G. B., Dziock, F., Rixen, C. & Lavorel, S. (2013) Linking traits between plants and invertebrate herbivores to track functional effects of land-use changes. *Journal of Vegetation Science*, 24, 949–962.

Moretti, M., Dias, A. T. C., de Bello, F., Altermatt, F., Chown, S. L., Azcarate, F. M., Bell, J. R., Fournier, B., Hedde, M., Hortal, J., Ibanez, S., Ockinger, E., Sousa, J. P., Ellers, J. & Berg, M. P. (2017) Handbook of protocols for standardized measurement of terrestrial invertebrate functional traits. *Functional Ecology*, 31, 558–567.

Mouchet, M., Guilhaumon, F., Villéger, S., Mason, N. W. H., Tomasini, J. A. & Mouillot, D. (2008) Towards a consensus for calculating dendrogram-based functional diversity indices. *Oikos*, 117, 794–800.

Mouillot, D., Bellwood, D. R., Baraloto, C., Chave, J., Galzin, R., Harmelin-Vivien, M., Kulbicki, M., Lavergne, S., Lavorel, S., Mouquet, N., Paine, C. E. T., Renaud, J. & Thuiller, W. (2013a) Rare species support vulnerable functions in high-diversity ecosystems. *PLOS Biology*, 11, 1001569.

Mouillot, D., Graham, N. A. J., Villéger, S., Mason, N. W. H. & Bellwood, D. R. (2013b) A functional approach reveals community responses to disturbances. *Trends in Ecology & Evolution*, 28, 167–177.

Mouillot, D., Mason, W. H. N., Dumay, O. & Wilson, J. B. (2005) Functional regularity: a neglected aspect of functional diversity. *Oecologia*, 142, 353–359.

Mouillot, D., Villéger, S., Scherer-Lorenzen, M. & Mason, N. W. (2011) Functional structure of biological communities predicts ecosystem multifunctionality. *PLOS ONE*, 6, e17476.

Mudrák, O., Doležal, J., Vitová, A. & Lepš, J. (2019) Variation in plant functional traits is best explained by the species identity: stability of trait-based species ranking across meadow management regimes. *Functional Ecology*, 33, 746–755.

Münkemüller, T., Lavergne, S., Bzeznik, B., Dray, S., Jombart, T., Schiffers, K. & Thuiller, W. (2012) How to measure and test phylogenetic signal. *Methods in Ecology and Evolution*, 3, 743–756.

Muñoz, M. C., Schaefer, H. M., Böhning-Gaese, K. & Schleuning, M. (2017) Importance of animal and plant traits for fruit removal and seedling recruitment in a tropical forest. *Oikos*, 126, 823–832.

Myhrvold, N. P., Baldridge, E., Chan, B., Sivam, D., Freeman, D. L. & Morgan Ernest, S. K. (2015) An amniote life-history database to perform comparative analyses with birds, mammals, and reptiles. *Ecology*, 96, 3109.

Naeem, S. & Wright, J. P. (2003) Disentangling biodiversity effects on ecosystem functioning: deriving solutions to a seemingly insurmountable problem. *Ecology Letters*, 6, 567–579.

Nakano, S. & Murakami, M. (2001) Reciprocal subsidies: dynamic interdependence between terrestrial and aquatic food webs. *Proceedings of the National Academy of Sciences of the United States of America*, 98, 166–170.

Nesshöver, C., Assmuth, T., Irvine, K. N., Rusch, G. M., Waylen, K. A., Delbaere, B., Haase, D., Jones-Walters, L., Keune, H., Kovacs, E., Krauze, K., Külvik, M., Rey, F., van Dijk, J., Vistad, O. I., Wilkinson, M. E. & Wittmer, H. (2017) The science, policy and practice of nature-based solutions: an interdisciplinary perspective. *Science of the Total Environment*, 579, 1215–1227.

Nguyen, N. H., Song, Z., Bates, S. T., Branco, S., Tedersoo, L., Menke, J., Schilling, J. S. & Kennedy, P. G. (2016) FUNGuild: an open annotation tool for parsing fungal community datasets by ecological guild. *Fungal Ecology*, 20, 241–248.

Nickel, H. & Remane, R. (2002) Artenliste der Zikaden Deutschlands, mit Angaben zu Nährpflanzen, Nahrungsbreite, Lebenszyklen, Areal und Gefährdung (Hemiptera). *Beiträge zur Zikandenkunde*, 5, 27–64.

Nunes, A., Köbel, M., Pinho, P., Matos, P., de Bello, F., Correia, O. & Branquinho, C. (2017) Which plant traits respond to aridity? A critical step to assess functional diversity in Mediterranean drylands. *Agricultural and Forest Meteorology*, 239, 176–184.

Nyffeler, M., Olson, E. J. & Symondson, W. O. C. (2016) Plant-eating by spiders. *Journal of Arachnology*, 44, 15–27.

Oh, H. J., Jeong, H. G., Nam, G. S., Oda, Y., Dai, W., Lee, E. H., Kong, D., Hwang, S. J. & Chang, K. H. (2017) Comparison of taxon-based and trophi-based response patterns of rotifer community to water quality: applicability of the rotifer functional group as an indicator of water quality. *Animal Cells and Systems*, 21, 133–140.

Olesen, J. M., Bascompte, J., Dupont, Y. L., Elberling, H., Rasmussen, C. & Jordano, P. (2011) Missing and forbidden links in mutualistic networks. *Proceedings of the Royal Society B – Biological Sciences*, 278, 725–732.

Oliveira, B. F., Sao-Pedro, V. A., Santos-Barrera, G., Penone, C. & Costa, G. C. (2017) Data Descriptor: AmphiBIO, a global database for amphibian ecological traits. *Scientific Data*, 4, 170123.

Ostenfeld, C. H. (1908) *The land-vegetation of the Færöes, with special reference to the higher plants*. In: Botany of the Faeroes, Part III (ed. Warming, E.), pp. 867–1026. Copenhagen and Christiania: Gylendalske Boghandel, Nordisk Forlag.

Ovaskainen, O., Roy, D. B., Fox, R., Anderson, B. J. & Orme, D. (2015) Uncovering hidden spatial structure in species communities with spatially explicit joint species distribution models. *Methods in Ecology and Evolution*, 7, 428–436.

Pagel, M. D. (1999) Inferring the historical patterns of biological evolution. *Nature*, 401, 877–884.

Paine, C. E. T., Baraloto, C. & Díaz, S. (2015) Optimal strategies for sampling functional traits in species-rich forests. *Functional Ecology*, 29, 1325–1331.

Pakeman, R. J. (2004) Consistency of plant species and trait responses to grazing along a productivity gradient: a multi-site analysis. *Journal of Ecology*, 92, 893–905.

Pakeman, R. J. (2011) Multivariate identification of plant functional response and effect traits in an agricultural landscape. *Ecology*, 92, 1353–1365.

Pakeman, R. J. (2014) Functional trait metrics are sensitive to the completeness of the species' trait data? *Methods in Ecology and Evolution*, 5, 9–15.

Pakeman, R. J. & Quested, H. M. (2007) Sampling plant functional traits: what proportion of the species need to be measured? *Applied Vegetation Science*, 10, 91–96.

Pakeman, R. J., Lennon, J. J. & Brooker, R. W. (2011) Trait assembly in plant assemblages and its modulation by productivity and disturbance. *Oecologia*, 167, 209–218.

Paliy, O. & Shankar, V. (2016) Application of multivariate statistical techniques in microbial ecology. *Molecular Ecology*, 25, 1032–1057.

Palkovacs, E. P., Marshall, M. C., Lamphere, B. A., Lynch, B. R., Weese, D. J., Fraser, D. F., Reznick, D. N., Pringle, C. M. & Kinnison, M. T. (2009) Experimental evaluation of evolution and coevolution as agents of ecosystem change in Trinidadian streams. *Philosophical Transactions of the Royal Society B – Biological Sciences*, 364, 1617–1628.

Palmer, M. A., Zedler, J. B. & Falk, D. A. (2016) *Foundations of restoration ecology*. 2nd ed. Washington, DC: Island Press.

Paradis, E. (2012) *Analysis of phylogenetics and evolution with R*. Berlin: Springer.

Pardo, I., Roquet, C., Lavergne, S., Olesen, J. M., Gómez, D. & García, M. B. (2017) Spatial congruence between taxonomic, phylogenetic and functional hotspots: true pattern or methodological artefact? *Diversity and Distributions*, 23, 209–220.

Parker, I. M., Simberloff, D., Lonsdale, W. M., Goodell, K., Wonham, M., Kareiva, P. M., Williamson, M. H., Von Holle, B., Moyle, P. B., Byers, J. E. & Goldwasser, L. (1999) Impact: toward a framework for understanding the ecological effects of invader. *Biological Invasions*, 1, 3–19.

Parmesan, C. & Yohe, G. (2003) A globally coherent fingerprint of climate change impacts across natural systems. *Nature*, 421, 37–42.

Parr, C. L., Dunn, R. R., Sanders, N. J., Weiser, M. D., Photakis, M., Bishop, T. R., Fitzpatrick, M. C., Arnan, X., Baccaro, F., Brandão, C. R. F., Chick, L., Donoso, D. A., Fayle, T. M., Gómez, C., Grossman, B., Munyai, T. C., Pacheco, R., Retana, J., Robinson, A., Sagata, K., Silva, R. R., Tista, M., Vasconcelos, H., Yates, M. & Gibb, H. (2017) GlobalAnts: a new database on the geography of ant traits (Hymenoptera: Formicidae). *Insect Conservation and Diversity*, 10, 5–20.

Pärtel, M., Szava-Kovats, R. & Zobel, M. (2011) Dark diversity: shedding light on absent species. *Trends in Ecology & Evolution*, 26, 124–128.

Pascual, M. & Dunne, J. A. (2006) *Ecological networks: linking structure to dynamics in food webs*. Oxford University Press. Oxford, UK.

Pataki, D. E., McCarthy, H. R., Gillespie, T., Jenerette, G. D. & Pincetl, S. (2013) A trait-based ecology of the Los Angeles urban forest. *Ecosphere*, 4, 1–20.

Pausas, J. G. & Verdú, M. (2010) The jungle of methods for evaluating phenotypic and phylogenetic structure of communities. *BioScience*, 60, 614–625.

Pavoine, S. & Bonsall, M. B. (2011) Measuring biodiversity to explain community assembly: a unified approach. *Biological Reviews*, 86, 792–812.

Pavoine, S., Gasc, A., Bonsall, M. B. & Mason, N. W. H. (2013) Correlations between phylogenetic and functional diversity: mathematical artefacts or true ecological and evolutionary processes? *Journal of Vegetation Science*, 24, 781–793.

Pavoine, S., Marcon, E. & Ricotta, C. (2016) 'Equivalent numbers' for species, phylogenetic or functional diversity in a nested hierarchy of multiple scales. *Methods in Ecology and Evolution*, 7, 1152–1163.

Pavoine, S., Vallet, J., Dufour, A. B., Gachet, S. & Daniel, H. (2009) On the challenge of treating various types of variables: application for improving the measurement of functional diversity. *Oikos*, 118, 391–402.

Pavoine, S., Vela, E., Gachet, S., de Bélair, G. & Bonsall, M. B. (2011) Linking patterns in phylogeny, traits, abiotic variables and space: a novel approach to linking environmental filtering and plant community assembly. *Journal of Ecology*, 99, 165–175.

Pearman, P. B., Lavergne, S., Roquet, C., Wüest, R., Zimmermann, N. E. & Thuiller, W. (2014) Phylogenetic patterns of climatic, habitat and trophic niches in a European avian assemblage. *Global Ecology and Biogeography*, 23, 414–424.

Peay, K. G. (2016) The mutualistic niche: mycorrhizal symbiosis and community dynamics. *Annual Review of Ecology, Evolution, and Systematics*, 47, 143–164.

Peco, B., Navarro, E., Carmona, C. P., Medina, N. G. & Marques, M. J. (2017) Effects of grazing abandonment on soil multifunctionality: the role of plant functional traits. *Agriculture Ecosystems & Environment*, 249, 215–225.

Pellissier, L., Albouy, C., Bascompte, J., Farwig, N., Graham, C., Loreau, M., Maglianesi, M. A., Melián, C. J., Pitteloud, C., Roslin, T., Rohr, R., Saavedra, S., Thuiller, W., Woodward, G., Zimmermann, N. E. & Gravel, D. (2018) Comparing species interaction networks along environmental gradients. *Biological Reviews*, 93, 785–800.

Pennell, M. W. & Harmon, L. J. (2013) An integrative view of phylogenetic comparative methods: connections to population genetics, community ecology, and paleobiology. *Annals of the New York Academy of Sciences*, 1289, 90–105.

Pennell, M. W., FitzJohn, R. G., Cornwell, W. K. & Harmon, L. J. (2015) Model adequacy and the macroevolution of angiosperm functional traits. *American Naturalist*, 186, E33–E50.

Penone, C., Davidson, A. D., Shoemaker, K. T., Di Marco, M., Rondinini, C., Brooks, T. M., Young, B. E., Graham, C. H. & Costa, G. C. (2014) Imputation of missing data in life-history trait datasets: which approach performs the best? *Methods in Ecology and Evolution*, 5, 961–970.

Peres-Neto, P. R., Dray, S. & ter Braak, C. J. F. (2017) Linking trait variation to the environment: critical issues with community-weighted mean correlation resolved by the fourth-corner approach. *Ecography*, 40, 806–816.

Pérez-Harguindeguy, N., Díaz, S., Garnier, E., Lavorel, S., Poorter, H., Jaureguiberry, P., Bret-Harte, M. S., Cornwell, W. K., Craine, J. M., Gurvich, D. E., Urcelay, C., Veneklaas, E. J., Reich, P. B., Poorter, L., Wright, I. J., Ray, P., Enrico, L., Pausas, J. G., de Vos, A. C., Buchmann, N., Funes, G., Quétier, F., Hodgson, J. G., Thompson, K., Morgan, H. D., ter Steege, H., Sack, L., Blonder, B., Poschlod, P., Vaieretti, M. V., Conti, G., Staver, A. C., Aquino, S. & Cornelissen, J. H. C. (2013) New handbook for standardised measurement of plant functional traits worldwide. *Australian Journal of Botany*, 61, 167–234.

Perez-Ramos, I. M., Matias, L., Gomez-Aparicio, L. & Godoy, O. (2019) Functional traits and phenotypic plasticity modulate species coexistence across contrasting climatic conditions. *Nature Communications*, 10, 2555.

Perović, D. J., Gïmez-Virués, S., Landis, D. A., Wäckers, F., Gurr, G. M., Wratten, S. D., You, M. S. & Desneux, N. (2018) Managing biological control services through multi-trophic trait interactions: review and guidelines for implementation at local and landscape scales. *Biological Reviews*, 93, 306–321.

Perring, M. P., Standish, R. J., Price, J. N., Craig, M. D., Erickson, T. E., Ruthrof, K. X., Whiteley, A. S., Valentine, L. E. & Hobbs, R. J. (2015) Advances in restoration ecology: rising to the challenges of the coming decades. *Ecosphere*, 6, art131.

Petchey, O. L. & Gaston, K. J. (2002) Functional diversity (FD), species richness and community composition. *Ecology Letters*, 5, 402–411.

Petchey, O. L. & Gaston, K. J. (2006) Functional diversity: back to basics and looking forward. *Ecology Letters*, 9, 741–758.

Petchey, O. L., Hector, A. & Gaston, K. J. (2004) How do different measures of functional diversity perform? *Ecology*, 85, 847–857.

Petersen, H. & Luxton, M. (1982) A comparative-analysis of soil fauna populations and their role in decomposition processes. *Oikos*, 39, 287–388.

Pey, B., Laporte, B. & Hedde, M. (2014) BETSI. https://portail.betsi.cnrs.fr/.

Pfennig, D. W. & McGee, M. (2010) Resource polyphenism increases species richness: a test of the hypothesis. *Philosophical Transactions of the Royal Society B – Biological Sciences*, 365, 577–591.

Pfestorf, H., Weiß, L., Müller, J., Boch, S., Socher, S. A., Prati, D., Schöning, I., Weisser, W., Fischer, M. & Jeltsch, F. (2013) Community mean traits as additional indicators to monitor effects of land-use intensity on grassland plant diversity. *Perspectives in Plant Ecology, Evolution and Systematics*, 15, 1–11.

Phillips, B. L., Brown, G. P. & Shine, R. (2010) Life-history evolution in range-shifting populations. *Ecology*, 91, 1617–1627.

Phillips, B. L., Brown, G. P., Webb, J. K. & Shine, R. (2006) Invasion and the evolution of speed in toad. *Nature*, 439, 803.

Piccini, I., Nervo, B., Forshage, M., Celi, L., Palestrini, C., Rolando, A. & Roslin, T. (2018) Dung beetles as drivers of ecosystem multifunctionality: are response and effect traits interwoven? *Science of the Total Environment*, 616, 1440–1448.

Pickett, S. T. A. & Bazzaz, F. A. (1978) Organization of an assemblage of early successional species on a soil moisture gradient. *Ecology*, 59, 1248–1255.

Pierce, S., Negreiros, D., Cerabolini, B. E. L., Kattge, J., Díaz, S., Kleyer, M., Shipley, B., Wright, S. J., Soudzilovskaia, N. A., Onipchenko, V. G., van Bodegom, P. M., Frenette-Dussault, C., Weiher, E., Pinho, B. X., Cornelissen, J. H. C., Grime, J. P., Thompson, K., Hunt, R., Wilson, P. J., Buffa, G., Nyakunga, O. C., Reich, P. B., Caccianiga, M., Mangili, F., Ceriani, R. M., Luzzaro, A., Brusa, G., Siefert, A., Barbosa, N. P. U., Chapin, F. S., III, Cornwell, W. K., Fang, J., Fernandes, G. W., Garnier, E., Le Stradic, S., Penuelas, J., Melo, F. P. L., Slaviero, A., Tabarelli, M. & Tampucci, D. (2017) A global method for calculating plant CSR ecological strategies applied across biomes world-wide. *Functional Ecology*, 31, 444–457.

Pillai, P. & Gouhier, T. C. (2019) Not even wrong: the spurious measurement of biodiversity's effects on ecosystem functioning. *Ecology*, 100, e02645.

Pistón, N., de Bello, F., Dias, A. T. C., Götzenberger, L., Rosado, B. H. P., de Mattos, E. A., Salguero-Gómez, R. & Carmona, C. P. (2019) Multidimensional ecological analyses demonstrate how interactions between functional traits shape fitness and life history strategies. *Journal of Ecology*, 107, 2317–2328.

Piton, G., Legay, N., Arnoldi, C., Lavorel, S., Clément, J. C. & Foulquier, A. (2020) Using proxies of microbial community-weighted means traits to explain the cascading effect of management intensity, soil and plant traits on ecosystem resilience in mountain grasslands. *Journal of Ecology*, 108, 876–893.

Pizzatto, L. & Dubey, S. (2012) Colour-polymorphic snake species are older. *Biological Journal of the Linnean Society*, 107, 210–218.

Poisot, T., Canard, E., Mouillot, D., Mouquet, N. & Gravel, D. (2012) The dissimilarity of species interaction networks. *Ecology Letters*, 15, 1353–1361.

Pollock, L. J., Morris, W. K. & Vesk, P. A. (2012) The role of functional traits in species distributions revealed through a hierarchical model. *Ecography*, 35, 716–725.

Post, D. M. & Palkovacs, E. P. (2009) Eco-evolutionary feedbacks in community and ecosystem ecology: interactions between the ecological theatre and the evolutionary play. *Philosophical Transactions of the Royal Society of London B – Biological Sciences*, 364, 1629–1640.

Powell, J. R. & Rillig, M. C. (2018) Biodiversity of arbuscular mycorrhizal fungi and ecosystem function. *New Phytologist*, 220, 1059–1075.

Prescott, C. E. & Zukswert, J. M. (2016) Invasive plant species and litter decomposition: time to challenge assumptions. *New Phytologist*, 209, 5–7.

Price, J. N. & Pärtel, M. (2013) Can limiting similarity increase invasion resistance? A meta-analysis of experimental studies. *Oikos*, 122, 649–656.

Prinzing, A. (2016) On the opportunity of using phylogenetic information to ask evolutionary questions in functional community ecology. *Folia Geobotanica*, 51, 69–74.

Prinzing, A., Reiffers, R., Braakhekke, W. G., Hennekens, S. M., Tackenberg, O., Ozinga, W. A., Schaminée, J. H. J. & Van Groenendael, J. M. (2008) Less lineages – more trait variation: phylogenetically clustered plant communities are functionally more diverse. *Ecology Letters*, 11, 809–819.

Puy, J., Dvorakova, H., Carmona, C. P., de Bello, F., Hiiesalu, I. & Latzel, V. (2018) Improved demethylation in ecological epigenetic experiments: testing a simple and harmless foliar demethylation application. *Methods in Ecology and Evolution*, 9, 744–753.

Pyron, R. A., Costa, G. C., Patten, M. A. & Burbrink, F. T. (2015) Phylogenetic niche conservatism and the evolutionary basis of ecological speciation. *Biological Reviews*, 90, 1248–1262.

Rafferty, N. E. & Ives, A. R. (2013) Phylogenetic trait-based analyses of ecological networks. *Ecology*, 94, 2321–2333.

Raunkiær, C. C. (1934) *The life forms of plants and statistical plant geography*. Oxford: Clarendon Press.

Read, J., Fletcher, T. D., Wevill, T. & Deletic, A. (2010) Plant traits that enhance pollutant removal from stormwater in biofiltration systems. *International Journal of Phytoremediation*, 12, 34–53.

Rees, M. (1995) EC-PC comparative analyses? *Journal of Ecology*, 83, 891.

Reich, P. B. (2014) The world-wide 'fast–slow' plant economics spectrum: a traits manifesto. *Journal of Ecology*, 102, 275–301.

Relyea, R. A. & Yurewicz, K. L. (2002) Predicting community outcomes from pairwise interactions: integrating density- and trait-mediated effects. *Oecologia*, 131, 569–579.

Renner, S. C. & van oesel, W. (2017) Ecological and functional traits in 99 bird species over a large-scale gradient in Germany. *Data*, 2, 12.

Richardson, S. J., Press, M. C., Parsons, A. N. & Hartley, S. E. (2002) How do nutrients and warming impact on plant communities and their insect herbivores? A 9-year study from a sub-Arctic heath. *Journal of Ecology*, 90, 544–556.

Ricklefs, R. E. (2004) A comprehensive framework for global patterns in biodiversity. *Ecology Letters*, 7, 1–15.

Ricotta, C. & Moretti, M. (2011) CWM and Rao's quadratic diversity: a unified framework for functional ecology. *Oecologia*, 167, 181–188.

Ricotta, C., de Bello, F., Moretti, M., Caccianiga, M., Cerabolini, B. E. L. & Pavoine, S. (2016) Measuring the functional redundancy of biological communities: a quantitative guide. *Methods in Ecology and Evolution*, 7, 1386–1395.

Riibak, K., Reitalu, T., Tamme, R., Helm, A., Gerhold, P., Znamenskiy, S., Bengtsson, K., Rosen, E., Prentice, H. C. & Pärtel, M. (2015) Dark diversity in dry calcareous grasslands is determined by dispersal ability and stress-tolerance. *Ecography*, 38, 713–721.

Rolo, V., Olivier, P. I. & van Aarde, R. (2017) Tree and bird functional groups as indicators of recovery of regenerating subtropical coastal dune forests. *Restoration Ecology*, 25, 788–797.

Rosado, B. H. P., Dias, A. T. C. & de Mattos, E. (2013) Going back to basics: importance of ecophysiology when choosing functional traits for studying communities and ecosystems. *Natureza & Conservação*, 11, 15–22.

Roscher, C., Schumacher, J., Gubsch, M., Lipowsky, A., Weigelt, A., Buchmann, N., Schmid, B. & Schulze, E.-D. (2012) Using plant functional traits to explain diversity–productivity relationships. *PLOS ONE*, 7, e36760.

Rosenfield, M. F. & Müller, S. C. (2017) Predicting restored communities based on reference ecosystems using a trait-based approach. *Forest Ecology and Management*, 391, 176–183.

Rosenthal, G. A. & Berenbaum, M. R. (2012) *Herbivores: their interactions with secondary plant metabolites: ecological and evolutionary processes.* New York: Academic Press.

Rota, C., Manzano, P., Carmona, C. P., Malo, J. E. & Peco, B. (2017) Plant community assembly in Mediterranean grasslands: understanding the interplay between grazing and spatio-temporal water availability. *Journal of Vegetation Science*, 28, 149–159.

Roucou, A., Violle, C., Fort, F., Roumet, P., Ecarnot, M. & Vile, D. (2018) Shifts in plant functional strategies over the course of wheat domestication. *Journal of Applied Ecology*, 55, 25–37.

Rudman, S. M., Kreitzman, M., Chan, K. M. A. & Schluter, D. (2017) Evosystem services: rapid evolution and the provision of ecosystem services. *Trends in Ecology and Evolution*, 32, 403–415.

Sabo, J. L. & Power, M. E. (2002) River–watershed exchange: effects of riverine subsidies on riparian lizards and their terrestrial prey. *Ecology*, 83, 1860–1869.

Salguero-Gómez, R., Jones, O. R., Archer, C. R., Bein, C., de Buhr, H., Farack, C., Gottschalk, F., Hartmann, A., Henning, A., Hoppe, G., Römer, G., Ruoff, T., Sommer, V., Wille, J., Voigt, J., Zeh, S., Vieregg, D., Buckley, Y. M., Che-Castaldo, J., Hodgson, D., Scheuerlein, A., Caswell, H. & Vaupel, J. W. (2016) COMADRE: a global data base of animal demography. *Journal of Animal Ecology*, 85, 371–384.

Salguero-Gómez, R., Jones, O. R., Archer, C. R., Buckley, Y. M., Che-Castaldo, J., Caswell, H., Hodgson, D., Scheuerlein, A., Conde, D. A., Brinks, E., de Buhr, H., Farack, C., Gottschalk, F., Hartmann, A., Henning, A., Hoppe, G., Römer, G., Runge, J., Ruoff, T., Wille, J., Zeh, S., Davison, R., Vieregg, D., Baudisch, A., Altwegg, R., Colchero, F., Dong, M., de Kroon, H., Lebreton, J.-D., Metcalf, C. J. E., Neel, M. M., Parker, I. M., Takada, T., Valverde, T., Vélez-Espino, L. A., Wardle, G. M., Franco, M., Vaupel, J. W. & Rees, M. (2015) The COMPADRE plant matrix database: an open online repository for plant demography. *Journal of Ecology*, 103, 202–218.

Sanford, G. M., Lutterschmidt, W. I. & Hutchison, V. H. (2002) The comparative method revisited. *BioScience*, 52, 830.

Sasaki, T., Katabuchi, M., Kamiyama, C., Shimazaki, M., Nakashizuka, T. & Hikosaka, K. (2014) Vulnerability of moorland plant communities to environmental change: consequences of realistic species loss on functional diversity. *Journal of Applied Ecology*, 51, 299–308.

Schaffers, A. P. & Sykora, K. V. (2000) Reliability of Ellenberg indicator values for moisture, nitrogen and soil reaction: a comparison with field measurements. *Journal of Vegetation Science*, 11, 225–244.

Scherer-Lorenzen, M., Palmborg, C., Prinz, A. & Schulze, E. D. (2003) The role of plant diversity and composition for nitrate leaching in grasslands. *Ecology*, 84, 1539–1552.

Schimper, A. F. W. (1903) *Plant-geography upon a physiological basis.* Oxford: Clarendon Press.

Schleuning, M., Fründ, J. & García, D. (2015) Predicting ecosystem functions from biodiversity and mutualistic networks: an extension of trait-based concepts to plant–animal interactions. *Ecography*, 38, 380–392.

Schlichting, C. D. & Levin, D. A. (1986) Phenotypic plasticity – an evolving plant character. *Biological Journal of the Linnean Society*, 29, 37–47.

Schloss, C. A., Nunez, T. A. & Lawler, J. J. (2012) Dispersal will limit ability of mammals to track climate change in the Western Hemisphere. *Proceedings of the National Academy of Sciences of the United States of America*, 109, 8606–8611.

Schluter, D. & McPhail, J. D. (1992) Ecological character displacement and speciation in sticklebacks. *American Naturalist*, 140, 85–108.

Schmera, D., Eros, T. & Podani, J. (2009) A measure for assessing functional diversity in ecological communities. *Aquatic Ecology*, 43, 157–167.

Schmera, D., Heino, J., Podani, J., Erös, T. & Dolédec, S. (2017) Functional diversity: a review of methodology and current knowledge in freshwater macroinvertebrate research. *Hydrobiologia*, 787, 27–44.

Schmidt-Kloiber, A. & Hering, D. (2015) www.freshwaterecology.info – An online tool that unifies, standardises and codifies more than 20,000 European freshwater organisms and their ecological preferences. *Ecological Indicators*, 53, 271–282.

Schmitz, O. J. (2008) Effects of predator grassland hunting mode on grassland ecosystem function. *Science*, 319, 952–954.

Schmitz, O. J. (2010) *Resolving ecosystem complexity*. Princeton University Press. Princeton, NJ.

Schmitz, O. J., Buchkowski, R. W., Burghardt, K. T. & Donihue, C. M. (2015) Functional traits and trait-mediated interactions. Connecting community–level interactions with ecosystem functioning. *Advances in Ecological Research*, 52, 319–343.

Schneider, F. D., Fichtmueller, D., Gossner, M. M., Güntsch, A., Jochum, M., König-Ries, B., Le Provost, G., Manning, P., Ostrowski, A., Penone, C. & Simons, N. K. (2019) Towards an ecological trait-data standard. *Methods in Ecology and Evolution*, 10, 2006–2019.

Schoener, T. W. (2011) The Newest Synthesis: understanding ecological dynamics. *Science*, 331, 426–429.

Schouw, J. F. (1823) *Grundzüge einer allgemeinen Pflanzengeographie*. De Gruyter, Incorporated.

Schumacher, J. & Roscher, C. (2009) Differential effects of functional traits on aboveground biomass in semi-natural grasslands. *Oikos*, 118, 1659–1668.

Schwarz, N., Moretti, M., Bugalho, M. N., Davies, Z. G., Haase, D., Hack, J., Hof, A., Melero, Y., Pett, T. J. & Knapp, S. (2017) Understanding biodiversity-ecosystem service relationships in urban areas: a comprehensive literature review. *Ecosystem Services*, 27, 161–171.

Schweiger, O., Settele, J., Kudrna, O., Klotz, S. & Kühn, I. (2008) Climate change can cause spatial mismatch of trophically interacting species. *Ecology*, 89, 3472–3479.

Schweitzer, J. A., Bailey, J. K., Rehill, B. J., Martinsen, G. D., Hart, S. C., Lindroth, R. L., Keim, P. & Whitham, T. G. (2004) Genetically based trait in a dominant tree affects ecosystem processes. *Ecology Letters*, 7, 127–134.

Seabloom, E. W., Harpole, W. S., Reichman, O. J. & Tilman, D. (2003) Invasion, competitive dominance, and resource use by exotic and native California grassland species. *Proceedings of the National Academy of Sciences of the United States of America*, 100, 13384–13389.

Sebastián-González, E. (2017) Drivers of species' role in avian seed-dispersal mutualistic networks. *Journal of Animal Ecology*, 86, 878–887.

Semper, C. (1881) *Animal life as affected by the natural conditions of existence.* New York: D. Appleton and Co.

Sengupta, S., Ergon, T. & Leinaaa, H. P. (2016) Genotypic differences in embryonic life history traits of *Folsomia quadrioculata* (Collembola: Isotomidae) across a wide geographical range. *Ecological Entomology*, 41, 72–84.

Sgrò, C. M., Terblanche, J. S. & Hoffmann, A. A. (2016) What can plasticity contribute to insect responses to climate change? *Annual Review of Entomology*, 61, 433–451.

Shipley, B. (2000) *Cause and correlation in biology: a user's guide to path analysis, structural equations and causal inference.* Cambridge University Press. Cambridge, UK.

Shipley, B. (2012) *From plant traits to vegetation structure.* Cambridge University Press. Cambridge, UK.

Shipley, B., Belluau, M., Kühn, I., Soudzilovskaia, N. A., Bahn, M., Penuelas, J., Kattge, J., Sack, L., Cavender-Bares, J., Ozinga, W. A., Blonder, B., van Bodegom, P. M., Manning, P., Hickler, T., Sosinski, E., Pillar, V. D. P., Onipchenko, V. & Poschlod, P. (2017) Predicting habitat affinities of plant species using commonly measured functional traits. *Journal of Vegetation Science*, 28, 1082–1095.

Shipley, B., de Bello, F., Cornelissen, J. H. C., Lalibertée, E., Laughlin, D. C. & Reich, P. B. (2016) Reinforcing loose foundation stones in trait-based plant ecology. *Oecologia*, 180, 923–931.

Shipley, B., Paine, C. E. T. & Baraloto, C. (2012) Quantifying the importance of local niche-based and stochastic processes to tropical tree community assembly. *Ecology*, 93, 760–769.

Shipley, B., Vile, D. & Garnier, E. (2006) From plant traits to plant communities: a statistical mechanistic approach to biodiversity. *Science*, 314, 812–814.

Shurin, J. B. (2000) Dispersal limitation, invasion resistance, and the structure of pond zooplankton communities. *Ecology*, 81, 3074–3086.

Siefert, A., Violle, C., Chalmandrier, L., Albert, C. H., Taudiere, A., Fajardo, A., Aarssen, L. W., Baraloto, C., Carlucci, M. B., Cianciaruso, M. V., Dantas, V. D., de Bello, F., Duarte, L. D. S., Fonseca, C. R., Freschet, G. T., Gaucherand, S., Gross, N., Hikosaka, K., Jackson, B., Jung, V., Kamiyama, C., Katabuchi, M., Kembel, S. W., Kichenin, E., Kraft, N. J. B., Lagerstrom, A., Le Bagousse-Pinguet, Y., Li, Y. Z., Mason, N., Messier, J., Nakashizuka, T., McC Overton, J., Peltzer, D. A., Perez-Ramos, I. M., Pillar, V. D., Prentice, H. C., Richardson, S., Sasaki, T., Schamp, B. S., Schob, C., Shipley, B., Sundqvist, M., Sykes, M. T., Vandewalle, M. & Wardle, D. A. (2015) A global meta-analysis of the relative extent of intraspecific trait variation in plant communities. *Ecology Letters*, 18, 1406–1419.

Silvertown, J., Dodd, M., Gowing, D. J. G., Lawson, C. S. & McConway, K. J. (2006a) Phylogeny and the hierarchical organization of plant diversity. *Ecology*, 87, S39–S49.

Silvertown, J., Poulton, P., Johnston, E., Edwards, G., Heard, M. & Biss, P. M. (2006b) The Park Grass Experiment 1856–2006: its contribution to ecology. *Journal of Ecology*, 94, 801–814.

Simberloff, D. S. (1970) Taxonomic diversity of island biotas. *Evolution*, 24, 23–47.

Šmilauer, P. & Lepš, J. (2014) *Multivariate analysis of ecological data using CANOCO 5.* 2nd ed. Cambridge University Press. Cambridge, UK.

Smith, A. B., Godsoe, W., Rodríguez-Sánchez, F., Wang, H. H. & Warren, D. (2018) Niche estimation above and below the species level. *Trends in Ecology & Evolution*, 34, 260–273.

Smith, T. M., Shugart, H. H. & Woodward, F. I. (1997) (eds.) *Plant functional types: their relevance to ecosystem properties and global change.* Cambridge University Press. Cambridge, UK.

Solé-Senan, X. O., Juárez-Escario, A., Robleño, I., Conesa, J. A. & Recasens, J. (2017) Using the response–effect trait framework to disentangle the effects of agricultural intensification on the provision of ecosystem services by Mediterranean arable plants. *Agriculture, Ecosystems & Environment*, 247, 255–264.

Soliveres, S., Lehmann, A., Boch, S., Altermatt, F., Carrara, F., Crowther, T. W., Delgado-Baquerizo, M., Kempel, A., Maynard, D. S., Rillig, M. C., Singh, B. K., Trivedi, P. & Allan, E. (2018) Intransitive competition is common across five major taxonomic groups and is driven by productivity, competitive rank and functional traits. *Journal of Ecology*, 106, 852–864.

Spasojevic, M. J. & Suding, K. N. (2012) Inferring community assembly mechanisms from functional diversity patterns: the importance of multiple assembly processes. *Journal of Ecology*, 100, 652–661.

Speight, M. C. D. (2014) *Syrph the Net: the database of European Syrphidae (Diptera)*. Species accounts of European Syrphidae (Diptera). Syrph the Net Publications.

Spielman, D., Brook, B. W. & Frankham, R. (2004) Most species are not driven to extinction before genetic factors impact them. *Proceedings of the National Academy of Sciences of the United States of America*, 101, 15261–15264.

Spitz, J., Ridoux, V. & Brind'Amour, A. (2014) Let's go beyond taxonomy in diet description: testing a trait-based approach to prey–predator relationships. *Journal of Animal Ecology*, 83, 1137–1148.

Stavert, J. R., Linan-Cembrano, G., Beggs, J. R., Howlett, B. G., Pattemore, D. E. & Bartomeus, I. (2016) Hairiness: the missing link between pollinators and pollination. *Peerj*, 4, 18.

Sterk, M., Gort, G., Klimkowska, A., van Ruijven, J., van Teeffelen, A. J. A. & Wamelink, G. W. W. (2013) Assess ecosystem resilience: linking response and effect traits to environmental variability. *Ecological Indicators*, 30, 21–27.

Stevenson, P. R. & Guzmán-Caro, D. C. (2010) Nutrient transport within and between habitats through seed dispersal processes by woolly monkeys in North-Western Amazonia. *American Journal of Primatology*, 72, 992–1003.

Storkey, J. (2006) A functional group approach to the management of UK arable weeds to support biological diversity. *Weed Research*, 46, 513–522.

Stubbs, W. J. & Wilson, J. B. (2004) Evidence for limiting similarity in a sand dune community. *Journal of Ecology*, 92, 557–567

Suding, K. N., Lavorel, S., Chapin, F. S., Cornelissen, J. H. C., Díaz, S., Garnier, E., Goldberg, D., Hooper, D. U., Jackson, S. T. & Navas, M. L. (2008) Scaling environmental change through the community-level: a trait-based response-and-effect framework for plants. *Global Change Biology*, 14, 1125–1140.

Sultan, S. E. (1996) Phenotypic plasticity for offspring traits in *Polygonum persicaria*. *Ecology*, 77, 1791–1807.

Sultan, S. E. (2000) Phenotypic plasticity for plant development, function and life history. *Trends in Plant Science*, 5, 537–542.

Swan, C. M. & Palmer, M. A. (2006) Composition of speciose leaf litter alters stream detritivore growth, feeding activity and leaf breakdown. *Oecologia*, 147, 469–478.

Swanson, H. K., Lysy, M., Power, M., Stasko, A. D., Johnson, J. D. & Reist, J. D. (2015) A new probabilistic method for quantifying n-dimensional ecological niches and niche overlap. *Ecology*, 96, 318–324.

Swenson, N. G. (2014a) *Functional and phylogenetic ecology in R*. Berlin: Springer.

Swenson, N. G. (2014b) Phylogenetic imputation of plant functional trait databases. *Ecography*, 37, 105–110.

Swenson, N. G. & Enquist, B. J. (2009) Opposing assembly mechanisms in a Neotropical dry forest: implications for phylogenetic and functional community ecology. *Ecology*, 90, 2161–2170.

Swenson, N. G., Enquist, B. J., Pither, J., Thompson, J. & Zimmerman, J. K. (2006) The problem and promise of scale dependency in community phylogenetics. *Ecology*, 87, 2418–2424.

Swenson, N. G., Enquist, B. J., Thompson, J. & Zimmerman, J. K. (2007) The influence of spatial and size scale on phylogenetic relatedness in tropical forest communities. *Ecology*, 88, 1770–1780.

Tacutu R., Thornton D., Johnson E., Budovsky A., Barardo D., Craig T., Diana E., Lehmann G., Toren D., Wang J., Fraifeld V. E., de Magalhaes J. P. (2018) Human Ageing Genomic Resources: new and updated databases. *Nucleic Acids Research* 46, D1083–D1090.

Tamme, R., Götzenberger, L., Zobel, M., Bullock, J. M., Hooftman, D. A. P., Kaasik, A. & Pärtel, M. (2014) Predicting species' maximum dispersal distances from simple plant traits. *Ecology*, 95, 505–513.

Taugourdeau, S., Villerd, J., Plantureux, S., Huguenin-Elie, O. & Amiaud, B. (2014) Filling the gap in functional trait databases: use of ecological hypotheses to replace missing data. *Ecology and Evolution*, 4, 944–958.

Tavşanoğlu, Ç. & Pausas, J. G. (2018) A functional trait database for Mediterranean Basin plants. *Scientific Data*, 5, 180135.

Taylor, A. R., Lenoir, L., Vegerfors, B. & Persson, T. (2019) Ant and earthworm bioturbation in cold–temperate ecosystems. *Ecosystems*, 22, 981–994.

Thomas, C. D., Bodsworth, E. J., Wilson, R. J., Simmons, A. D., Davies, Z. G., Musche, M. & Conradt, L. (2001) Ecological and evolutionary processes at expanding range margins. *Nature*, 411, 577–581.

Thompson, K., Askew, A. P., Grime, J. P., Dunnett, N. P. & Willis, A. J. (2005) Biodiversity, ecosystem function and plant traits in mature and immature plant communities. *Functional Ecology*, 19, 355–358.

Thornhill, I. A., Biggs, J., Hill, M. J., Briers, R., Gledhill, D., Wood, P. J., Gee, J. H. R., Ledger, M. & Hassall, C. (2018) The functional response and resilience in small waterbodies along land-use and environmental gradients. *Global Change Biology*, 24, 3079–3092.

Thorpe, R. S., Reardon, J. T. & Malhotra, A. (2005) Common garden and natural selection experiments support ecotypic differentiation in the Dominican anole (*Anolis oculatus*). *American Naturalist*, 165, 495–504.

Thuiller, W., Guéguen, M., Georges, D., Bonet, R., Chalmandrier, L., Garraud, L., Renaud, J., Roquet, C., Van Es, J., Zimmermann, N. E. & Lavergne, S. (2014) Are different facets of plant diversity well protected against climate and land cover changes? A test study in the French Alps. *Ecography*, 37, 1254–1266.

Thuiller, W., Lavorel, S. & Araujo, M. B. (2005) Niche properties and geographical extent as predictors of species sensitivity to climate change. *Global Ecology and Biogeography*, 14, 347–357.

Tilman, D., Wedin, D. & Knops, J. (1996) Productivity and sustainability influenced by biodiversity in grassland ecosystems. *Nature*, 379, 718–720.

Tobner, C. M., Paquette, A., Gravel, D., Reich, P. B., Williams, L. J. & Messier, C. (2016) Functional identity is the main driver of diversity effects in young tree communities. *Ecology Letters*, 19, 638–647.

Trochet, A., Moulherat, S., Calvez, O., Stevens, V. M., Clobert, J. & Schmeller, D. S. (2014) A database of life-history traits of European amphibians. *Biodiversity Data Journal*, 2, e4123. https://doi.org/10.3897/BDJ.2.e4123

Turcotte, M. M., Reznick, D. N. & Hare, J. D. (2011) The impact of rapid evolution on population dynamics in the wild: experimental test of eco-evolutionary dynamics. *Ecology Letters*, 14, 1084–1092.

Umana, M. N., Mi, X. C., Cao, M., Enquist, B. J., Hao, Z. Q., Howe, R., Iida, Y., Johnson, D., Lin, L. X., Liu, X. J., Ma, K. P., Sun, I. F., Thompson, J., Uriarte, M., Wang, X. G., Wolf, A., Yang, J., Zimmerman, J. K. & Swenson, N. G. (2017) The role of functional uniqueness and spatial aggregation in explaining rarity in trees. *Global Ecology and Biogeography*, 26, 777–786.

USDA & NRCS (2018) *The PLANTS Database*. Greensboro, NC: National Plant Data Team.

Uyeda, J. C., Zenil-Ferguson, R. & Pennell, M. W. (2018) Rethinking phylogenetic comparative methods. *Systematic Biology*, 67, 1091–1109.

Valencia, E., Gross, N., Quero, J. L., Carmona, C. P., Ochoa, V., Gozalo, B., Delgado-Baquerizo, M., Dumack, K., Hamonts, K., Singh, B. K., Bonkowski, M. & Maestre, F. T. (2018) Cascading effects from plants to soil microorganisms explain how plant species richness and simulated climate change affect soil multifunctionality. *Global Change Biology*, 24, 5642–5654.

Valencia, E., Maestre, F. T., Le Bagousse-Pinguet, Y., Quero, J. L., Tamme, R., Börger, L., García-Gómez, M. & Gross, N. (2015) Functional diversity enhances the resistance of ecosystem multifunctionality to aridity in Mediterranean drylands. *New Phytologist*, 206, 660–671.

van Kleunen, M., Weber, E. & Fischer, M. (2010) A meta-analysis of trait differences between invasive and non-invasive plant species. *Ecology Letters*, 13, 235–245.

van Klink, R., Lepš, J., Vermeulen, R. & de Bello, F. (2019) Functional differences stabilize beetle communities by weakening interspecific temporal synchrony. *Ecology*, 100, e02748.

Vandewalle, M., de Bello, F., Berg, M. P., Bolger, T., Dolédec, S., Dubs, F., Feld, C. K., Harrington, R., Harrison, P. A. & Lavorel, S. (2010) Functional traits as indicators of biodiversity response to land use changes across ecosystems and organisms. *Biodiversity and Conservation*, 19, 2921–2947.

Vellend, M. (2010) Conceptual synthesis in community ecology. *Quarterly Review of Biology*, 85, 183–206.

Vellend, M. (2016) *The theory of ecological communities*. Princeton University Press. Princeton, NJ.

Verdú, M., Rey, P. J., Alcántara, J. M., Siles, G. & Valiente-Banuet, A. (2009) Phylogenetic signatures of facilitation and competition in successional communities. *Journal of Ecology*, 97, 1171–1180.

Vesk, P. A. & Westoby, M. (2001) Predicting plant species' responses to grazing. *Journal of Applied Ecology*, 38, 897–909.

Via, S. (1999) Reproductive isolation between sympatric races of pea aphids. I. Gene flow restriction and habitat choice. *Evolution*, 53, 1446–1457.

Via, S. (2001) Sympatric speciation in animals: the ugly duckling grows up. *Trends in Ecology & Evolution*, 16, 381–390.

Vilà, M., Espinar, J. L., Hejda, M., Hulme, P. E., Jarošík, V., Maron, J. L., Pergl, J., Schaffner, U., Sun, Y. & Pyšek, P. (2011) Ecological impacts of invasive alien plants: a meta-analysis of their effects on species, communities and ecosystems. *Ecology Letters*, 14, 702–708.

Villéger, S., Grenouillet, G. & Brosse, S. (2013) Decomposing functional diversity reveals that low functional diversity is driven by low functional turnover in European fish assemblages. *Global Ecology and Biogeography*, 22, 671–681.

Villéger, S., Mason, N. W. H. & Mouillot, D. (2008) New multidimensional functional diversity indices for a multifaceted framework in functional ecology. *Ecology*, 89, 2290–2301.

Violle, C., Enquist, B. J., McGill, B. J., Jiang, L., Albert, C. H., Hulshof, C., Jung, V. & Messier, J. (2012) The return of the variance: intraspecific variability in community ecology. *Trends in Ecology & Evolution*, 27, 244–252.

Violle, C., Navas, M. L., Vile, D., Kazakou, E., Fortunel, C., Hummel, I. & Garnier, E. (2007) Let the concept of trait be functional! *Oikos*, 116, 882–892.

Violle, C., Nemergut, D. R., Pu, Z. C. & Jiang, L. (2011) Phylogenetic limiting similarity and competitive exclusion. *Ecology Letters*, 14, 782–787.

Violle, C., Thuiller, W., Mouquet, N., Munoz, F., Kraft, N. J. B., Cadotte, M. W., Livingstone, S. W. & Mouillot, D. (2017) Functional rarity: the ecology of outliers. *Trends in Ecology & Evolution*, 32, 356–367.

Visser, M. E. & Holleman, L. J. M. (2001) Warmer springs disrupt the synchrony of oak and winter moth phenology. *Proceedings of the Royal Society B – Biological Sciences*, 268, 289–294.

Visser, M. E., Noordwijk, A. V., Tinbergen, J. M. & Lessells, C. M. (1998) Warmer springs lead to mistimed reproduction in great tits (*Parus major*). *Proceedings of the Royal Society B – Biological Sciences*, 265, 1867–1870.

Volterra, V. (1926) Variazioni e fluttuazioni del numero d'individui in specie animali conviventi. Memoria della Reale Accademia. *Nazionale dei Lincei*, 2, 31–113.

von Döhren, P. & Haase, D. (2015) Ecosystem disservices research: a review of the state of the art with a focus on cities. *Ecological Indicators*, 52, 490–497.

von Liebig, J. (1843) Die Wechselwirthschaft. *Justus Liebigs Annalen DerChemie*, 46, 58–97.

Vos, V. C. A., van Ruijven, J., Berg, M. P., Peeters, E. & Berendse, F. (2011) Macro-detritivore identity drives leaf litter diversity effects. *Oikos*, 120, 1092–1098.

Walker, B., Kinzig, A. & Langridge, J. (1999) Plant attribute diversity, resilience, and ecosystem function: the nature and significance of dominant and minor species. *Ecosystems*, 2, 95–113.

Walther, G. R., Post, E., Convey, P., Menzel, A., Parmesan, C., Beebee, T. J. C., Fromentin, J. M., Hoegh-Guldberg, O. & Bairlein, F. (2002) Ecological responses to recent climate change. *Nature*, 416, 389–395.

Wardle, D. A., Bardgett, R. D., Klironomos, J. N., Setala, H., van der Putten, W. H. & Wall, D. H. (2004) Ecological linkages between aboveground and belowground biota. *Science*, 304, 1629–1633.

Warming, E. (1909) *Oecology of plants – an introduction to the study of plant-communities*. Oxford: Clarendon Press.

Watson, H. C. (1847–1859) *Cybele Britannica: or British plants and their geographical relations*. London: Longman & Company.

Webb, C. O., Ackerly, D. D., McPeek, M. A. & Donoghue, M. J. (2002) Phylogenies and community ecology. *Annual Review of Ecology and Systematics*, 33, 475–505.

Weiher, E. & Keddy, P. A. (1995) Assembly rules, null models, and trait dispersion: new questions from old patterns. *Oikos*, 74, 159.

Weiher, E., Clarke, G. D. P. & Keddy, P. A. (1998) Community assembly rules, morphological dispersion, and the coexistence of plant species. *Oikos*, 81, 309–322.

Westoby, M. (1998) A leaf-height-seed (LHS) plant ecology strategy scheme. *Plant and Soil*, 199, 213–227.

Westoby, M., Falster, D. S., Moles, A. T., Vesk, P. A. & Wright, I. J. (2002) Plant ecological strategies: some leading dimensions of variation between species. *Annual Review of Ecology and Systematics*, 33, 125–159.

Westoby, M., Leishman, M. R. & Lord, J. M. (1995) On misinterpreting the 'phylogenetic correction'. *Journal of Ecology*, 83, 531.

Whitham, T. G., Bailey, J. K., Schweitzer, J. A., Shuster, S. M., Bangert, R. K., Leroy, C. J., Lonsdorf, E. V., Allan, G. J., DiFazio, S. P., Potts, B. M., Fischer, D. G., Gehring, C. A., Lindroth, R. L., Marks, J. C., Hart, S. C., Wimp, G. M. & Wooley, S. C. (2006) A framework for community and ecosystem genetics: from genes to ecosystems. *Nature Reviews Genetics*, 7, 510–523.

Whittaker, R. H. (1956) Vegetation of the Great Smoky Mountains. *Ecological Monographs*, 26, 1–80.

Whittaker, R. J. & Fernández-Palacios, J. M. (2007) *Island biogeography: ecology, evolution, and conservation*. Oxford University Press. Oxford, UK.

Wilby, A. (2002) Ecosystem engineering: a trivialized concept? *Trends in Ecology and Evolution*, 17, 307–308.

Wildi, O. (2016) Why mean indicator values are not biased. *Journal of Vegetation Science*, 27, 40–49.

Wilman, H., Belmaker, J., Simpson, J., de la Rosa, C., Rivadeneira, M. M. & Jetz, W. (2014) EltonTraits 1.0: species-level foraging attributes of the world's birds and mammals. *Ecology*, 95, 2027.

Wilson, J. B. (2007) Trait-divergence assembly rules have been demonstrated: limiting similarity lives! A reply to Grime. *Journal of Vegetation Science*, 18, 451–452.

Winfrey, B. K., Hatt, B. E. & Ambrose, R. F. (2018) Biodiversity and functional diversity of Australian stormwater biofilter plant communities. *Landscape and Urban Planning*, 170, 112–137.

Wisz, M. S., Pottier, J., Kissling, W. D., Pellissier, L., Lenoir, J., Damgaard, C. F., Dormann, C. F., Forchhammer, M. C., Grytnes, J. A., Guisan, A., Heikkinen, R. K., Høye, T. T., Kühn, I., Luoto, M., Maiorano, L., Nilsson, M. C., Normand, S., Öckinger, E., Schmidt, N. M., Termansen, M., Timmermann, A., Wardle, D. A., Aastrup, P. & Svenning, J.-C. (2013) The role of biotic interactions in shaping distributions and realised assemblages of species: implications for species distribution modelling. *Biological Reviews*, 88, 15–30.

Wittmann, M. E., Barnes, M. A., Jerde, C. L., Jones, L. A. & Lodge, D. M. (2016) Confronting species distribution model predictions with species functional traits. *Ecology and Evolution*, 6, 873–879.

Wolf, A., Doughty, C. E. & Malhi, Y. (2013) Lateral diffusion of nutrients by mammalian herbivores in terrestrial cosystems. *PLOS ONE*, 8, 1–10.

Wong, B. B. M. & Candolin, U. (2015) Behavioral responses to changing environments. *Behavioral Ecology*, 26, 665–673.

Wright, I. J. & Westoby, M. (2002) Leaves at low versus high rainfall: coordination of structure, lifespan and physiology. *New Phytologist*, 155, 403–416.

Wright, I. J., Dong, N., Maire, V., Prentice, I. C., Westoby, M., Díaz, S., Gallagher, R. V., Jacobs, B. F., Kooyman, R., Law, E. A., Leishman, M. R., Niinemets, U., Reich, P. B., Sack, L., Villar, R., Wang, H. & Wilf, P. (2017) Global climatic drivers of leaf size. *Science*, 357, 917–921.

Wright, I. J., Reich, P. B., Westoby, M., Ackerly, D. D., Baruch, Z., Bongers, F., Cavender-Bares, J., Chapin, T., Cornelissen, J. H. C., Diemer, M., Flexas, J., Garnier, E., Groom, P. K., Gulias, J., Hikosaka, K., Lamont, B. B., Lee, T., Lee, W., Lusk, C., Midgley, J. J., Navas, M. L., Niinemets, U., Oleksyn, J., Osada, N., Poorter, H., Poot, P., Prior, L., Pyankov, V. I., Roumet, C., Thomas, S. C., Tjoelker, M. G., Veneklaas, E. J. & Villar, R. (2004) The worldwide leaf economics spectrum. *Nature*, 428, 821–827.

Yang, L. H. & Rudolf, V. H. W. (2010) Phenology, ontogeny and the effects of climate change on the timing of species interactions. *Ecology Letters*, 13, 1–10.

Yang, N., Butenschoen, O., Rana, R., Kohler, L., Hertel, D., Leuschner, C., Scheu, S., Polle, A. & Pena, R. (2019) Leaf litter species identity influences biochemical composition of ectomycorrhizal fungi. *Mycorrhiza*, 29, 85–96.

Yu, Q. & Pulkkinen, P. (2003) Genotype–environment interaction and stability in growth of aspen hybrid clones. *Forest Ecology and Management*, 173, 25–35.

Zelený, D. (2018) Which results of the standard test for community-weighted mean approach are too optimistic? *Journal of Vegetation Science*, 29, 953–966.

Zelený, D. & Schaffers, A. P. (2012) Too good to be true: pitfalls of using mean Ellenberg indicator values in vegetation analyses. *Journal of Vegetation Science*, 23, 419–431.

Zimmer, M., Pennings, S. C., Buck, T. L. & Carefoot, T. H. (2002) Species-specific patterns of litter processing by terrestrial isopods (Isopoda: Oniscidea) in high intertidal salt marshes and coastal forests. *Functional Ecology*, 16, 596–607.

Ziter, C. (2016) The biodiversity–ecosystem service relationship in urban areas: a quantitative review. *Oikos*, 125, 761–768.

Zizzari, Z. V. & Ellers, J. (2014) Rapid shift in thermal resistance between generations through maternal heat exposure. *Oikos*, 123, 1365–1370.

Zobel, M. (1997) The relative role of species pools in determining plant species richness: an alternative explanation of species coexistence? *Trends in Ecology & Evolution*, 12, 266–269.

Zobel, M. (2016) The species pool concept as a framework for studying patterns of plant diversity. *Journal of Vegetation Science*, 27, 8–18.

Index

Abiotic and biotic factors, 6, 9, 12, 62–67, 111, 114, 138–139, 147, 151, 183
   biotic interactions, 39–43, 59, 63–64, 73, 110, 120, 129, 132–140, 146, 168, 189, 240
Abiotic and biotic filters. See Abiotic and biotic factors
Abundance. See Species abundance
Abundance ranking. See Species relative abundance
Adaptation, 6, 37, 39, 41, 57–59, 110–115, 153, 157
Agroecosystems, 199, 231, 237–240
Alien species. See Invasion
Alpha diversity, 98
ANOVA, 120–122, 158, 163

Beta diversity, 98–103, 126, 226
Between-species trait variability, 32–33, 36, 119–121
Biodiversity, 1–2, 4, 171, 178, 199, 232, 238
   biodiversity indicators, 232, 234–237
   biodiversity monitoring, 232–237, 248
Biodiversity and ecosystem functions (BEF), 179, 181–185, 192
Biodiversity effects
   additive partitioning, 187
   idiosyncratic species effects, 183, 234
   mass ratio hypothesis, 182, 184
   net diversity effect, 187, 189
   niche complementarity. See Niche: Niche complementarity
   non-additive effects, 182
   selection effect, 187
Bioindicators. See Biodiversity: biodiversity indicators
Bipartite networks, 203
Brownian model of evolution. See Brownian motion
Brownian motion, 153–160, 163–164

Categorical traits, 21, 29, 42, 45, 48, 51, 69, 78, 80, 159, 164–165, 171
CATS model, 147, 246
Character displacement, 133, 173, 175
Circular traits, 31, 45, 50
Clustering. See Community assembly: convergence

Coefficient of variation (CV), 119, 121
Coexistence, 132–140, 168, 226
Common garden, 58, 117–119
Community, 2–3, 58, 63, 65, 69, 76, 81, 98–99, 110, 211
Community assembly, 65, 67, 110, 115, 120, 124, 129–143, 147, 167–170, 172, 205
   chequerboard patterns, 130
   convergence, 6, 134, 139–143, 147, 168
   divergence, 138–143, 148, 168
   environmental filtering, 64–67, 70, 110, 130–131, 137, 140–141, 147, 168, 205, 214
   limiting similarity, 94, 120, 130, 133–137, 139, 142, 148, 167, 245
Community level analyses, 67, 69, 72, 199
Community phylogenetic structure, 168
Community trait variance. See Functional diversity (FD) and Variance
Community weighted mean (CWM), 69, 78–86, 121–124, 147, 181–189, 199–201, 227, 233–234, 246–247
Competition. See Abiotic and biotic factors: biotic interactions
Context dependency, 5, 23
Convergent evolution, 157
Convex hull, 88, 93
Critical upper thermal limit, 12
C-S-R, 41–43, 61, 147, 239

Dark diversity. See Species pool
Decoupling functional and phylogenetic diversity, 172–173
Density-dependence, 40, 135, 137, 248
Differentiation between species. See Dissimilarity between species
Dispersal, 18, 65, 73, 110, 113–115, 130–131, 140, 147, 196, 205, 245, 247
Dissimilarity between species, 36, 44–50, 86, 90, 98, 124, 133, 137, 143, 170, 173, 201, 214, 223, 226
   differences in competitive ability, 65, 115, 136
   differences in intrinsic growth rate, 136
   fitness differences, 136, 139

Gower distance, 53
  phylogenetic distance, 164–165, 167
  phylogenetic relatedness, 161, 166, 173
Divergent evolution, 111, 175, 228
Diversification, 154, 174
Domestication, 237
Dominance, 3, 83, 92, 100, 132, 138, 182, 184
Drift, 107, 110, 130
Dummy variable, 45, 48, 164
Dynamic box ranges, 126

Ecological literacy, 248
Ecosystem function, 1, 3–4, 12–14, 16, 18, 38, 85, 97, 117, 148, 192, 201, 205, 214, 232, 244–248
  ecosystem services trade-offs and clusters, 179, 244
Ecosystem functioning. *See* Ecosystem function
Ecosystem process. *See* Ecosystem function
Ecosystem service. *See* Ecosystem function
Effect trait, 14, 18–21, 38, 55, 85, 117, 177–180, 186, 190–192, 194, 197–201, 205, 238, 247
Ellenberg indicator values, 9
Environmental preference. *See* Niche: realized niche
Equalizing mechanisms. *See* Community assembly: convergence
Equivalent numbers, 100
Evenness, 3, 83, 94, 96
Evolutionary distance between species. *See* Dissimilarity between species: phylogenetic distance
Evolutionary dynamics, 105, 153
Exotic species. *See* Invasion

Facilitation. *See* Abiotic and biotic factors: biotic interacions
Feeding guild, 3–4, 208
Filters. *See* Abiotic and biotic factors
Fitness, 6–16, 19, 21, 55, 70–72, 107, 113, 130, 137, 139, 167, 178
Functional diversity (FD), 4, 70, 77–78, 86–104, 124–128, 139, 143–147, 149, 167–173, 182, 185–188, 199, 227, 233–234, 246
  functional attribute diversity (FAD), 88, 91, 102
  functional dispersion (FDis), 91, 228
  functional divergence (FDiv), 93, 96, 189
  functional evenness, 4, 94–96, 126
  functional redundancy, 55, 97, 126, 244
  functional regularity (FRO), 94
  functional richness, 4, 87, 95–96, 99, 126, 144–146, 228
  mean neighbour taxon distance (MNTD), 95, 99
  mean pairwise dissimilarity (MPD), 91–94, 97–99, 103
  Rao diversity, 94, 96–104, 120, 126, 143, 199
  Trait Even Distribution (TED), 95
  Trait Onion Peeling (TOP), 93

Functional groups, 4, 36, 53–56, 61, 76, 86, 89, 97, 184, 215, 234, 237
  clustering dendrograms, 53, 55, 89
  functional effect groups, 38, 97
  functional response groups, 38
Functional marker, 10
Functional trait, definition, 7–14
Functional types. *See* Functional groups
Fuzzy coding, 30, 45, 48, 50

Gamma diversity, 98
Genetic diversity, 112–113
Genetic variation, 107, 118–119, 210
Genotype, 106–112, 118, 201
Genotype–environment interaction, 109
Geographical scale, 58, 64, 105, 111, 117, 175, 210, 223
Gradient, 6, 9, 37, 55, 58, 60, 64, 67–74, 83, 111, 114, 118, 121–124, 144, 147, 161, 171, 211, 218, 221–226
  direct gradient, 62
  indirect gradient, 62
  resource gradient, 62
Green urban infrastructure, 240–244

Habitat preference. *See* Niche: realized niche
Higher-order interactions, 137
Hypervolume, 6, 87, 126

Interaction chain, 137, 207
Interaction networks, 202–207
Interspecific trait variability. *See* Between-species trait variability
Intransitive interactions, 137
Intraspecific trait variability (ITV), 23, 36, 67, 74, 105–128, 189, 201, 210, 218, 221, 224–225, 232, 246
  aITV, 122
  wITV, 120–121, 225
Invasion, 114–115, 148–149, 246–248

Leaf economic spectrum, 42, 51, 172, 180, 238
L-H-S, 41–43, 50, 74, 90
Life history, 111–112, 115, 191, 208
Log transformation, 52, 83, 86, 88, 90, 121, 228
Lotka–Volterra model, 132

Matching traits, 197, 204
Mean field theory, 105
Missing trait data, 31, 50, 80, 166, 173, 226–230
  imputation, 28, 31, 228
Modular networks, 203
Molecular operational taxonomic units (MOTUs), 3
Multi-trait dissimilarity. *See* Dissimilarity between species

Multitrophic interactions, 195–207
Mutualism. *See* Abiotic and biotic factors: biotic interactions

NAs. *See* Missing trait data
Neutral theory, 110, 120, 130, 139, 206
Niche
  alpha niche, 63
  beta niche, 63
  ecological niche, 6, 9, 40, 57, 69, 111, 115, 160
  fundamental niche, 63, 138, 175
  niche breadth, 63
  niche complementarity, 182, 187, 238, 244
  niche conservatism, phylogenetic, 156, 228
  niche differences. *See* niche differentiation
  niche differentiation, 94, 111, 130, 133–134, 139, 146. *See also* Community assembly
  niche divergence, phylogenetic, 228
  niche modelling, 175
  niche optimum, 63
  niche partitioning. *See* niche differentiation
  niche shift, 28, 189, 208
  niche space, 9
  niche width. *See* niche breadth
  realized niche, 63
  species ditribution modelling, 72
Nominal traits. *See* Categorical traits
Null model, 84, 88, 95, 130, 141, 143–147, 157, 168

Ordinal traits, 21, 30, 44
Overdispersion. *See* Community assembly; Divergent evolution

Paradox of the plankton, 132
Performance, 10–12, 15, 19, 232
Performance traits. *See* Performance
Permanova, 120
Phenotype, 7, 107, 109, 112–114, 156, 237
Phenotypic diversity, 77, 113, 117–118
Phenotypic plasticity, 28, 49, 107, 118, 202, 223
  epigenetic variation, 59, 109
  transgenerational plasticity, 109
Phylogenetic community structure. *See* Community phylogenetic structure
Phylogenetic comparative method, 160–166
Phylogenetic diversity, 22, 86, 99, 103, 167–173
Phylogenetic eigenvectors, 165
Phylogenetic independent contrasts (PICs), 166
Phylogenetic signal, 157–159, 166, 168–170, 173
Phylogenetic tree, 152–153, 157, 160, 163, 166, 168, 173, 175
Plot. *See* Sampling unit
Polymorphism, 28, 111, 113
Polyploidy, 114
Population, 2, 6, 9, 23, 28, 32, 40, 58, 67, 69, 77, 105–106, 110, 112–113, 118, 216, 221, 248

Principal component analysis (PCA), 42, 51, 180
Principal coordinate analyses (PCoA), 51, 88, 103, 165
Probability density, 77, 126, 225
Proxy, trait, 10, 21, 167

Quadratic entropy. *See* Functional diversity (FD): Rao diversity

r/K selection, 39–41
Randomizations, 84, 120, 143–147
Range size, 9, 114
Reaction norm, 108
Reproduction, asexual, 114
Response trait, 12, 19–20, 55, 57, 84, 117, 177, 190, 198, 205, 238
Response–effect trait framework, 16, 178, 190–192, 197–201, 205, 209
Restoration, 148, 244–246
Ruderal species, 41, 239

Sampling unit, 60, 76, 78, 81, 98, 100, 121, 146, 164, 220, 224
Selection pressure, 19, 39, 58, 106–107, 110, 118, 130, 138, 153, 202
Simpson diversity. *See* Rao diversity
Spatial scale, 17, 27, 29, 32, 64, 66, 99, 110, 124, 126, 142, 146, 169, 218, 221, 223
Speciation, 39, 65, 110–112, 130, 153–157, 174
Species abundance, 60–70, 74, 76, 83, 86, 96, 103, 121, 126, 135, 144, 146–147, 182, 187, 205–207, 214, 226–230, 232, 237, 245
Species coexistence. *See* Coexistence
Species composition, 3, 5, 55, 70, 121, 132, 148, 190, 212, 214, 222, 232, 234, 244
Species distribution, 9, 12, 37, 57, 63–64, 72–74, 113–114, 173–175
Species distribution modelling (SDM). *See* Niche: species distribution modeling
Species diversity, 3, 92, 115, 226, 235
Species effects, 4, 55, 183, 188, 234, 247
Species pool, 64–67, 115, 132, 139, 142, 146–147, 185, 236
Species relative abundance, 3, 78–83, 100
Species replacements. *See* Turnover in species composition
Species richness, 3–4, 74, 89, 92, 96–97, 99–100, 120, 143, 145–146, 185, 188, 226, 233, 235
Species-level analyses, 67–70, 85, 160–166
Stability, 180, 183, 197, 204, 207, 237
Standardized trait protocols, 28, 32–34
  acclimation, 10, 33
  handbook of protocols, 32, 34, 218
  precision and accuracy, 216, 219, 221

Taxon-free sampling, 221, 226, 230
Taxonomy, 4, 17, 27, 92, 98, 167, 226

Trait convergence. *See* Community assembly: convergence; Convergent evolution
Trait database, 20, 22–28, 32, 42, 166, 192, 210, 219, 223, 226
Trait divergence. *See* Community assembly; Divergent evolution
Trait evolution. *See* Brownian motion; Convergent evolution; Divergent evolution
Trait measurement. *See* Standardized trait protocols
Trait probability density (TPD), 126
Trait sampling strategies, xii, 32, 106, 124, 210–230
Trait syndromes, 21, 43, 179–180, 199, 238
Trait trade-offs, 39–42, 50, 160
Trait-based ecology, 2, 4, 7
Trait–environment relationships, 57, 67
Traitspace model, 147, 246

Trophic effect trait, 198–199, 202
Trophic interactions, 2. *See also* Multitrophic interactions
Trophic response trait, 197–201
TRY. *See* Trait database
Turnover in species composition, 85, 98, 102, 121–124, 207, 211, 214, 221, 223, 244

Underdispersion. *See* Community assembly: convergence

Variance, 77, 88, 90–92, 94, 96, 99, 103, 119, 121, 124, 158, 224

Within-species analyses, 58, 67
Within-species trait variability. *See* Intraspecific trait variability (ITV)

£38.99

Milton Keynes UK
Ingram Content Group UK Ltd.
UKHW031303160224
437934UK00020B/170